SUICIDE
Guidelines for Assessment, Management, and Treatment

edited by

Bruce Bongar

New York   Oxford
OXFORD UNIVERSITY PRESS
1992

Oxford University Press

Oxford   New York   Toronto
Delhi   Bombay   Calcutta   Madras   Karachi
Kuala Lumpur   Singapore   Hong Kong   Tokyo
Nairobi   Dar es Salaam   Cape Town
Melbourne   Auckland   Madrid

and associated companies in
Berlin   Ibadan

Published by Oxford University Press, Inc.
200 Madison Avenue, New York, New York 10016

Oxford is a registered trademark of Oxford University Press

Library of Congress Cataloging-in-Publication Data
Suicide : guidelines for assessment, management, and treatment /
edited by Bruce Bongar. p.   cm.
Includes bibliographical references and index.
ISBN 0-19-506846-7
1. Suicide—Risk factors.   2. Suicidal behavior—Diagnosis.
3. Suicidal behavior—Prevention.   4. Suicide—Prevention.
I. Bongar, Bruce Michael.
[DNLM: 1. Personality Assessment. 2. Suicide—prevention & control.
3. Suicide—psychology. HV 6546 S94825]
RC569.S9345   1992   616.83′8445—dc20
DNLM/DLC   92-1065

9 8 7 6 5 4 3 2 1

Printed in the United States of America
on acid-free paper

Foreword

The relatively new field of suicidology as a formal discipline is served well by the availability of this book. Unlike some of the lengthy compendia on suicide, this book manages to provide both the "data" of suicide and the necessary information for managing the risk of suicide in clinical practice.

Almost twenty years have elapsed since the publication by Oxford University Press of an earlier text, which also systematically reviewed information on suicide and suicide prevention from a multidisciplinary perspective. In 1967, funded by a grant from the Center for Studies of Suicide Prevention at the National Institute of Mental Health, the first formal postgraduate Fellowship Program in Suicidology was offered by the Department of Psychiatry and Behavioral Sciences at The Johns Hopkins University School of Medicine. The multidisciplinary curriculum was edited by me as *A Handbook for the Study of Suicide*, published in 1975, to provide the "basics" relevant to the understanding of suicidal behavior. My plan was to survey certain disciplines and present their contributions, recognizing the limitations inherent in the choices made. Nevertheless, that program, that text, and those fellows who learned and subsequently taught others have been significant in setting standards for training in the study of suicide.

This new work reviews, updates and expands the literature of this multifaceted field. Proper attention is given to the priorities of what is needed for the clinician and the clinician-in-training. This guidebook succeeds in doing justice to complexity, to what is known and also to what is unknown. Happily, this scenario on suicide provides creative integration of the information on this subject with application to everyday practice and concerns.

The editor has brought together a "Who's Who" of contemporary authorities in suicidology and has constructed a set of training materials that compose a curriculum. These experts serve as a core faculty for the reader of whatever age, discipline, or expertise who wishes a distillation of theoretical, empirical, and clinical insights.

For many years, I have had special concerns about the hit-or-miss approach to suicide within the training programs of mental health care or primary care specialties. Many disciplines and many offerings provide context, but not the focused information necessary to understand, assess, manage, and treat the suicidal individual. As the leading empirical researcher on training needs in our specialty, Dr. Bongar has been sensitive in developing a model text that is both broad based yet especially helpful in responding effectively to the "cry for help."

It is my hope that *Suicide: Guidelines for Assessment, Management, and Treatment* will underscore the need for new training programs in suicidology. In the meantime, I congratulate Dr. Bongar in prospectively providing a text which disseminates the subject matter in a most illuminating manner.

Seymour Perlin, M.D.

Acknowledgments

I would like to acknowledge the efforts of the staff at Oxford University Press, and to mention specifically the marvelous assistance, experienced editorial advice, and enthusiasm that my editor Joan Bossert provided for this work. I would also like specifically to thank Larry Olson and his assistant Geoff Feinberg for their help with logistics on this project, and to thank Joan's assistant, Allison Sole, for her help with details of coordination. Also, thanks to Berta Steiner and her staff for their patience and fine copy editing work.

To my wife Debbora and to our young son Brandon, thank you for bearing with me through yet another lengthy and difficult project. Your deep love and support are the joyful foundations of my life.

I would also like to take a moment to thank all of the distinguished contributors to this volume who unselfishly gave of their time and effort in crafting their chapters—all focused on the common goal of workable and useful guidelines for the practitioner and practitioner-in-training.

Last, this book is dedicated to my students, interns, and residents who over the years repeatedly asked me to construct a set of workable and concrete basic training guidelines for understanding the suicidal scenario. I hope that through the collective voice of the contributors in this volume we have satisfied your request, and provided you and the next generation of clinicians with some of the didactic resources needed to help safeguard and improve the quality of care for your patients.

Contents

Contributors

Robert J. Berchick, Ph.D.
Assistant Professor of Psychology in
  Psychiatry
Senior Consultant
Center for Cognitive Therapy
Department of Psychiatry
University Pennsylvania Medical School
Philadelphia, Pennsylvania

Alan L. Berman, Ph.D.
Director
National Center for the Study and
  Prevention of Suicide
Washington School of Psychiatry
Past-President, American Association of
  Suicidology
Private Practice of Psychotherapy and
  Consultation
Washington Psychological Center
Washington, D.C.

Bruce Bongar, Ph.D.
Associate Professor
Clinical Psychology Program
Pacific Graduate School of Psychology
Palo Alto, California; and
Adjunct Associate Professor
Department of Psychiatry
University of Massachusetts Medical
  School
Worcester, Massachusetts

David C. Clark, Ph.D.
Associate Professor of Psychology and
  Psychiatry
Rush Medical College; and
Director Center for Suicide
  Research and Prevention
Department of Psychiatry
Rush-Presbyterian-St. Luke's Medical
  Center
Chicago, Illinois
Past-President, American Association of
  Suicidology

Betsy S. Comstock, M.D.
Professor of Clinical Psychiatry
Baylor College of Medicine
Houston, Texas
Past-President, American Association of
  Suicidology

Larry E. Dumont, M.D.
Director
Adolescent Dual Diagnosis Unit
Fair Oaks Hospital
Summit, New Jersey

Edward J. Dunne, Ph.D.
Clinical Instructor in Psychology
Department of Psychiatry
Columbia University's College of
    Physicians and Surgeons
Adjunct Professor of Psychiatry
State University of New York Health
    Sciences Center at Brooklyn (formerly
    Downstate Medical Center)
New York, New York

James R. Eyman, Ph.D.
Director, Suicide Research Program
C. F. Menninger Memorial Hospital
Topeka, Kansas

Susanne Kohn Eyman, Ph.D.
Mental Health Associates
Manhattan, Kansas

Jan Fawcett, M.D.
Professor of Psychiatry
Rush Medical College; and
Chair, Department of Psychiatry
Rush-Presbyterian-St. Luke's Medical
    Center
Medical Director
Center for Suicide Research and
    Prevention
Rush-Presbyterian-St. Luke's Medical
    Center
Chicago, Illinois

Mort Harmatz, Ph.D.
Professor
Department of Psychology
University of Massachusetts at Amherst
Amherst, Massachusetts

David A. Jobes, Ph.D.
Associate Professor
Department of Psychology
Catholic University; and
Assistant Director
National Center for the Study and
    Prevention of Suicide
Washington School of Psychiatry
Washington, D.C.

James W. Lomax, M.D.
Professor of Psychiatry
Baylor College of Medicine
Houston, Texas

Christopher G. Lovett, Ph.D.
Candidate, Boston Psychoanalytic
    Institute
Clinical Supervisor
McLean Hospital and Beth Israel
    Hospital
Harvard Medical School
Boston, Massachusetts

John T. Maltsberger, M.D.
President, American Association of
    Suicidology
Senior Consultant in the Psychosocial
    Program and Associate Psychiatrist
McLean Hospital
Harvard Medical School
Member, Boston Psychoanalytic Society
Boston, Massachusetts

Ronald W. Maris, Ph.D.
Professor of Preventive Medicine
Director, Center for the Study of Suicide
University of South Carolina, Columbia
Columbia, South Carolina
Editor-in-Chief, Suicide and Life
    Threatening Behavior
Past-President, American Association of
    Suicidology

John L. McIntosh, Ph.D.
Professor
Department of Psychology
Indiana University at South Bend
South Bend, Indiana
President-Elect, American Association of
   Suicidology

Jerome A. Motto, M.D.
Professor of Psychiatry
University of California, San Francisco
San Francisco, California
Past-President, American Association of
   Suicidology

Seymour Perlin, M.D.
Professor of Psychiatry and Behavioral
   Sciences
George Washington School of Medicine
Washington, D.C.; and
Former Professor of Psychiatry
The Johns Hopkins University School of
   Medicine
Baltimore, Maryland
Past-President, American Association of
   Suicidology

Cynthia R. Pfeffer, M.D.
Professor of Psychiatry
Cornell University Medical College
White Plains, New York
Past-President, American Association of
   Suicidology

Edwin S. Shneidman, Ph.D.
Professor of Thanatology Emeritus
Neuropsychiatric Institute and Hospital
   Center for Health Sciences
University of California at Los Angeles
Los Angeles, California
Founding President, American
   Association of Suicidology

Andrew Edmund Slaby, M.D., Ph.D.,
   MPH
Medical Director
Fair Oaks Hospital
Psychiatrist-in-Chief
The Regent Hospital
Summit, New Jersey; and
Clinical Professor of Psychiatry
New York University
Adjunct Clinical Professor of Psychiatry
Brown University and New York
   Medical College
Past-President, American Association of
   Suicidology

Mae S. Sokol, M.D.
Assistant Professor of Psychiatry
Cornell University Medical College
White Plains, New York

Fred D. Wright, Ed.D.
Director of Education
Assistant Professor of Psychology in
   Psychiatry
Center for Cognitive Therapy
Department of Psychiatry
University of Pennsylvania Medical
   School
Philadelphia, Pennsylvania

Robert I. Yufit, Ph.D.
Associate Professor of Clinical Psychology
Northwestern University Medical School;
   and Director,
Suicide Intervention Program
Department of Psychiatry and the
   Behavioral Sciences
Division of Clinical Psychology
Northwestern University Medical Center
Chicago, Illinois
Past-President, American Association of
   Suicidology

Introduction and Rationale

Suicide is the human act of self-inflicted, self-intended cessation—i.e., the willed permanent stopping of consciousness. The phenomenon is enormously complicated, encompassing a wide variety (and different ranges) of dysphoria, disturbance, and self-abnegation, resignation, despair, hopelessness, and so many other complex emotions, thoughts and behaviors (Shneidman, 1981). Suicide is the most frequently encountered emergency situation for mental health professionals (Schein, 1976), and clinicians consistently rank working with suicidal patients as the most stressful of all their clinical activities (Deutsch, 1984). This is a reflection of the fact that the relationship between psychopathology, suicide attempting, and completed suicide is dynamic and not yet well understood (Bongar, 1991; Maris, 1981; Peterson & Bongar, 1989; Shneidman, 1989). Recent studies show changes in the identity of high risk groups (Boyd, 1983; Evans & Farberow, 1988; Weissman, Klerman, Markowitz, & Ouelette, 1989; Maris, 1981, 1989; Roy, 1986; Shneidman, 1989), and follow-up studies demonstrate that risk factors among attempters may be significantly different from those for the general population (Frederick, 1978; Maris, 1981, 1989). Also, studies suggest that rates of suicide among the mentally ill (Hirschfeld & Davidson, 1988; Maris, 1981; Pokorny, 1964a, 1964b; Roy, 1986; Simon, 1988), and physically ill (Abram, Moore, & Westervelt 1971; Hirschfeld & Davidson, 1988; Maris, 1981; Roy, 1986) far exceed that of the population as a whole. Gutheil, Bursztajn, Brodsky, and Alexander (1991), citing the work of Lester Havens, referred to suicide as "the final outcome of a web of cause and chance" (p. 189).

Mental health and health care professionals need to understand that completed and attempted suicide represents a very serious public health problem in the United States today. Each year almost 30,000 individuals take their own lives, making suicide the eighth leading cause of death in the country (Alcohol, Drug Abuse, and Mental Health Administration 1989; Hirschfeld & Davidson, 1988). The death of a patient from suicide is a relatively common occurrence in the professional lives of mental health professionals; the average professional psychologist involved in direct patient care has better than a 20% chance of losing a patient to suicide at some time during his or her professional career, with the odds climbing to better than 50% for psychiatrists (Chemtob, Hamada, Bauer, Torigoe & Kinney 1988; Chemtob, Hamada, Bauer, Kinney & Torigoe, 1988). Even psychology trainees have approximately a 15% chance of losing a patient to suicide at some time during their training sequence (Brown, 1987; Kleespies, Smith, & Becker, 1990). One recent study found that the trainees who had experienced the suicide of a patient reported levels of stress that were equivalent to that found in "patient samples with bereavement and higher than that found with professional clinicians who had patient suicides" (Kleespies, Smith, & Becker, 1990, p. 257). The trainees who had lost a patient to suicide responded (in order of frequency) with feelings of shock, guilt or shame, denial or disbelief, feelings of incompetence, anger, depression, and a sense of being to blame.

Thus, as Chemtob et al. (1989) pointed out, it is vitally important that training programs in psychology and psychiatry

not convey, either explicitly or implicitly, the impression that patient suicides are an unlikely event (Brown, 1987). . . . It may be time that psychiatrists and psychologists explicitly acknowledge patient suicide as an important occupational hazard. (Chemtob et al., 1989, p. 299)

Fremouw, de Perczel, and Ellis (1990) recently pointed out that working effectively with the suicidal patient entails therapeutic activism (e.g., the delaying of the patient's suicidal impulses, the restoring of hope, environmental intervention, and consideration of hospitalization). Yet over fifty years ago Karl Menninger called for clinicians actively to use their intelligence and training in the battle against suicide—to strengthen the will to live against the wish to die (Menninger, 1938).

Suicide: Guidelines for Assessment, Management, and Treatment was created to address the current lacunae in our training efforts in the study of suicide. For the reality is that our training efforts in this area for the core mental health disciplines and for primary health care providers may well be inadequate for the purpose of preparing the next generation of practitioners to handle this often-encountered clinical crisis (Bongar, 1991, Bongar & Harmatz, 1989, 1991). We also know that the average practitioner has had relatively little formal training in working with suicidal patients (Bongar, 1991). Thus, in examining this all too real occupational hazard, the present book explicitly conveys the message that the management of suicidal patients may be a higher risk professional endeavor than most other clinical situations, and that the situation requires extensive clinical and didactic specialty training (Bongar, 1991; Bongar et al., 1989; Maltsberger, 1986; Roy, 1986).

The following chapters represent a response to this problem by distilling fundamentals of assessment, management, and treatment of suicidal patients, and proposing sensible and useful guidelines for clinical practice. The guiding principle for the construction of guidelines was to provide the reader with a foundation of basic and necessary clinical knowledge drawn from the contributors' accumulated clinical wisdom, from their review of recent empirical findings, and from their extensive personal experience in dealing with the suicidal scenario. As editor, I have had the pleasure of selecting and working with a group of contributors who include some of our most distinguished thinkers in the field of suicidology (presidents of the American Association of Suicidology, editors of journals, directors of training programs, chairs of departments, etc.). When I approached them, they enthusiastically agreed to construct their chapters in a way that would first and foremost address the immediate basic clinical concerns, and typical legal and ethical dilemmas, of the average practitioner and practitioner-in-training. One result is that the book presents a rich overview of contemporary developments in the clinical assessment, management, and treatment of suicidal patients, within the framework of a basic textbook of suicidology for clinical practice. The mosaic of knowledge contained in the book's chapters ranges across the theoretical, empirical, and epidemiological, to applied clinical approaches to working with special populations such as youth and the elderly, to specific information on psychological assessment, pharmacotherapy, dynamic and cognitive approaches, emergency care, hospitalization and its alternatives, postvention, to the realm of training, supervision, and ethical and forensic considerations. Yet each of the chapters shares in its concluding section a commitment to present essential and workable guidelines for practice. For at its heart, this book is an effort to provide a clear picture of the overall knowledge base in the study of suicide for the next generation of practitioners, and to update the existing skill and knowledge level of current mental health and health care professionals.

For supervisors and course instructors, it is also important to note that each of the chapters was designed as a module that covers a specific didactic element in a basic curriculum on suicidology. Each was constructed for use in a practicum, professional practice seminar, or internship/residency setting as the didactic component in teaching fundamental skills in the assessment, treatment, and management of suicidal patients.

The approach that should be taken to the study of suicide is obviously open to debate (Bongar & Harmatz, 1991). However, clinical competence in managing suicidal patients does not mean informational competence alone (Berman, 1986). Indeed, there is research to support the contention that a knowledge of risk factors and the capacity to respond in an effective way to those patients who present as an imminent risk of suicide may be two independent areas of clinical competency (Inman, Bascue, Kahn, & Shaw, 1984). Thus, we must be careful to guard against the conception of training in the study of suicide as a mechanical dissemination of factual knowledge. Those involved in teaching the subject must understand the "crucial importance of consultation and supervision as methods of training and education in suicidology" (Maris, Dorpat, Hathorne, Heilig, Powell, Stone, & Ward, 1973, p. 34). Therefore, a formal course of study in the care of the suicidal patient will need to include clinical supervision, in addition to the sort of didactic material contained in the present volume.

While there are relatively few true emergency situations in mental health practice, of these emergencies, suicide is probably the most compelling. Its unsuccessful assessment or management can be a source of great anguish to the practitioner (Lomax, 1986). Most practicing mental health professionals will see patients in their professional practice activities who meet the profile of an elevated risk for attempted or completed suicide. Clinical wisdom among mental health practitioners "admonishes that it is not a matter of whether one of their clients will someday commit suicide, but of when" (Fremouw, de Perczel, & Ellis, 1990, p. 129). Patient suicide is not a tragedy that exclusively confronts the mental health professions, but "the incidence of its occurrence is such a frequent issue, both clinically and sometimes legally, that it demands special consideration" from mental health professionals (Smith, 1986, p. 62).

In order to meet the challenge of the suicidal patient, clinicians need to approach this population with confidence rather than with fear (Berman & Cohen-Sandler, 1983). Confidence building begins with high quality training, knowledgeable clinical supervision, and the formulation of essential professional guidelines and standards aimed at moving clinical practice beyond merely defensive care (Bongar, 1991). In addition, it is critical that each clinician heed Pope's insight: "perhaps most importantly communicate that you care" (Pope, 1986, p. 20). Although individual practitioners may differ in the ways they demonstrate such caring, an inextricable element is that they truly extend themselves in their management efforts and demonstrate their commitment to quality by insuring that their knowledge of suicidology is current.

While the material contained in this book was designed to help practitioners effectively maximize their efforts to prevent suicide in clinical practice, we also must acknowledge that mental health and health care professionals are neither omniscient nor omnipotent (Bongar, 1991; Hoge & Appelbaum, 1989). When our best efforts as clinicians fail, we share with other suicide survivors a "despair born of death too soon" (Slawson, Flinn, & Schwartz, 1974, p. 63). We dedicate this book to the prevention of that despair.

Stanford, California   B.B.
May, 1992

REFERENCES

Abram, H. S., Moore, G. I., & Westervelt, F. B. (1971). Suicidal behavior in chronic dialysis patients. *American Journal of Psychiatry, 127,* 1199.

Alcohol, Drug Abuse, and Mental Health Administration (1989). *Report of the Secretary's Task Force on Youth Suicide Vols. I–IV.* DHSS Pub. No. (ADM) 89-1621-1624. Washington, D.C.: U.S. Government Printing Office.

Berman, A. (1986). A critical look at our adolescence: Notes on turning 18 (and 75). *Suicide and Life Threatening Behavior, 16*(1), 1–12.

Berman, A. L., & Cohen-Sandler, R. (1983). Suicide and malpractice: Expert testimony and the standard of care. *Professional Psychology: Research and Practice, 14*(1), 6–19.

Bongar, B. (1991). *The suicidal patient: Clinical and legal standards of care.* Washington, DC: American Psychological Association.

Bongar, B. & Harmatz, M. (1989). Graduate training in clinical psychology and the study of suicide. *Professional Psychology: Research and Practice, 20*,4, 209–213.

Bongar, B., & Harmatz, M. (1991). Clinical psychology graduate education in the study of suicide: Availability, resources, and importance. *Suicide & Life-Threatening Behavior, 21*(3), 231–244.

Bongar, B., Peterson, L. G., Harris, E. A., & Aissis, J. (1989). Clinical and legal considerations in the management of suicidal patients: An integrative overview. *Journal of Integrative and Eclectic Psychotherapy, 8*(1), 53–67.

Boyd, J. (1983). The increasing rate of suicide by firearms. *New England Journal of Medicine, 308*, 872–898.

Brown, H. N. (1987). The impact of suicides on therapists in training. *Comprehensive Psychiatry, 28*(2), 101–112.

Chemtob, C. M., Bauer, G. B., Hamada, R. S., Pelowski, S. R., & Muraoka, M. Y. (1989). Patient suicide: Occupational hazard for psychologists and psychiatrists. *Professional Psychology: Research and Practice, 20*(5), 294–300.

Chemtob, C. M., Hamada, R. S., Bauer, G. B., Kinney, B. & Torigoe, R. Y. (1988). Patient suicide: Frequency and impact on psychiatrists. *American Journal of Psychiatry, 145*, 224–228.

Chemtob, C. M., Hamada, R. S., Bauer, G. B., Torigoe, R. Y., & Kinney, B. (1988). Patient suicide: Frequency and impact on psychologists. *Professional Psychology: Research and Practice, 19*(4), 421–425.

Deutsch, C. J. (1984). Self-report sources of stress among psychotherapists. *Professional Psychology: Research and Practice, 15*, 833–845.

Evans, G., & Farberow, N. L. (1988). *The encyclopedia of suicide.* New York: Facts on File.

Frederick, C. J. (1978). Current trends in suicidal behavior. *American Journal of Psychotherapy, 32*(2), 172–200.

Fremouw, W. J., de Perczel, M., & Ellis, T. E. (1990). *Suicide risk: Assessment and response guidelines.* New York: Pergamon Press.

Gutheil, T. G., Bursztajn, H., Brodsky, A., & Alexander, V. (1991). *Decision making in psychiatry and the law.* Baltimore: Williams and Wilkins.

Hirschfeld, R., & Davidson, L. (1988). Risk factors for suicide. In A. J. Frances & R. E. Hales (Eds.) *American Psychiatric Press Review of Psychiatry Vol. 7* (pp. 307–333). Washington, D.C.: American Psychiatric Press

Hoge, S. K., & Appelbaum, P. S. (1989). Legal issues in outpatient psychiatry. In A. Lazare (Ed.) *Outpatient psychiatry* (pp. 605–621). Baltimore: Williams and Wilkins.

Inman, D., Bascue, J., Kahn, W. & Sharp, P. (1984). The relationship between suicide knowledge and suicide interviewing skills. *Death Education, 8*, 179–184.

Kleespies, P. M., Smith, M. R., & Becker, B. R. (1990). Psychology interns as patient suicide survivors: Incidence, impact, and recovery. *Professional Psychology: Research and Practice, 21*(4), 257–263.

Lomax, J. W. (1986). A proposed curriculum on suicide for psychiatric residency. *Suicide and Life-Threatening Behavior, 16*(1), 56–64.

Maltsberger, J. T. (1986). *Suicide risk: The formulation of clinical judgment.* New York: New York University Press.

Maris, R. W. (1981). *Pathways to suicide: A survey of self-destructive behaviors.* Baltimore: Johns Hopkins University Press.

Maris, R. W. (1988). Preface: Overview and discussion. In R. W. Maris (Ed.) (pp. vii–xxiii) *Understanding and preventing suicide: Plenary papers of the first combined meeting of the AAS and IASP.* New York: Guilford.

Maris, R. W. (1989). Preface: Strategies for studying suicide and suicidal behavior. *Suicide and Life Threatening Behavior, 19*(1), ix–x.

Maris, R. W., Dorpat, T. L., Hathorne, B. C., Heilig, S. M., Powell, W. J., Stone, H., & Ward, H. P. (1973). Education and training in suicidology for the seventies. In H. L. P. Resnick

& B. C. Hathorne (Eds.) *Suicide prevention in the seventies* (pp. 23–44). Washington, D.C.: National Institute of Mental Health and Government Printing Office.

Menninger, K. (1938). *Man against himself*. New York: Harcourt Brace Jovanovich.

Peterson, L. G. & Bongar, B. (1989). The suicidal patient. In A. Lazare (Ed.), *Outpatient psychiatry: diagnosis and treatment* (2nd ed.) (pp. 569–584). Baltimore: Williams and Wilkins.

Pokorny, A. D. (1964a). A follow-up study of 618 suicidal patients. *American Journal of Psychiatry, 122*, 1109–1116.

Pokorny, A. D. (1964b). Suicide rates in various psychiatric disorders. *Journal of Nervous and Mental Disease, 139*(6), 499–506.

Pope, K. (1986). Assessment and management of suicidal risks: Clinical and legal standards of care. *Independent Practitioner*, January 1986, 17–23.

Roy, A. (1986). Preface. In A. Roy (Ed.) *Suicide* (p. vii). Baltimore: Williams and Wilkins.

Schein, H. M. (1976). Obstacles in the education of psychiatric residents. *Omega, 7*, 75–82.

Shneidman, E. S. (1981). Psychotherapy with suicidal patients. *Suicide and Life-Threatening Behavior, 11*(4), 341–348.

Shneidman, E. S. (1989). Overview: A multidimensional approach to suicide. In D. G. Jacobs & H. N. Brown (Eds.) *Suicide: Understanding and responding: Harvard Medical School perspectives on suicide* (pp. 1–30). Madison, CT: International Universities Press.

Simon, R. I. (1988). *Concise guide to clinical psychiatry and the law*. Washington, D.C.: American Psychiatric Press.

Slawson, P. F., Flinn, D. E., Schwartz, D. A. (1974). Legal responsibility for suicide. *Psychiatric Quarterly, 42*, 50–64.

Smith, J. (1986). *Medical malpractice psychiatric care*. Colorado Springs, CO: Shepards/McGraw-Hill.

Weissman, M. M., Klerman, G. L., Markowitz, J. S., Ouelette, R. (1989). Suicidal ideation and suicide attempts in panic disorder and attacks. *New England Journal of Medicine, 321*, 1209–1214.

PART I
Theory and Research

1

What Do Suicides Have in Common?
Summary of the Psychological Approach

EDWIN S. SHNEIDMAN

Among Mill's canons for inductive inference, the "method of difference" is probably the most powerful for conducting experiments to generate new knowledge. Even so, a consideration of *similarities* among comprehensive phenomena grouped under a common label (in this case "suicide") can be particularly illuminating. This view permits us legitimately to ask, "What, in common sense terms, is common in suicide?"*

Let us begin by briefly citing two (of several) commonly discussed disparities among suicidal phenomena.

1. *Demographic disparities.* In the history of the study of suicide, the demographic approach (especially since Durkheim) has appeared to be perhaps the most useful and powerful approach. Thus we have innumerable studies of suicide in relation to sex, age, ethnicity, locale—countries, states of the United States, cities, neighborhoods, census tracts—socioeconomic status, occupation, and (the current favorite) discrete psychiatric diagnostic categories. In my view, there is an unseemly sibling rivalry latent in these studies: Norway loses to Italy, men lose to women, San Francisco loses to Los Angeles, adolescents and oldsters lose to the middle-aged, schizophrenics lose to depressives. It is a long and dreary list, as though one party in each pairing were somehow morally inferior to the other. For someone to know the demographic truth that the modal suicide in the United States is a not-currently-married Protestant Caucasian man in his forties with a history of suicide attempts is an interesting fact but of questionable usefulness in its wide-ranged application. Toward the end of my own suicidological career, I now muse that when one cannot pursue phenomena in detailed depth one counts them in large number. It is not a thoroughly adequate substitute.

2. *Disparities of approach.* In the preceding paragraph I have implied two approaches: the nomothetic (tabular, statistical, demographic, arithmetic) approach, and the idiographic (clinical, case-study, personal document, historical,

---

*Portions of this chapter have previously appeared in other publications (Shneidman, 1973, 1985, 1987, 1990a,b, 1991).

3

anamnestic) approach. (The terms *idiographic* and *nomothetic* are those of German philosopher Wilhelm Windelband [1848–1915] introduced to American readers by Gordon Allport [1897–1967].) There are several other approaches however, with perhaps a dozen *ways* to address and attempt to understand suicidal phenomena. The goal, *of course*, is to prevent these phenomena and treat them. The point to be made is that there are *several* avenues to the desired goals, and one should take umbrage only if a partisan of a particular view claims that his or her view is the *only* (or true or basic) way to provide "fundamental data" for understanding suicidal phenomena. Thirteen possible legitimate approaches are listed in Table 1.1. If these approaches can be visualized as a "circle of the etiology of suicide," the reader can understand that the remainder of this chapter focuses on but one sector—the psychological sector—of the total etiological pie. I should make it clear that personally I advocate research in *all* of the sectors, but in this setting I limit myself to the psychological commonalities and leave it to others more qualified to explicate those other legitimate approaches.

In previous publications (Shneidman, 1985, 1987, 1990a,b), I have referred to "the ten commonalities of suicide," specifically, the common psychological features in human self-destruction. They are outlined in Table 1.2 and are discussed briefly below.

1. The common *purpose* of suicide is to *seek a solution*. First, suicide is not a random, pointless, or purposeless act. To the sufferer it seems to be the only available answer to a real puzzler: What to do to get out of this unbearable situation? The purpose is to seek a solution to a perceived crisis—in my view the problem of overwhelming pain—that is generating intense suffering. The French social scientist Baechler (1975) wrote an extended book-length essay on suicide that focused on the assertion that to understand what a suicide is about one must know the psychological problems it was intended to solve. Suicide is problem-solving behavior; it is a practical act. Murray (1938) wrote that suicide, in general, is intended to reduce tension. He stated: "Suicide does not have *adaptive* (survival) value but it does have *adjustive* value for the organism. Suicide is *functional* because it abolishes painful tension." In every case, the "genotypic" (fundamental) purpose of suicide is just as Murray described it; in each case, the "phenotypic" purposes must be determined directly or indirectly from the victim or indirectly by means of an extended "psychological autopsy"—close interviewing with a number of knowledgeable survivors. The general question to a suicidal person is "What is going on?"

2. The common *goal* of suicide is *cessation* of consciousness. Suicide is both a moving toward and a moving away. The common practical goal of suicide is to stop the painful flow of consciousness. Suicide is best understood not so much as a movement toward a reified Death, as it is in terms of "cessation"—the complete (and irreversible) stopping of one's consciousness of unendurable pain. Murray (1967) stated: "Their intention was an urgently felt need to stop unbearable anguish . . . an action to put an end to intolerable affects." Furthermore, as Litman (personal communication, 1990) pointed out, each suicide

Table 1.1. Contemporary Approaches to the Study of Suicidal Phenomena

Sector	Key references	Principal assertion: suicide is best understood
Life history	Allport (1942) Murray (1967) Runyon (1982)	As an episode in a long life history and that precursors of the suicidal death can be seen in previous patterns of response to comparable life situations
Personal documents	Allport (1942) Shneidman & Farberow (1957) Leenaars (1988)	Through the analysis of such personal documents as letters, diaries, autobiographies, and especially, suicide notes
Demographic; epidemiological	Graunt (1662) Sussmilch (1741) Dublin (1963) Hollinger & Offer (1986)	In terms of "census" data, e.g., statistics for sex, age, race, religious affiliation, marital status, socioeconomic status
Systems theoretical	Blaker (1972) Miller (1978) Tyler (1984)	As an act within a living system, both the individual and the individual-within-society as living systems
Philosophical; theological	Pepper (1942) Fowles (1964) Choron (1972) Battin (1982)	In answers to questions such as: What is the purpose of life? Are there forces beyond ourselves? What is our relation to the universe?
Sociocultural	Hendin (1964) Lifton et al. (1979) Iga (1986)	In terms of sociocultural data, such as knowledge of various countries and cultures (e.g., Sweden, Japan)
Sociological	Durkheim (1897) Douglas (1967) Maris (1981)	In terms of an individual's relationship to the society (e.g., estrangement from it, ties to it)
Dyadic and family	Pfeffer (1986) Richman (1987)	In terms of the stressful interaction between two people or within a family nexus
Psychiatric	Kraepelin (1883) APA *DSM*-3-R (1987)	In terms of mental illnesses (e.g., depression, schizophrenia, alcoholism)
Psychodynamic	Freud (1910) Zilboorg (1937) Menninger (1938) Litman (1967)	In terms of unconscious conflicts, especially unconscious hostility toward the father; suicide seen as unconscious murder
Psychological	Murray (1967) Shneidman (1985)	In terms of psychological pain, produced by the frustration of psychological needs
Constitutional	Roy (1986) Kety (1986)	As an expression of inborn (constitutional or genetic) factors.
Biological; biochemical	Bunney & Fawcett (1965) Asberg et al. (1986)	As a result of biochemical imbalances in the body fluids (blood) or organs (brain)

Table 1.2. Ten Commonalities of Suicide

1. Common *purpose* of suicide is to seek a *solution*
2. Common *goal* of suicide is *cessation of consciousness*
3. Common *stimulus* in suicide is *intolerable psychological pain*
4. Common *stressor* in suicide is *frustrated psychological needs*
5. Common *emotion* in suicide is *hopelessness-helplessness*
6. Common *cognitive state* in suicide is *ambivalence*
7. Common *perceptual state* in suicide is *constriction*
8. Common *action* in suicide is *egression*
9. Common *interpersonal act* in suicide is *communication of intention*
10. Common *consistency* in suicide is with *life-long coping patterns*

Source: *Definition of Suicide*, E. Shneidman, 1985, New York: John Wiley & Sons. By permission.

involves not only pain but the individual's *unwillingness* to tolerate that pain, the decision not to endure it, and the active will to stop it. It means that in psychotherapy one must focus on the amelioration of the pain as well as the "character" which (who) has chosen not to tolerate it.

3. The common *stimulus* in suicide is unendurable psychological *pain*. In any close analysis, suicide is best understood as a combined movement toward cessation *and* a movement away from intolerable, unendurable, unacceptable anguish. It is psychological pain of which we are speaking: metapain, the pain of feeling pain. From a traditional psychodynamic view, hostility, shame, guilt, fear, protest, and longing to join a deceased loved one have singly and in combination been identified as root factors in suicide. It is, in fact, none of these factors; rather, it is the pain involved with any or all of them together with the unwillingness to endure that idiosyncratically defined pain. Psychological pain is the center of suicide and constitutes the chief hurdle that must be lowered before any kind of therapy can be life-saving. There is a basic clinical rule: Reduce the level of suffering, often just a bit, and the individual can chose to live.

It should be clear by now that in my view the key to the psychological understanding of suicide is the psychological pain the individual idiosyncratically experiences in that situation in life. Kelly (1961) wrote that each suicidal act reflects an individual's own private (unique, idiosyncratic) epistemology—his or her *personal construct* of the world—and that a suicide is an individual's effort to validate his or her life. Kelly (1961) wrote:

Under what circumstances of outlook would anyone seek death rather than life? By stating our question in this way, we shall be inquiring into a psychological principle rather than a simple situational element. . . . To understand this act we must know what he died for. What personal constructions of life and truth, what anticipations formulated from his structured world seemed to be validated by so drastic an act . . . a suicide attempt can be regarded as an act designed to validate one's life.

What words other than "psychological pain" might we use? We might think of *experiential* pain or *circumstantial* pain, i.e., the experiencing of unbearably painful circumstances. I avoid the confusing term *existential* pain, which has become an intellectually fashionable word relating to a variety of dour philosophical orientations. When I speak of psychological, experiential, or circumstantial pain I do not mean to assert that there is some abstract unbearable philosophical pain of Being but, rather, to state the more modest and obvious fact that *some* relatively *rare* circumstances, experiences, or states of being appear—especially to a perturbed and constricted mind—to be hopelessly unbearable, and that suicide, through escape, can occur to such a mind as an available solution. No dentist or physician can find the somatic root for this kind of pain, but it can be as real as the most excruciating toothache and 10,000 times more dangerous. In clinical practice, suicide prevention is psychological pain management.

4. The common *stressor* in suicide is frustrated psychological *needs*. The psychological pains that are central to suicide are driven, created by, and sustained by frustrated, blocked, or thwarted psychological needs. Suicide is best understood not so much as an unreasonable act—every suicide seems logical to the person who commits it (given that person's major premises, styles of syllogizing, and the constricted focus)—as it is a reaction to unfulfilled psychological needs.

To understand suicide in this context, we must ask a broader question, which I believe is the key: What purposes do most human acts (including suicide) intend to accomplish? The best nondetailed answer to that question is that, in general, human acts are intended to satisfy a variety of human *needs*. There is no compelling, a priori reason why a typology (or classification or taxonomy) of suicidal acts might not parallel a general classification of human needs. Such a classification of psychological needs can be found in Murray's *Explorations in Personality* (1938). There, in a discussion of human needs, is a ready-made, viable taxonomy of the essential underpinnings of suicidal behaviors. Table 1.3 presents a partial listing of psychological needs.

Most suicides represent combinations of various needs, so a particular suicide might properly be understood in terms of two or more need categories. There are many pointless deaths but never a needless suicide. When one addresses the frustrated needs, the pain they cause is lessened; and the suicide does not occur. The therapist's function is to decrease the patient's acute discomfort. One way to operationalize this task is to focus on the thwarted needs. Questions such as "Where do you hurt?" can be useful for clarifying the suicidal picture.

The assessment of needs related to suicide may be a more complicated epistemological task than identification of the other commonalities of suicide. Psychological needs are more conceptual, more of an abstraction, and there is more inference necessary to name them. One requires access to a detailed anamnestic record or case history for the raw data from which inferences can reasonably be drawn. The use of the language of psychological needs immediately and radically changes the way in which one conceptualizes an individual's

Table 1.3. Partial Listing of the Murray Psychological Needs

Abasement: To submit passively to external force. To accept injury, blame, criticism, punishment. To surrender. To become resigned to fate. To admit inferiority, error, wrongdoing, or defeat. To confess and atone. To blame, belittle, or mutilate the self. To seek and enjoy pain, punishment, illness, and misfortune.

Achievement: To accomplish something difficult. To master, manipulate, or organize physical objects, humans, or ideas. To do this as rapidly and independently as possible. To overcome obstacles and attain a high standard. To excel oneself. To rival and surpass others. To increase self-regard by the successful exercise of talent.

Affiliation: To draw near and enjoyably cooperate or reciprocate with an allied other (who resembles the subject or who likes the subject). To please and win affection of a cathected object. To adhere and remain loyal to a friend.

Aggression: To overcome opposition forcefully. To fight. To attack or injure another. To oppose forcefully or punish another.

Autonomy: To get free, shake off restraint, break out of social confinement. To resist coercion and restriction. To avoid or quit activities prescribed by domineering authorities. To be independent and free to act according to desire. To defy convention.

Counteraction: To master or make up for a failure by restriving. To obliterate a humiliation by resumed action. To overcome weakness; to repress fear. To efface a dishonor by action. To seek for obsctacles and difficulties to overcome. To maintain self-respect and pride on a high level.

Defendance: To defend the self against assault, criticism, blame. To conceal or justify a misdeed, failure, or humiliation. To vindicate the ego.

Deference: To admire and support a superior. To praise, honor, or eulogize. To yield eagerly to the influence of an allied other. To emulate an exemplar. To conform to custom.

Dominance: To control one's human environment. To influence or direct the behavior of others by suggestion, seduction, persuasion, or command. To dissuade, restrain, or prohibit.

Exhibition: To make an impression. To be seen and heard. To excite, amaze, fascinate, entertain, shock, intrigue, amuse, or entice others.

Harmavoidance: To avoid pain, physical injury, illness, and death. To escape from a dangerous situation. To take precautionary measures.

Infavoidance: To avoid humiliation. To quit embarrassing situations. To avoid conditions that lead to scorn, derision, or indifference of others. To refrain from action because of fear of failure.

Inviolacy: To protect the self. To remain separate. To resist attempts of others to intrude upon or invade one's own psychological space. To maintain a psychological distance. To be isolated, reticent, concealed; immune from criticism.

Nurturance: To give sympathy and gratify the needs of another person, especially an infant or someone who is weak, disabled, tired, inexperienced, infirm, defeated, humiliated, lonely, rejected, sick, mentally confused. To feed, help, support, console, protect, comfort, nurse, heal; to nurture.

Order: To put things or ideas in order. To achieve arrangement, organization, balance, tidiness, and precision among things in the outer world or ideas in the inner world.

Play: To act for "fun" without further purpose. To like to laugh and make jokes. To enjoy relaxation of stress. To participate in pleasurable activities for their own sake.

Rejection: To separate oneself from a negatively cathected object. To exclude, abandon, expel, or remain indifferent to an inferior object. To snub or jilt another.

Sentience: To seek and enjoy sensuous experience. To give an important place to creature comforts of taste and touch and the other senses.

Succorance: To have one's needs gratified by the sympathetic aid of another person. To be nursed, supported, sustained, protected, loved, advised, guided, indulged, forgiven, consoled, taken care of. To remain close to a devoted protector. To always have a supporter.

Understanding: To ask and answer questions. To be interested in theory. To speculate, formulate, analyze, and generalize. To want to know the answers to general questions.

Source: *Explorations in Personality* by H. A. Murray, New York: Oxford Press, 1938, Chapter 3, pp. 142–242. Adapted by permission.

suicidal behavior and the ways in which one can think to help the suicidal person.

Furthermore, one must distinguish between (1) the disposition of needs with which that individual ordinarily lives, which can be rated simply by assigning 100 points among the 20 needs, the use of a "constant sum" (in this case 100); and (2) the disposition of needs causing the frustration regarding which that individual is willing (indeed believes it necessary) to die. It is precisely when these two (omnipresent) lists diverge that an individual is suicidal. In any event, it is those (transiently) blocked or thwarted psychological needs that beg to be addressed, even if only partially.

5. The common *emotion* is *helplessness-hopelessness*. In the suicidal state there is a pervasive feeling of helplessness-hopelessness: "There is nothing I can do [except to commit suicide], and there is no one who can help me [with the pain I am suffering]." Underlying all of the emotions—hostility, guilt, shame—is the emotion of active, impotent ennui, the feeling of helplessness-hopelessness. The most effective way to reduce elevated lethality is to reduce the elevated perturbation made by the feelings of helplessness-hopelessness. One person who committed suicide said: "All things cease to shine [and] the rays of hope are lost."

6. The common *cognitive state* toward suicide is *ambivalence*. The (non-Aristotelian) accommodation to the psychological realities of mental life—simultaneous contradictory feelings (such as love and hate toward the same person)—is called *ambivalence*. It is the common internal attitude toward suicide: to believe that one has to do it and at the same time to yearn (even to plan) for rescue and intervention. The therapist uses this ambivalence and plays for time so the affect rather than the bullet can be discharged.

7. The common *perceptual* state is *constriction*. Suicide is not best understood as a psychosis, a neurosis, or a character disorder (100% of suicides are highly perturbed). It is more accurately seen as a more-or-less transient psychological constriction of affect and intellect. Synonyms for constriction are a tunneling, focusing, or narrowing of the range of options usually available to that individual's consciousness when the mind is not panicked into dichotomous thinking: Either some specific (almost magical) total solution *or* cessation: *Caesar aut nihil* (all or nothing). The range of life choices has narrowed to two—not much of a range. The usual life-sustaining images of loved ones are not even within the mind. The fact that suicide is committed by individuals who are in a special constricted condition suggests that no one should ever commit suicide while disturbed. It takes a mind capable of scanning a range of options (comprising more than two) to make a decision as important as taking one's life. At the outset, it is vital to counter the suicidal person's constriction of thought by widening the mental blinders and increasing the number of options beyond only two dichotomous options of either a magical resolution or being dead. The word for which the therapist should be alert is "only." "There's only one thing to do . . . is to jump off something good and high."

8. The common *action* in suicide is *escape* (egression). Egression is a person's

intended departure from a region of distress. Suicide is the ultimate egression, beside which all others (running away from home, quitting a job, deserting an army, leaving a spouse) pale. We need to give the person other exits, other ways of doing something. (We also need to close certain lethal exits, such as removing a gun.) This notion is related to expanding the constriction.

9. The common *interpersonal act* in suicide is *communication* of intent. Perhaps the most interesting finding from large numbers of psychological autopsies of suicidal deaths is that in most cases (80%) there were clear verbal or behavioral clues to the impending lethal event. Individuals intent on committing suicide, albeit ambivalent about it, consciously or unconsciously emit signals of distress, wimpers of helplessness, or pleas for response, or they provide opportunities for rescue in the usually dyadic interplay that is an integral part of the suicidal drama. The common interpersonal act of suicide is paradoxical communication of intent with the usual verbal and behavioral clues.

10. The common *consistency* in suicide is with lifelong *coping patterns*. With suicide we are initially thrown off the scent because suicide is an act that, by definition, the individual has never done before, so there seem to be no precedents. Yet there are some deep consistencies with lifelong coping patterns. We must look to previous episodes of deep perturbation, distress, duress, threat, and the capacity to endure psychological pain in order to find paradigms of egression in that person's life.

It is now possible to combine these ten items into a more succinct theoretical model. Imagine a cube (Figure 1.1) made up of 125 cubelets: 25 squares on a plane and 5 cubelets in each row and each column. I call these three faces of the cube (and the three components of this theoretical model) pain, perturbation, and press, defined as follows.

1. *Pain.* Pain is represented on the front of the cube. It refers to the psychological pain resulting from thwarted psychological needs. The left column of cubelets represents little or no pain, the next column some bearable pain, and so on until the right-most column represents intolerable psychological pain.

2. *Perturbation.* Perturbation, assigned to the side plane of the large cube, is a general term meaning the state of being upset or perturbed. Ranked here on a 5-point rating scale, perturbation includes *everything* in *DSM-III-R*. In relation to suicide, perturbation includes (1) perceptual constriction, and (2) a proclivity for precipitous self-harming or inimical action. *Constriction* refers to a reduction of the individual's perceptual and cognitive range. At its worst, the individual reduces the viable options to two and then to one. On this plane in the cube, the bottom row reflects open-mindedness, wide mentational scope, and relatively clear thinking. The top row reflects constriction of thought, tunnel vision, and a narrowing of focus to few options, with cessation, death, and egression as one (and ultimately the only) solution to the problem of pain and frustrated needs. *Penchant for action* refers to something akin to impulsiveness: a tendency to get things over with, to bring them inappropriately to a quick resolution, to have little patience and low tolerance for open and stressful situations, to jump to conclusions, and to grasp at opportunities for more

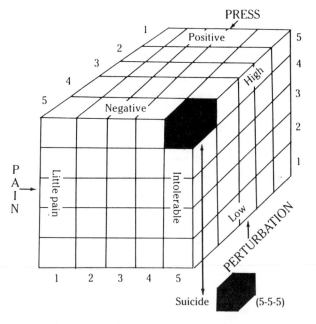

Figure 1.1 Theoretical cubic model of suicide.

rapid resolution. The lowest column of cubelets reflects a high tolerance for ambiguity, a capacity for patience and waiting. As we move upward on this plane, we come closer to the tendency for precipitous and consumatory action, this lethal impulsivity, which is what the topmost column on the side plane represents.

3. *Press.* Murray (1938) may have had the word *pressure* in mind when he decided to use the word *press* (pl. press) to represent those aspects of the inner and outer world, or environment, that impinge on, move, touch, or affect the individual and make a difference. "Press" refers to the events done to the individual (and the way they are incorporated and interpreted) to which he or she reacts. There is positive press (good genes and happy fortune) and negative press (conditions or events that perturb, threaten, stress, or harm the individual). It is the latter that are relevant to suicide. Press includes both actual and imagined events in the sense that everything is mediated by the mind. In this cubic model, press ranges from positive (back row of the top plane) to negative (front row of the top plane).

In this theoretical cube of 125 cubelets there is only one cubelet that is numbered 5–5–5. Not every individual who is in a 5–5–5 cubelet commits suicide: He or she may commit homicide, "go crazy," become amnesic, or destroy a career; but in this conceptual model no one commits suicide except those in the 5–5–5 cubelet: maximum pain, maximum perturbation, maximum negative press. The implications for therapy and life-saving redress seem obvious: Reduce any of these three dimensions from a 5 to a 4 (or less). Preferably, reduce all three. Individuals can lead long, unhappy lives as 4s and 3s, but the specific goal in suicide prevention is to remove the individual from the 5–5–5 cubelet, to save the life. All else—demographic variables, family history, suicidal history—is peripheral except as those and other factors bear on the presently felt pain, perturbation, and negative press.

To place this presentation of the psychological approach to suicide in a wider perspective we can speak of a variety of fields of knowledge that have a legitimate interest in suicidology (Table 1.1). Each of these approaches can contribute to our knowledge of suicidal phenomena. In this same sense, any one of them errs when it claims that its sector represents the whole circle of etiology or that it is the principal way to understand suicide. Furthermore, proponents of all the approaches must be vigilant that they do not confuse *concomitance*— events (in the brain, the blood, society, interpersonal relationships, within the psyche) that occur at the same time as heightened suicidality—with *causality*, events that necessarily precede the onset of a suicidal state and because of their causal connection present us with reasonable avenues to effective prevention.

Arguably, the best documented contemporary case of suicide currently in the public domain is contained in the diary of Arthur Crew Inman, as edited by Harvard professor Daniel Aaron (1985). Aaron reduced Inman's 160 typed volumes, covering the years 1895 to 1963, to two volumes, a mere 1600 pages. Inman, an independently wealthy Boston recluse, was an addicted, dictatorial, loathsome, hypochondriacal, politically reactionary (he wrote about presidents from Wilson to Kennedy and particularly hated FDR), anti-Semitic, adolescent-molesting, difficult son and husband—a death-bound man with unusual insights and capacities for social commentary and self-revelation. The point is not that the contents are unfailingly unattractive; the point is that the record is full. The diary materials are supplemented by Inman's physician's reports and other materials. All in all, the published Inman diaries constitute a gold mine of sociological and psychological ore and are a major resource for serious students of suicide, and they are available to everyone with access to an adequate library or from Harvard University Press. Given the above, earnest suicidologists are in the position to conduct studies of their own devising.

Let me suggest one example: Using the set of Murray psychological needs, one could, for any particular time in Inman's life, rate his psychological need constellation. This rating could be accomplished by distributing 100 points among the 20 needs. There are needs with which any individual, Inman in this case, lives; and there are needs, specifically in suicidal individuals toward the end of their lives, for which they are willing to die. This disposition of psychological needs within an individual is what, in large part, defines his "personality" and is therefore related to his suicide. With every suicide there are certain critical psychological needs, the frustration of which is not tolerated.

The assessment of the need-scope at various points throughout Inman's life would permit us to see the development, ebb and flow, growths, and regressions of his personality over time, actually from the beginning of his diary (in 1919) to the time of his suicide by gunshot in 1963. In addition—because we know from the beginning that there is going to be a suicidal outcome—we can study something of the psychological genesis of this particular suicide and, by reasonable extrapolation, suicide in general.

If several scholars could, under a general, mutually agreed upon framework, independently undertake such a task, their subsequent meetings and discussions of their ratings and understandings would provide a heuristically dynamic symposium. The possibilities for new understandings would be heightened if the prepared participants represented a variety of disciplines: psychology, internal medicine, psychiatry, sociology, epidemiology, history, and political science. It would also be most desirable for the raters to have done successive Q sortings (Block, 1961; Shneidman, 1984 a,b; Stephenson, 1953) of Inman's general personality characteristics. If this were done, it would permit the differences among successive need evaluations by different raters to be related to the differences of those raters' general perceptions of Inman's personality. Such exercises would permit the researchers to focus on both the substrata and the acute drama of suicide-within-a-life—on both the suicidal career and the acute penultimate suicidal drama.

The alert reader can recognize the suggestion outlined above as borrowed directly from Murray's "assessment council" (1938), which involves the coming together of several senior experts representing different scientific disciplines who have examined the "same" corpus of material on the *same* case. It is an obviously powerful technique, underused largely because of the practical and logistical difficulties involved in arranging it, but these factors, in my opinion, do not constitute sufficient reasons for earnest investigators to eschew it. I hope that it is one general avenue of investigation that future suicidological activities will someday explore.

REFERENCES

Aaron, D. (Ed.) (1985). *The Inman diary: public and private confession.* (Vol. 1 & 2). Cambridge, MA: Harvard University Press.

Allport, G. (1942). *The use of personal documents as psychological science.* New York: Social Science Research Council.

American Psychiatric Association (1987). *Diagnostic and statistical manual, third edition, revised.* Washington, DC: American Psychiatric Association.

Asberg, M., Nordstrom, P., & Traskman-Bendz, L. (1986). Biological factors in suicide. In A. Roy (Ed.), *Suicide.* Baltimore: Williams & Wilkins.

Baechler, J. (1975/1979). *Suicides.* New York: Basic Books (originally published as *Les Suicides,* 1975).

Battin, M. P. (1982). *Ethical issues in suicide.* Englewood Cliffs, NJ: Prentice-Hall.

Blaker, K. P. (1972). Systems theory and self-destructive behavior. *Perspectives in Psychiatric Care, 10,* 168–172.

Block, J. (1961). *The Q-sort method in personality assessment and psychiatric research.* Springfield, IL: Charles C. Thomas.

Bunney, W. E., & Fawcett, J. A. (1965). Possibility of a biochemical test for suicide potential. *Archives of General Psychiatry, 13,* 232–239.

Choron, J. (1972). *Suicide.* New York: Scribner.

Douglas, J. D. (1967). *The social meaning of suicide*. Princeton: Princeton University Press.

Dublin, L. I. (1963). *Suicide: a sociological and statistical study*. New York: Ronald Press.

Durkheim, (1897/1951). *Suicide*. Glencoe, IL: Free Press.

Fowles, J. (1964). *The aristos*. Boston: Little, Brown.

Freud, S. (1910/1967). Comments on suicide. In P. Friedman (Ed.), *On suicide*. New York: International Universities Press.

Graunt, J. (1662). *Natural and political annotations . . . upon the Bills of Mortality*. London.

Hendin, H. (1964). *Suicide in Scandinavia*. Orlando, FL: Grune & Stratton.

Hollinger, P. C., & Offer, D. (1986). *Sociodemographic, epidemiologic and individual attributes of youth suicides*. Bethesda: Department of Health and Human Services.

Iga, M. (1986). *The thorn in the chrysanthemum: suicide and economic success in modern Japan*. Berkeley: University of California Press.

Kelly, G. A. (1961). Suicide: the personal construct point of view. In N. Farberow & E. Shneidman (Eds.), *The cry for help*. New York: McGraw-Hill.

Kety, S. (1986). Genetic factors in suicide. In A. Roy (Ed.), *Suicide*. Baltimore: Williams & Wilkins.

Kraepelin, E. (1883/1915). *Textbook of psychiatry*. Berlin.

Leenaars, A. (1988). *Suicide notes*. New York: Behavioral Sciences Press.

Lifton, R. J., Kato, S., & Reich, M. R. (1979). *Six lives/six deaths*. New Haven: Yale University Press.

Litman, R. E. (1967). Sigmund Freud on suicide. In E. Shneidman (Ed.), *Essays in self-destruction*. New York: Science House.

Maris, R. (1981). *Pathways to suicide*. Baltimore: Johns Hopkins University Press.

Menninger, K. A. (1938). *Man against himself*. New York: Harcourt Brace.

Miller, J. G. (1978). *Living systems*. New York: McGraw-Hill.

Murray, H. A. (1938). *Explorations in personality*. New York: Oxford University Press.

Murray, H. A. (1967). Dead to the world: the passions of Herman Melville. In E. Shneidman (Ed.), *Essays in self-destruction*. New York: Science House.

Pepper, S. (1942). *World hypotheses*. Berkeley: University of California Press.

Pfeffer, C. (1986). *The suicidal child*. New York: Guilford Press.

Richman, J. (1987). *Family therapy for suicidal individuals*. New York: Springer-Verlag.

Roy, A. (Ed.), (1986). *Suicide*. Baltimore: Williams & Wilkins.

Runyon, W. M. (1982). *Life histories and psychobiography: explorations in theory and method*. New York: Oxford University Press.

Shneidman, E. (1973). Suicide. In *Encyclopaedia Britannica*. Chicago: William Benton.

Shneidman, E. (1980). *Voices of death*. New York: Harper & Row.

Shneidman, E. (1984a). Suicide and suicide prevention. In R. J. Corsini & B. D. Ozaki (Eds.), *Encyclopaedia of psychology*. New York: Wiley.

Shneidman, E. (1984b). Personality and "success" among a selected group of lawyers. *Journal of Personality Assessment, 48*, 609–616.

Shneidman, E. (1985). *Definition of suicide*. New York: John Wiley & Sons.

Shneidman, E. (1987). A psychological approach to suicide. In G. R. VandenBos & B. K. Bryant (Eds.), *Cataclysms, crises and catastrophes: psychology in action*. Washington, DC: American Psychological Association.

Shneidman, E. (1990a). A life in death: notes of a committed suicidologist. In C. E. Walker (Ed.), *A history of clinical psychology in autobiography*. Pacific Grove, CA: Brooks/Cole.

Shneidman, E. (1990b). The commonalities of suicide across the life span. In A. Leenaars (Ed.), *Life span perspectives of suicide*. New York: Plenum.

Shneidman, E. (1991). A conspectus of the suicidal scenario. In R. Maris, A. Berman, J. Maltsberger, & R. Yufit (Eds.), *Assessment and prediction of suicide*. New York: Guilford Press.

Shneidman, E., & Farberow, N. (Eds.), (1957). *Clues to suicide*. New York: McGraw-Hill.

Stephenson, W. (1953). *The study of behavior: Q-technique and its methodology*. Chicago: University of Chicago Press.

Sussmilch, J. (1741). *Die Gottliche Ordnung in den Veranderungen des Menschlichen Geschlechts, aus der Gerburt, dem Tode, und der Fortsflanzug desselben erwiesen.* Berlin: J. C. Spenser.

Tyler, L. (1984). *Thinking creatively.* San Francisco: Jossey-Bass.

Zilboorg, G. (1937). Considerations in suicide with particular reference to that of the young. *American Journal of Orthopsychiatry, 7,* 15–31.

2

Review of Empirical Risk Factors for Evaluation of the Suicidal Patient

DAVID C. CLARK AND JAN FAWCETT

This chapter is devoted to a consideration of strategies for assessing risk for *completed* suicide, in contrast to risk for *nonfatal* suicide attempts, and to application of empirical knowledge about suicide to estimate the severity of suicide risk. As we show, persons at risk for imminent suicide may differ in important respects from those at high risk to die by suicide in the more distant future, for example, 1 to 5 years later. Because one of the primary tasks of the clinician is to distinguish *emergent* suicide risk from other problems so the clinician can focus attention, time, and therapeutic efforts on keeping that patient alive for another few days or weeks until the suicidal crisis has passed or the risk of suicide has been reduced, we emphasize the difference between profiles associated with *imminent* and *long-term* suicide risk as gleaned from retrospective and prospective studies.

Basic Epidemiological Patterns

Suicide is the eighth leading cause of death in the United States as of 1989, accounting for more than 30,000 deaths each year (1.4% of all deaths) (National Center for Health Statistics, 1992). This statistic is grounds for concluding that suicide is a major public health problem. Because only 12.2 of every 100,000 persons dies by suicide each year, however, "suicide is an infrequent event in anyone's clinical practice, so the danger is easily forgotten" (Murphy, 1986b, p. 572). Annual suicide rates and numbers for 1989 by sex, white/nonwhite groups, and young/elderly groups are outlined in Table 2.1.

Three times more men than women die by suicide in the United States annually, and the suicide rate for whites is two times higher than that for nonwhites (i.e., African-Americans and Hispanics). When these rates are applied to the actual number of white and nonwhite persons in the United States, one finds that 72% of those dying by suicide each year are white males, 19% are white females, 7% are nonwhite males, and 2% are nonwhite females (National Center for Health Statistics, 1992).

Whereas suicide rates for adolescents have increased threefold since 1955,

16

Table 2.1. 1989 U.S. Data on Completed Suicide

Population	Annual rate per 100,000	Actual number
Nation as a whole	12.2	30,232
Males	19.9	24,102
Females	4.8	6,130
Whites	13.1	27,424
Nonwhites	7.1	2,808
Young (15–24 years)	13.3	4,870
25 Years to 64 years	14.9	19,134
Elderly (65+ years)	20.1	6,228

Source: National Center for Health Statistics, 1992. Advance report of final mortality statistics, 1989. *NCHS Monthly Vital Statistics Report 40*, suppl 2.

the rates for the elderly (those 65 years and over) have decreased just as dramatically over four decades. Despite the dramatic extent of these changes by age group, it continues to remain true that those 15 to 24 years old have a lower suicide rate and those over age 60 have a higher suicide rate than those between the ages of 25 and 55. The age group differences in suicide rate apparent during the 1980s are slight compared to the pronounced age group differences during the 1930 to 1950 era, when the elderly evidenced rates three times higher than those for other age groups. Today the age of the person being evaluated contributes little useful information to the estimation of suicide risk (Holinger, 1987).

Kreitman (1988) showed that young persons who have been widowed evidence the highest suicide rate of any marital status group, but that this higher rate associated with young widowed persons decays with age, so that in older adult widowed cohorts the suicide rate falls to a rate equivalent to that for single persons. The death of a spouse may be a superordinate disruption for the young, who have been preparing for the psychological expansion of life, not its abrupt termination, and who are least prepared for the burden of grieving.

Apart from the narrow category of young widowed persons, parental status is probably a better demographic marker for increased suicide risk than marital status. Veevers (1973) was the first to demonstrate that the variable *marital status* functions as a proxy for the variable *responsibility for children under age 18 years*, which is not generally codified in vital statistics data. According to this theory, the presence of children under age 18 years in a home imputes some parental responsibilities, and these responsibilities (not marital status) are associated with a low risk of death by suicide. The few replication studies reported comparing the relations among marital status, parental status, and suicide risk support Veever's hypothesis (Fawcett et al., 1987; Lindesay, 1989; Smith et al., 1988).

Nonfatal suicide attempt rates, on the other hand, reveal a different pattern (Table 2.2). About 2.9% of the general adult population have made a suicide attempt. Suicide attempt rates are three times higher for women than men.

Table 2.2. Lifetime Prevalence of
Suicide Attempts in the United States

Population	Prevalence (%)
Adults aged 18 years and over	2.9
Males	1.5
Females	4.2
18–24 Years	3.4
25–44 Years	4.0
45–64 Years	2.1
65+ Years	1.1
Non-Black/non-Hispanic	3.0
Black	2.3
Hispanic	3.3
Married	2.0
Never married	2.9
Separated/divorced	8.5
Widowed	2.0

Source: Suicide attempts in the epidemiologic catchment area study by E. K. Moscicki et al., 1988, *Yale Journal of Biology and Medicine, 61*, 259–268. Reprinted by permission.

Suicide attempt rates are higher for those aged 25–44 years and lower after age 45 years when compared to persons in the 18- to 24-year-old age range. The rates are four times higher for persons separated or divorced than persons in other marital status categories, and the rates show little difference by race or ethnicity (Table 2.2) (Moscicki et al., 1988).

Completed Suicide in the Absence of a Psychiatric Disorder is Rare

Suicides rarely occur in the absence of major psychopathology. Community-based psychological autopsy studies of 100 or more suicides in the United States, United Kingdom, Sweden, and Australia using structured symptom interviews to elicit data converge on the conclusion that a recent major psychiatric disorder is implicated in no less than 93% of cases of adult suicide (Barraclough, Bunch, Nelson, & Sainsbury, 1974; Beskow, 1979; Chynoweth, Tonge, & Armstrong, 1980; Dorpat & Ripley, 1960; Fowler, Rich, & Young, 1986; Robins, Gasser, Kaye, Wilkinson, & Murphy, 1959a). The diagnoses implicated most often have been major depression (40–60% of cases), chronic alcoholism (20%), and schizophrenia (10%) (Clayton, 1985; Murphy, 1986). Major depression and alcoholism account for 57 to 86% of all suicides. These findings are summarized in Table 2.3. These diagnostic rates may be conservative underestimates because most psychological autopsy studies used a hierarchically organized diagnostic scheme that permits a maximum of one psychiatric diagnosis per subject. One study (Dorpat & Ripley, 1960) found that 9% of persons dying by suicide evidenced character pathology but no psychiatric disorder.

Table 2.3. Rates of Psychopathology Identified in Community-Based Psychological Autopsy Studies of Completed Suicide.

	Robins et al., 1959	Dorpat & Ripley, 1960	Barraclough et al., 1974	Beskow, 1979	Chynoweth et al., 1980	Rich, Fowler, & Young, 1986
Sample size	134	114	100	271	135	283
Diagnostic classification						
Affective disorder	45%	30%	70%	28%	55%	44%
Alcoholism	23%	27%	15%	31%	20%	54%
(Affective disorder or substance use disorder)	(69%)	(57%)	(86%)	(65%)	(89%)	(~69%)
Schizophrenia	2%	12%	3%	~3%	4%	3%
Organic brain syndrome	4%	4%	0%	2%	not reported	4%
Character pathology & no other diagnosis	0%	9%	0%	4%	3%	not reported
Mentally ill but unspecified diagnosis	19%	16%	6%	0%	not reported	3%
Mentally well	2%–6%	0%	1%	3%	12%	5%

To consider the relation between major psychiatric disorder and suicide from another vantage point, follow-up mortality studies of patients afflicted with major affective disorder, alcoholism, and schizophrenia consistently demonstrate these patients are at elevated risk for death by suicide (Allebeck, 1989; Black, Warrack, & Winokur, 1985; Drake, Gates, Whitaker, & Cotton, 1985; Martin, Cloninger, Guze, & Clayton, 1985; Miles, 1977; Murphy & Wetzel, 1990; Tsuang, Woolson, & Fleming, 1980) often at rates higher than the lifetime risk for those with a history of suicide attempts (Cullberg, Wasserman, & Stefansson, 1988; Pokorny, 1966). Thus the lifetime risk for death by suicide in the U.S. population as a whole is 1.4%, and that for persons treated for a suicide attempt is approximately 7%, whereas the rates for persons with a diagnosis of major depression, schizophrenia, or alcoholism are estimated to be 15%, 10% and 3%, respectively. There is also good evidence to suggest that children and adolescents with a history of psychiatric inpatient care die by suicide at higher rates than children and adolescents in the general community (Kuperman, Black, & Burns, 1988).

Although it would be simplistic (and erroneous) to think that these specific psychiatric disorders "cause" suicide in and of themselves, there is empirical justification for thinking that a suicide rarely if ever occurs in the *absence* of a major psychiatric disorder. This theory, of course, runs contrary to lay theories that suicide is caused by acute life stress and problems (Sawyer & Sobal, 1987) and highlights the fact that mental health professionals have not done a good job of educating the public about the nature of psychopathology or the relation between psychopathology and suicide.

If we begin by accepting the premise that suicide risk is relatively low among persons free of psychiatric disorder and relatively high among persons meeting diagnostic criteria for a psychiatric diagnosis, two important implications for assessing suicide risk follow. First, the most efficient strategy for detecting persons at high risk for completed suicide is to screen for psychopathology and then concentrate more closely on those persons with major depression, alcoholism or drug use disorders, schizophrenia, organic brain syndromes, and high-risk psychiatric disorders. Second, *all persons meeting diagnostic criteria for a psychiatric diagnosis should be handled as persons liable to attempt suicide* until a more refined, second-stage evaluation clarifies whether the risk posed is high or low, and whether it is imminent or long term. We argue later that the constituent risk factors of high and imminent suicide risk may vary from one psychiatric diagnosis to another. Thus psychodiagnostic and suicide risk evaluations should necessarily be considered interrelated and complementary undertakings.

It is important to remember, however, that two-thirds of all persons meeting diagnostic criteria for a psychiatric disorder never make contact with a health professional and never seek health care services (Myers et al., 1984). Psychological autopsy studies have confirmed the important observation that most persons at high risk for suicide by virtue of major depression, alcoholism, drug abuse/dependence, or schizophrenia do not pursue evaluation or treatment by a mental health professional; in fact, 43 to 48% of persons who died by suicide in community-based studies had never seen a mental health professional in their entire lifetime (Barraclough et al., 1974; Beskow, 1979; Hagnell & Rorsman, 1978, 1979). Only 25 to 30% of persons who die by suicide are under the care of a mental health professional at the time of their death.

We now recommend ways to explore suicidal tendencies and estimate suicide risk in the context of a clinical interview by an experienced mental health professional.

Detecting Suicidal Thoughts and Gauging Their Intensity

The most fruitful assessment interviews are conducted in a comfortable, private, quiet location at a time when neither the interviewer nor the patient feels rushed or pressed for time. Ideally, the interviewer begins by asking the patient to provide an overview of the current dilemma. The interviewer listens quietly without interrupting while the patient explains his or her greatest concerns in his or her own words. By setting aside the first 5 to 10 minutes of the interview to grasp the patient's chief complaints, his or her conceptualization of the problem (e.g., situational, interpersonal, or intrapsychic), and his or her word choices, the interviewer often elicits an overview of the dilemma that helps to structure inquiry for the remainder of the interview and creates an opportunity to study the patient's pace and style of talking. Interviewers should slow down their own talking speed (a little) with patients who talk slowly in order to be sure the patients fully understand all the questions and interpretations posed.

Because most patients evidencing psychopathology are at elevated risk for suicide attempts or death by suicide, all mental status examinations and psychodiagnostic evaluations should routinely cover questions about suicidal thinking and behavior. In particular, patients who meet diagnostic criteria for a major depressive disorder, alcoholism, drug abuse or dependence, or schizophrenia should be asked about suicidal thinking and behavior each time an assessment is warranted. Routine questioning about suicidal phenomena serves two purposes: It often elicits examples of recent suicidal ideation or behavior by the patient, and it alerts the patient that the interviewer continues to be interested in hearing about suicidal thoughts and behavior over time.

One of the most effective ways to introduce the topic of suicide in a way that helps patients reveal and expand on their experiences is to progress from the patients' characteristic dysphoric states to symptoms of hopelessness and then to morbid thinking and suicide. Thus for the patient who has complained of feeling "empty" and "anguished," the next series of questions might be: "Have you been discouraged/felt hopeless? What kind of future do you see for yourself? Can you see yourself or your situation getting any better?" (working from the SADS interview) (Endicott & Spitzer, 1978). After gauging the patient's degree, frequency, and duration of hopelessness, the interviewer might then ask: "When persons feel empty or anguished, they may think about dying or even killing themselves. Have you?" (Endicott & Spitzer, 1978).

The interviewer who has a matter-of-fact approach to questions about morbid and suicidal thinking makes it easier for the patient to volunteer information and probably elicits more complete data. Some interviewers harbor anxieties about posing these questions for a variety of reasons: Some are inexperienced and are not confident they know what to do if the patient should announce an intention to die by suicide; some do not want to hear suicidal symptoms for fear they must bear legal responsibility for the patient; some become overstimulated or personally frightened by the patient's suicidal thoughts; and some simply do not like to work with suicidal patients. For this reason it is crucial that clinical training curricula and internships address the issue of suicide crisis response and familiarize interviewers with their private responses to a patient's expressed suicidal wish. The interviewer who is unprepared to hear that wish or who is too frightened by that wish cannot assess the degree of suicide risk posed and is unable to implement a therapeutic intervention.

When exploring the quality and severity of suicidal thoughts, the interviewer's first task is to establish whether the patient has had any thoughts of death or suicide. Patients who respond "no" too quickly or firmly may be embarrassed by the question, may be eager to establish (in their own mind as well as that of the interviewer) that they are not having suicidal thoughts *now* or may be eager to establish that they would never *act* on such thoughts. If interviewers suspect any of these types of response, they should disarm the patient by explaining that there are *degrees* of suicidal thoughts and behaviors, so it is important to gauge the severity of the patient's own thinking on a scale of relativity. For example, the interviewer begins by asking: "When persons feel empty or anguished, they may think about dying or even killing themselves. Have you?" The patient stiffens and replies curtly: "No, I'm Catholic." The interviewer then might say reassuringly: "Sometimes even Catholic people, when they are feeling *this* empty and anguished, have thoughts of death or suicidal ideas. The thoughts may come and go. Have you had any?"

If the patient has had thoughts of death, the interviewer's second task is to establish whether they have been limited to passive thoughts of death (e.g., "I would be better off dead," "I wish I were dead"). If the patient has had thoughts about taking his or her own life or envisioned potentially suicidal acts, the third step is to determine if the patient has ever considered a specific method of suicide.

If the patient has *not* considered a specific method, the fourth step is to ask the patient point-blank if he or she intends to follow through with the suicidal ideas by implementing them in the near future, and why or why not. At this point in the evaluation it is a mistake to inject any note or moral persuasion against suicide. The object is to sample the patient's own thinking processes without trampling on the phenomena of interest. The risk of "muddying the water" with premature or overzealous admonitions against suicide is great and serves only to defeat a thorough evaluation. Patients do not tend to interpret this simple probing of their intentions as incitement or encouragement to suicide. It is particularly valuable to listen to patients list reasons why they believe suicide is unlikely, as these reasons often constitute a psychologically meaningful catalogue of their inhibitory mechanisms opposing the suicide option. Reasons patients list, when asked, why they believe they will not take their own life sometimes include fear of self-injury or pain, responsibility for dependent children, reluctance to expose family members to a suicide, and a wish to be present at some meaningful future event. The influence of these inhibitory mechanisms or "counterweights" should not be overestimated, but they should be figured into the equation the clinician is developing about the patient's status to estimate suicide risk.

If the patient *has* considered a specific method, on the other hand, the fourth step is to inquire about all methods of suicide the patient has considered, listening for or asking about how frequently the plan is considered, the degree to which the patient has mentally rehearsed the plan, and if the patient has translated any of these thoughts into action, however tentative. It is important to ask patients about what *other* methods of suicide they have considered each time they finish answering the question until the interviewer is certain the patient has described every last plan and method. It is not unusual to find that a cagey patient will produce the most carefully rehearsed (in thought and deed) plan *last*, willing to discuss the most serious plan only because the interviewer's diligent and thorough inquiry has won the patient's begrudging respect and trust.

The line distinguishing suicidal thoughts from behavior is often an ambiguous one but must be considered carefully when evaluating suicide risk. The following are examples of patients who have crossed a line separating mental consideration/rehearsal from action: (1) the man who confesses that he picked up all his dry cleaning on a day not ordinarily set aside for that errand "so his affairs would all be in order if he killed himself;" (2) the woman who visited the open roof of a tall building so she could peer over the ledge, envisioning where she would fall if she jumped; and (3) the inpatient who was found still in his bed and grinning foolishly during morning rounds with a shoelace tied neatly around his neck in the fashion of a bow tie. As is made clear later in this chapter, we do not believe that persons who think and talk about suicide in the absence of suicidal behavior are *unlikely* to attempt suicide, nor do we believe that most suicidal deaths come from the ranks of those who have shown some prior suicidal behavior. We do believe, however, that persons who bridge the gap from suicidal thoughts to suicidal behavior evidence less psychological ability to inhibit mental impulses.

For these reasons, the concept of the *suicidal gesture* is a misleading one. The phrase implies that some suicidal behaviors are more innocuous than others by virtue of the fact that they were more tentative, more symbolic, more communicative, or less potentially lethal. We believe this perspective is a dangerous fallacy because it implied that people who "dabble" in suicide will continue to "dabble" without really hurting themselves. This assumption is dangerous because it encourages clinicians to lower their vigilance about suicide risk at a time when the patient is actively experimenting with suicidal behaviors. Many patients intent on killing themselves make a relatively harmless "gesture" first; and at first blush it is difficult or impossible to know whether their intention was to attract help or to build courage for the "real thing."

The clinician should be cautious about inferring subconscious or unconscious suicidal tendencies from physically reckless behavior, however dangerous or repetitive. Though death is too often the final common pathway for both suicide and self-endangering behavior, there is evidence—at least among adolescents—that physical recklessness and suicidal behavior represent different phenomena (Clark, Sommerfeldt, Schwarz, & Watel, 1990).

As a fifth and final step, the interviewer should always explore whether the patient's thinking about death or suicide has encompassed other persons. It is important to remember that about 5% of all those who kill another person in the United States each year die by suicide or make a suicide attempt soon after the homicide (Rosenbaum, 1990; Wolfgang, 1958), about 3 to 4% of all suicides occur in the context of a combination murder-suicide (Beskow, 1979; Dorpat & Ripley, 1960; Robins et al., 1959a), and about 20% of all those who kill a child subsequently die by suicide or make a suicide attempt (Adelson, 1961; Rodenburg, 1971). In cases of combination murder-suicide, the perpetrator is most commonly a husband who kills his wife and children before taking his own life, or a wife who kills her children before taking her own life. Severely depressed persons are often surprised and relieved to find that the interviewer inquires about the degree to which suicidal thinking has also implicated other persons or family members and generally discuss the vicissitudes of their homicidal ideas in a forthright manner. A surprisingly large proportion of parents develop homicidal ideas about their children in parallel with their own suicidal ideas, though most can successfully suppress the homicidal ideas in order to protect their children.

The clinical assumption underlining the five steps outlined above is that risk of imminent suicide varies in direct proportion to the step reached by inquiry: Persons with no thought of death are a lowest risk, those with nonsuicidal thoughts of death are at low risk, those having suicidal thoughts without a specific method are at elevated risk, and those with a method for suicide in mind are at highest risk. Among those who have considered a specific method for suicide, degrees of mental rehearsal and penetration of the thought/action barrier help define increments of high risk.

It should be unnecessary to add that whenever a patient verbalizes suicidal wishes and whenever the clinician believes that suicide risk is more than negligible, the clinician should determine if instruments for suicide (e.g., guns, medications, poisons) are readily available to the patient and, if so, have them removed from the patient's environs. This step ordinarily requires the clinician to discuss suicide risk and accessible instruments of suicide with family members as well as the patient. By bringing family members into the evaluation, the clinician is able to obtain more complete and reliable information and to educate family members about the nature and meaning of suicide risk.

This general strategy for gauging the intensity of suicidal thoughts ia good beginning point for assessing suicide risk, and it has been outlined for convenience in Table 2.4. This strategy should not be considered sufficient or reliable in isolation. *The interviewer should never be lulled into a false sense of security when a patient denies having suicidal ideas or when the patient describes mild suicidal thoughts.* In one prospective study of patients with major affective disorders, for example, more than half the patients who died by suicide within 6 months of a thorough and structured assessment reported mild or no suicidal thoughts to an experienced clinical interviewer (Fawcett et al., 1987, 1990).

Suicidal Communications to Family Members and Friends

More than a half-dozen community-based psychological autopsy studies of completed suicide in different parts of the world involving large samples have

Table 2.4. Detecting and Gauging Suicidal Ideation by Steps

Step 1: Listen.	Listen to patient's conceptualization of his/her problem.
Step 2: Pinpoint dysphoric affects.	Inquire about sad, apathetic, anhedonic, irritable, anxious (etc.) mood; learn the patient's vocabulary for these states; assess the frequency/intensity/duration of these states.
Step 3: Evaluate hopelessness.	Inquire whether the dysphoric mood states are associated with discouragement, pessimism, or hopelessness. Assess the frequency/intensity/duration of hopelessness.
Step 4: Probe for morbid thoughts. [If there are none, skip to Step 10. If there are some, then . . .]	Inquire about any thoughts of death or suicide.
Step 5: Passive thoughts of death only? [If yes, skip to Step 10. If no, then . . .]	Are symptoms limited to *passive* thoughts of death?
Step 6: Specific methods of suicide contemplated? If no, continue. If yes, skip to step 8 . . .	Has the patient considered any *specific* methods of suicide? Continue to probe for different methods.
Step 7: Likelihood of suicide? [Then skip to Step 10.]	Ask the patient if he/she intends to follow through on the suicidal ideas in the near future, and why or why not.
Step 8: Assess all suicidal plans.	For each method of suicide considered, find out how often the plan is contemplated, the degree to which the patient has mentally rehearsed the plan, and if the plan has been translated into any actions.
Step 9: Probe for homicidal thoughts.	Inquire whether the patient's thinking about death or suicide has encompassed other persons.
Step 10: Interview family members and intimates for evidence of morbid or suicidal thinking.	Educate family members and intimates about patient's illness and about suicide risk when appropriate. Inquire about any morbid or suicidal communications, however vague or indirect.

agreed that more than 40% communicated their suicidal intent in clear, specific terms, and another 30% communicated a wish to die or their preoccupations with death to several persons over a course of weeks to months immediately preceding their death (Barraclough et al., 1974; Beskow, 1979; Chynoweth et al., 1980; Dorpat & Ripley, 1960; Fowler, Rich, & Young, 1986; Hagnell & Rorsman, 1978, 1979, 1980; Rich, Fowler, Fogarty, & Young, 1988; Rich, Young, & Fowler, 1986; Robins, 1981; Robins et al., 1959a; Robins, Murphy, Wilkinson, Gassner, & Kayes, 1959b).

Although these findings underline a nonimpulsive quality of the final suicidal act and a willingness to communicate, it must be remembered that this information was gleaned in retrospect. A surprising or odd verbalization about "being better off dead" may have had a different meaning to a surviving family member interviewed 2 weeks before than it does 2 weeks after the suicide. A research finding serves to amplify this point: In a psychological autopsy study of completed suicide by 72 persons aged 65 years and over just concluded, we found that 48% of the suicide victims expressed explicit thoughts of suicide and another 26% repeatedly expressed thoughts of death during the last months of life to family members and close friends; yet in 89% of the cases all informants reported being completely surprised by the act of suicide.

What are some possible reasons that so many family members and intimates could hear and remember suicidal communications after the fact, yet did not respond to them as genuine or emergent at the time they were spoken? A person may not always find it possible to consciously recognize a loved one's suicidal anguish, especially in situations where the potential responder feels ill-equipped to intervene or access help. A family member may not be able to reconcile the facts that a previously healthy loved one is at the present moment expressing suicidal wishes that sound truly alien or weird, so it becomes simpler to ignore or dismiss the incongruous observations. Sometimes conscious recognition of a loved one's suicidal tendencies gives rise to feelings of personal inadequacy or guilt in the significant other, so that person maintains his or her own psychological equilibrium by not pursuing clues or fully realizing the facts. Moreover, sometimes a family member or intimate friend harbors ambivalent or mixed feelings about a person who becomes increasingly demanding, dependent, irritable, thoughtless, selfish, or withholding as his or her depressive and suicidal state intensifies, so there is some real temptation to call the sick person's bluff and dare him or her to commit suicide. Whatever the reason, it seems clear that it is difficult for family members or close friends to recognize symptoms of a suicidal crisis or respond to suicidal communications.

How does one reconcile the facts that patients who die by suicide within months of a structured psychiatric assessment often report little or no suicidal ideation (Fawcett et al., 1987, 1990), but that more than two-thirds of all suicide completers talked with family members and close friends about their suicidal or morbid thoughts? Our clinical experience suggests that patients at risk for imminent suicide are much more likely to discuss suicidal impulses and plans openly with those closest, that is, spouses, parents, children, lovers, and close friends (Fawcett, 1969). The same patients are much less likely to discuss suicidal impulses and plans with other family members, casual friends, coworkers, and health professionals. Thus we recommend that *the interviewer conducting a suicide risk assessment routinely contact and interview several family members and close friends*, inquiring specifically about examples of morbid preoccupations and suicidal thoughts.

Routine questioning of relatives and friends serves four purposes: It provides an opportunity to educate them about suicide risk in general and the content of suicidal communications in particular; it often elicits examples of recent suicidal ideation or behavior by the patient; it conveys the message that the interviewer is interested in the informants' observations; and it alerts the informants that the interviewer will continue to be interested in hearing about suicidal thoughts and behavior over time.

When eliciting the history of a young person for purposes of a suicide risk assessment, however, parents may not always be knowledgeable or reliable informants about prior suicide attempts. Walker, Moreau, and Weissman (1990) examined the reliability of mothers as informants about the past suicidal behavior of their offspring aged 6 to 23 years by conducting independent face-to-face structured interviews. Mothers reported that their child had made a suicide attempt in only one-third of the cases where the child reported having made a suicide attempt; rarely did a mother report an attempt not volunteered by the child. Cases of suicide attempts reported by children but not by their mothers tended to be more severe than cases reported by both in terms of age at first attempt, number of lifetime attempts, and the deliberateness of the attempt.

Smoke Screen Posed by Situational or "Appropriate" Depressions

Murphy (1975) showed that in a group of 60 suicide victims who had recently been under the care of physicians, the patient's depressive mood was recognized before suicide occurred in most cases. In 80%, however, depressive symptoms were interpreted as a "natural reaction" to situational life events of an adverse nature and not as an illness that required specific treatment. Thus the physicians tended to deal with piecemeal symptoms (e.g., insomnia) and failed to recognize the depressive syndrome as an illness requiring comprehensive management.

Community-based psychological autopsy studies have consistently shown that although more than half of all suicide victims had seen a physician during the last month of their life, about 60% had never seen a mental health professional in their entire life, not even once (Barraclough et al., 1974; Beskow, 1979; Chynoweth et al., 1980; Dorpat & Ripley, 1960; Fowler et al., 1986; Robins et al., 1959b). The implication is that *persons about to die by suicide do not recognize their own psychological symptoms, nor do they appreciate the psychological nature of their dilemma, focusing instead on physical symptoms and physical health concerns.* The problem is compounded by severely depressed patients' tendency to lose insight about their condition as one common feature of the illness: Many severely depressed patients are unaware that they are depressed and complain only about their somatic or anxiety symptoms.

The wish to die is rarely the result of a clear-headed decision based on the quality of life one faces in the future (Brown, Henteleff, Barakat, & Rowe, 1986; Whitlock, 1986). Community-based psychological autopsies of suicide have consistently demonstrated that only 2 to 4% of all suicide victims were terminally ill at the time (Barraclough et al., 1974; Beskow, 1979; Chynoweth et al., 1980; Robins et al., 1959b). Clinicians familiar with severely or terminally ill medical patients know that the small number who express suicidal thoughts or the wish to die change their minds many times; it would be a mistake to allow the patient to make an irreversible decision to die, just as it would be a mistake to accept the patient's premise of helplessness and hopelessness without an adequate assessment for the presence of treatable depressive illness.

Furthermore, many believe that the elderly have other understandable "reasons" for contemplating or implementing suicide (e.g., deteriorating health, severe illness, outliving family members and peers, loneliness, isolation). These beliefs lead some professionals to question the utility of spending time or health care resources on persons so temporally close to death by natural causes in the absence of suicide. The findings from community studies, however, consistently show that these attitudes and beliefs are not based in fact (Younger, Clark, Oehmig-Lindroth, & Stein, 1990). Social isolation, poverty, deteriorating health, and severe illness rarely if ever stand alone (i.e., in the absence of a psychiatric illness) as "causes" of suicidal thinking among the elderly. The great bulk of elderly persons in the general population are not depressed, and most do not develop suicidal thoughts (Moscicki et al., 1988; Myers et al., 1984).

The common tendency for clinicians and laypersons to overlook the diagnosis of clinical depression when "reasons" for experiencing a depressed mood are present often leads to the omission of psychotherapeutic and psychopharmacological treatment measures that might alleviate the severity of a depressive illness, alleviate functional impairment, and reduce suicide risk. Although the importance of precipitating stressful events cannot be ignored in the assessment and treatment of a suicidal patient, the importance of these events is too often permitted to overshadow the fact that *patients in episodes of "situational" depressive illnesses are no less likely to die by suicide than patients in episodes of "endogenous" or "vegetative" depressive illnesses* (Black, Winokur, & Nasrallah, 1987; Fawcett et al., 1987, 1990).

Given that primary care health providers treat a far greater proportion of persons with psychiatric disorders in the community than mental health professionals (Blacker & Clare, 1987; Regier, Goldberg, & Taube, 1978; Shapiro, Skinner, Dramer, Steinwachs, & Regier, 1985) and given that primary care providers fail to recognize a substantial proportion of the psychopathology posed by their patients (Blacker & Clare, 1987; Marks, Goldberg, & Hillier, 1979; Skuse & Williams, 1984; Verhaak, 1988), it seems that continued efforts to educate the general internist and family physician concerning the authenticity of psychiatric disorders, the importance of diagnosis, the importance of assessing suicide risk, the appropriate treatment of depressive syndromes, and the importance of making referrals to mental health specialists are warranted.

Our warning is that stressful life events impinging on the patient in recent months or interpreted as precipitants of the current disorder or crisis have an enormous potential to distract the evaluator from recognizing the authenticity of a depressive illness, the severity of a patient's depressive symptoms, and the acuity of suicide risk. The fact is that depression is an illness and can be treated even though the precipitant is obvious or understandable. In this regard Fawcett's aphorism (1972) is useful to remember: "The presence of a *reason* for depression does not constitute a reason for ignoring its presence."

"Siren Song" Quality of Severe Suicidal Preoccupations and Implications for Verbal Interventions

Most patients experiencing an episode of major depression develop morbid thoughts or suicidal ideas, however fleeting or tentative. Ideas about death and suicide are simply one of the ubiquitous features of a depressive illness. The subset of depressed patients who develop persistent, compelling, or convincing suicidal thoughts, on the other hand, are difficult to characterize with any uniformity. Some suicidal persons are placed in jeopardy by the severity of their illness state or their sense of acute hopelessness. Others may have less native capacity to inhibit violent or self-destructive impulses that are fanned or magnified by their illness state. Still others may have a long history of preoccupations with matters of death and afterlife that colors their bout of acute psychiatric illness.

Whatever the explanation or underlying cause, patients who experience suicidal thoughts with a persistent, compelling, or convincing quality can be conceptualized as persons responding to a "siren song," one with deceptively alluring appeal. By evoking the mythic siren song, we mean to emphasize both the seductive and the irresistible qualities of the "suicide solution" for the most severely ill patients, who experience nothing but intolerable and unrelieved pain from horizon to horizon. Ulysses, mythical king of Ithaca and leader of the Greeks in the Trojan War, had the foreknowledge to strap himself to the mast of his ship and press wax into his oarsmen's ears in order to resist responding to the siren song luring mariners to shipwreck on the rocky shoals surrounding the sea nymphs' island. An important aspect of the story is that he did not possess the capacity to resist their sweet singing—not even the hero Ulysses.

The siren song wove a web of illusions and false promises hinting of bliss but delivering sure death. The siren song drove good judgment and all other practical, common-sense thoughts out of a sailor's head, leaving him intent on one purpose: to embrace the sea nymph. The siren song turned the listening sailor into an ardent, heartsick suitor rushing toward the arms of Death with no ability to hear any other voice.

By the same token, many persons in acute suicidal crisis (including those with considerable psychological depth and resources when not in an episode of acute psychiatric illness) do not have the capacity to resist the internal logic, the emotional tug, and the pain relief offered by the solution of suicide. These patients temporarily lose all sense of the impact their death would have on beloved others. Sometimes listening to patients describe their experience of suicidal thoughts, the irresistible siren song quality is evident. More often, the siren song effect can be inferred from patients' strenuous efforts to rationalize, romanticize, or otherwise justify their suicidal preoccupations by donning a cloak of intellectualism, philosophical detachment, existentialism, cynicism, religious fervor, or even pop culture, as when adolescents latch onto a song lyric that seems to "explain" or "justify" their own suicidal or self-destructive impulses.

The point is that once suicidal preoccupations have reached a level of intensity here portrayed metaphorically as a siren song, there is great danger that patients have become irrefutably convinced of the sense of value of their suicidal ideas, so that all verbal interventions (i.e., interpretations, discussions, negotiations, pleas) are rendered ineffective. We do not think one can reason reliably with persons in acute suicidal crisis any more than one can reason with a person who believes God is sending personal messages via advertising billboards or one who believes (medical evidence to the contrary) that a tumor is causing death. In this kind of acute and severe crisis, the patients' verbal assurances are not sufficient to convince us that they can resist the siren song. Our skepticism about the momentary value of verbal interventions or psychotherapy is greater yet in situations where the patient, already swaying to the beat of the siren song, has also been using alcohol or other psychoactive drugs (any quantity or frequency) that have the potential to disinhibit. In these situations the clinical equivalent of "lashing the sailor to his mast" would be to ensure unremitting 24-hour supervision, usually in the form of psychiatric hospitalization, and to institute those psychotherapeutic, psychopharmacological, or other treatment measures that might be expected to alleviate the severity of depressive symptoms or other underlying illness states.

Sometimes patients who need to be hospitalized for their own protection oppose the decision to hospitalize. Although it is preferable to appeal to the family for help in persuading the patient to check into the hospital voluntarily or for help with involuntary commitment procedures, they are often reluctant or side with the patient. In these situations clinicians who have strong and demonstrable reasons for believing that suicide risk is appreciable must do everything within their professional and legal powers to protect the patient's life. It means that sometimes involuntary commitment proceedings must be instituted when there has been no explicit suicidal threat, when the patient is opposed to hospitalization, or when the patient or family members threaten legal retribution. In retrospect after a patient death by suicide, failure to hospitalize or explore every avenue for hospitalization exposes the clinician to far more legal jeopardy than a well-reasoned decision to implement involuntary commitment. If the patient cannot be hospitalized for a short period for whatever reason, the patient should be seen every day and the clinician should involve the patient's family in daily conferences and treatment planning. With luck and time, the emergency psychological and psychopharmacological treatment measures instituted on admission to the hospital protect patients from their own suicidal impulses and gradually reduce symptom severity, diminishing acute suicide risk to the point where verbal psychotherapy can be resumed.

Significance of a History of Nonfatal Suicide Attempts

The interviewer should systematically ask about all prior suicide attempts during the course of a suicide risk evaluation. The object is to discern if the patient has any history of translating suicidal thoughts into palpable behavior and to understand how the patient has managed similar crises in the past. A history of nonfatal suicide attempts is associated with a higher risk for subsequent nonfatal attempts (Buglass & Horton, 1974; Clark, Gibbons, Fawcett, & Scheftner, 1989; Kreitman & Casey, 1988; Morgan, Barton, Pottle, Pocock, & Burns-Cox, 1976) but is not a reliable index of risk for completed suicide in and of itself.

Psychological autopsy studies of community-based populations have consistently shown that only 18 to 38% of those who die by suicide had ever attempted suicide on a prior occasion (Barraclough et al., 1974; Dorpat & Ripley, 1960; Rich et al., 1986; Robins, 1981). Turning this observation around, long-term follow-up studies of persons who have made nonfatal suicide attempts show that only 7 to 10% eventually die by suicide (Cullberg et al., 1988; Ettlinger, 1964; Motto, 1965; Weiss & Scott, 1974). Because 1.4% of the U.S. population dies by suicide in the course of a lifetime (National Center for Health Statistics, 1992), the 7 to 10% risk associated with a history of nonfatal suicide attempts is clearly five to six times greater than that for the general population. Thus a history of nonfatal attempts constitutes one definition of a group at elevated risk for completed suicide, although 90 to 93% of all attempters never go on to die by suicide during their lifetime.

We do not intend to minimize the plight of the attempter or the danger of completed suicide among those with a history of nonfatal attempts. Instead, we want to emphasize the fact that there are many demographic and clinical differences between completers and nonfatal attempters (Linehan, 1986). Men die by suicide three to four times more often than women, but women make nonfatal attempts three to four times more often than men (Murphy, 1986b). Prevalence rates for major psychiatric disorder are much higher among suicide completers, and prevalence rates of personality disorder are much higher among nonfatal attempters (Murphy, 1986b). Living conditions, social circumstances, and acute life upsets appear to play a much greater role in shaping nonfatal attempts than death by suicide (Murphy, 1986b; Robins, 1986). The appropriate conclusion to draw is one of caution about generalizing from one group to the other.

Failures of Empirical Risk Prediction Models

Murphy (1972, 1983, 1984), MacKinnon and Farberow (1975), Pokorny (1983), Motto, Heilbron and Juster (1985), and Clark, Young, Scheftner, Fawcett, and Fogg (1987) have demonstrated that when trying to predict a behavior as infrequent in the general population as death by suicide, it is inevitable that a large number of false positives and false negatives are generated under the best of circumstances, effectively precluding the clinician from decisive emergency intervention in specific cases. MacKinnon and Farberow (1975) demonstrated that even if one began with a suicide prediction formula that showed 99% sensitivity, something far beyond present capabilities, still only 20% of suicides could be correctly predicted. Furthermore, no one has ever demonstrated that any standardized suicide risk prediction scale can pick out persons who go on to die by suicide in samples beyond the sample that generated the scale (Clark et al., 1987). Pokorny (1983) and Motto and colleagues (1985) suggested that the best one can do is to identify *groups* of persons at elevated risk without being able to pinpoint precisely the persons who are likely to die, reminding us that there will always be persons not identified as being at risk by screening criteria who will nonetheless go on to die by suicide.

Proposal for a Diagnostic Hierarchy to Guide Empirical Suicide Risk Evaluations

An alternative way to bring optimal precision to the task of suicide risk prediction is to tailor risk prediction formulas to identify suicide completers narrowly defined and not the wider band of completers and nonfatal attempters considered together. We have already suggested that risk assessment formulas meant to identify persons likely to make nonfatal attempts, multiple nonfatal attempts (Kreitman & Casey, 1988; Linehan, 1986), or medically serious nonfatal attempts differ qualitatively from those characterizing suicide completers.

A second way to bring optimal precision to the task of suicide risk prediction is to tailor risk assessment formulas to the principal psychiatric diagnosis implicated (Clark, 1990). In this regard, Pokorny (1983) may have been first to propose that a diagnostically organized hierarchy has the potential to refine clinical thinking and improve our ability to predict suicide. We conceptualize the final behavioral outcome of suicide as the product of a complex multivariate equation and hypothesize that one necessary variable in that equation is a major psychiatric disorder. The remainder of the equation defining death by suicide may take different forms, depending on whether the critical diagnostic variable is major depression, alcoholism, schizophrenia, or some other disorder. Studies that shed light on clinical features associated with greater risk for death by suicide within a specific diagnosis include ones focusing on major affective disorders (Black et al., 1987, 1988; Fawcett et al., 1987, 1990; Scheftner et al., 1988), alcoholism (Berglund, 1984; Motto, 1980; Murphy & Wetzel, 1990; Murphy, Armstrong, Hermele, Fischer, & Clendenin, 1979), and schizophrenia (Allebeck, 1989; Allebeck, Varla, Kristjansson, & Wistedt, 1987; Breier & Astrachan, 1984; Drake & Cotton, 1986; Drake et al., 1984, 1985; Westermeyer & Harrow, 1989).

Although there is likely to be some overlap among these diagnosis-specific suicide risk profiles, they may be substantively more different than alike, in which case it would be premature or erroneous to screen all patients using a single all-purpose risk profile. In its ambition to serve all situations, an all-purpose risk scale might well fail to be useful in any situation.

Assessing Suicide Risk in the Context of a Major Affective Disorder

Approximately 15% of persons treated for major depression ultimately die by suicide (Guze & Robins, 1970; Miles, 1977), as opposed to a U.S. general population rate of only 1.0 to 1.4%. Noncycling manic disorder is associated with a relatively low risk for suicide (Black et al., 1988). Note that the lifetime risk associated with a diagnosis of major depression is higher than that characterizing persons who have ever made a suicide attempt (7–10%, cited above).

Needle-in-a-Haystack Dilemma

Suicidal ideation is a common symptom among those afflicted with a major depression. If the interviewer works in a clinical setting that tends to receive severely depressed or impaired persons, such as an emergency room or inpatient psychiatric unit, most of the patients in need of evaluation may have suicidal thoughts and as many may have made a recent suicide attempt. In addition to a depressive illness, many of these patients may evidence a history of substance abuse or be acutely intoxicated at intake. In these settings, how does the clinician decide which high-risk patients are at *extreme* risk for imminent suicide requiring immediate attention and a concentration of treatment resources, and which high-risk patients need not be supervised as closely? The problem is made more difficult by the facts that imminent suicide risk is not directly associated with the severity of the depression by traditional measures (e.g., the Hamilton Depression Scale) or with global indices of functional impairment (Goodwin & Guze, 1989).

Anxious Depressions

In a prospective study of 954 psychiatric patients entering treatment for an affective disorder at five university medical centers, we and our colleagues

(Fawcett, 1988; Fawcett et al., 1990) found that most affective diagnostic categories (e.g., bipolar type, schizoaffective type, endogenous subtype depression, psychotic subtype depression), symptom features (e.g., suicidal ideation, history of suicide attempts, medical seriousness of previous attempts, severity of depressed mood, low self-esteem, guilt, weight loss), and psychosocial circumstances (e.g., sex, age, marital status, acute life stress) had little value for "predicting" which patients would die by suicide over the next year. These subjects had been selected from routine admissions to the participating psychiatric hospitals and received the usual care available; subjects were not recruited for a treatment study, and the investigators did not supervise their care. Within this sample of patients already at high risk for suicide by virtue of a major depressive illness and suicidal thoughts, it therefore appeared that the clinical markers commonly thought to define greater suicide risk were no longer discriminating.

Furthermore, our data showed that symptom features associated with death by suicide within 1 year of assessment (*short term*) tended to be different from the features associated with death by suicide 1 to 10 years later (*long term*). This finding is consistent with the oft-demonstrated finding that the incidence of completed and attempted suicide (both) is highest during the first 6 months following a hospital admission (Clark et al., 1989; Clayton, 1985), then levels off after 1 year to assume a steady, level rate.

In the sample of patients with major affective disorder, the following symptoms were associated with short-term suicide risk (Table 2.5): severe psychic anxiety, severe anhedonia, global insomnia, diminished concentration, indecision, acute overuse of alcohol, panic attacks, and obsessive-compulsive features (the latter two symptoms in the context of a major depression and not as an independent diagnosis) (Fawcett et al., 1990). Absence of responsibility for children under the age of 18 years, fewer lifetime episodes of major depression, fewer friendships during adolescence, and a current episode of cycling affective illness (i.e., cycling between depression and mania/hypomania without

Table 2.5. Risk Factors Among Persons with Major Depressive Disorder

Short-term risk for suicide (i.e., within 6–12 months)	Long-term risk for suicide (i.e., 1–10 years later)
Severe psychic anxiety	Acute suicidal ideation
Severe anhedonia	History of suicide attempts
Global insomnia	Severe hopelessness
Diminished concentration	Diminished concentration
Indecision	Indecision
Acute overuse of alcohol	
Panic attacks	
Obsessive-compulsive features	
Current episode of cycling affective illness	
One to three lifetime episodes of depression	
Absence of children under age 18 years in the home	
Absence of friendships during adolescence	

periods of recovery between) also characterized those who died by suicide within 12 months of assessment. The anxiety-spectrum symptoms (severe psychic anxiety, panic attacks, obsessive-compulsive features, overuse of alcohol) could be interpreted as an index of depression severity or as an index of what Shneidman (1985) felicitously called "perturbation." The emergence of panic and anxiety as short-term risk factors may converge with the findings of Coryell and colleagues (Coryell, Noyes, & Clancy, 1982; Coryell, Noyes, & House, 1986) of high rates of suicide in patients with panic disorders and with the findings of Weissman and colleagues (Weissman, Klerman, Markowitz, & Ouellette, 1989; Johnson, Weissman, & Klerman, 1990) of a positive relation between high rates of suicide attempts and panic disorder.

Some of the short-term risk predictors, particularly psychic anxiety and panic attacks, may be more amenable to modification than others. Thus the clinician managing a severely anxious patient with major depression may want to target the potentially modifiable anxiety symptoms for immediate pharmacological intervention, in addition to beginning the indicated regimen of antidepressant therapy (which may well take 2 to 3 weeks to yield evidence of clinical improvement), as one strategy for lowering the degree of imminent risk.

The following symptoms were associated with suicide 1 to 10 years (i.e., long-term suicide) after the initial evaluation: acute suicidal ideation, history of suicide attempts, and severe hopelessness. Symptoms of diminished concentration and indecision were associated with completed suicide in both the short and long term. The marked differences between these short- and long-term risk profiles for depressed patients have led us to recommend that clinicians systematically distinguish between the profile of symptoms defining imminent suicide risk (measured in terms of hours, days, or weeks) and that defining more chronic risk, insofar as imminent risk constitutes the clinical priority of the moment.

Finally, the findings suggest that clinicians should not be lulled into a false sense of security when a patient denies having suicidal ideas. More than half of the patients who died by suicide within 6 months of a comprehensive evaluation reported mild or no suicidal thoughts to an experienced clinical interviewer. Suicidal communications may be heard more often by family members and intimates than by casual friends, coworkers, or health professionals.

Mood Improvement Immediately Preceding Suicide

Observations that patients hospitalized for depression show an elevated risk of death by suicide during the first 6 to 12 months after hospital discharge, and that the mood of many depressed patients improves distinctly for 1 to 3 days preceding a suicide, have led to a variety of explanatory theories. One theory often repeated in clinical settings states that the return of normal levels of energy before other symptoms improve sometimes mobilizes or enables suicidal tendencies that have lain dormant or have not yet completely dissipated. Another prevalent clinical theory takes the position that patients who surrender to suicidal impulses after a long and arduous struggle may manifest a pervading sense of inner calm or peace for several days before taking their own life. Murphy (1986a) suggested that "mood fluctuations that frequently characterize the recovery phase of depression may be interpreted by the patient as relapse . . . [so that] in the absence of close contact with the physician [after hospital discharge] . . . despair may ensue, and then suicide." Schweizer, Dever, and Clary (1988) believed that when a severely depressed patient abruptly switches into a hypomanic state the change is sometimes misinterpreted as an initial sign of recovery, though the patient remains liable to sudden switches back into depression. They hypothesized that the period immediately following a switch from mania or hypomania back into depression may be one of particularly elevated suicide risk for patients with affective disorders.

Although there is no consensus about which theory is correct, observations of paradoxical clinical improvement during the days immediately preceding a death by suicide have been reported by many investigators. Thus it is important for the clinician to remain clinically vigilant through the early months of recovery following a depressive illness and following hospital discharge. These times are clearly periods of elevated suicide risk. The best available estimates are that 4 to 10% of all suicides occur inside hospitals (Beskow, 1979; Crammer, 1984; Robins et al., 1959b), and that 12 to 29% occur within a year after hospital discharge (Beskow, 1979; Robins et al., 1959b).

Bipolar and Cycling Affective Disorder

The frequent claim that suicide risk is higher among patients with a bipolar, in contrast to a unipolar, affective disorder is probably a presumption based on evidence that the bipolar type is a more severe form of affective disorder. Although suicide *attempt* rates may be higher for bipolars (Johnson & Hunt, 1979; Woodruff, Clayton, & Guze, 1972), evidence from several long-term follow-up studies of large samples of patients with major affective disorders has consistently demonstrated equivalent rates of completed suicide among unipolar and bipolar subjects (Black et al., 1987; Fawcett et al., 1987, 1990).

How is it possible to reconcile our positions that a cycling state is associated with a high risk for imminent suicide, but a bipolar history is not? Bipolar patients have a history of both depressive and manic (or hypomanic) states but do not necessarily have a history of cycling directly from one state into the other with no symptomatic recovery between depressive and manic phases. Our research shows that patients in a current episode of cycling affective disorder are the ones at elevated risk for imminent suicide.

Delusions, Hallucinations, and Paranoid Features

Delusions or hallucinations are observed in cases of major affective disorder more often than any other diagnostic group. Delusions or hallucinations accompany the depressive episodes of approximately 13 to 18% of all patients with major depression (Goodwin & Guze, 1989). In a 2- to 13-year follow-up mortality study of 1593 psychiatric inpatients admitted with a major affective disorder, Black and colleagues (1988) reported no difference in suicide rates when comparing psychotic and nonpsychotic patients by unipolar, bipolar, or combined subtypes. These findings are consistent with other retrospective (Coryell & Tsuang, 1982) and prospective (Fawcett et al., 1987) reports. It does not mean, of course, that *all* specific delusional or hallucinatory symptoms are unrelated to suicide risk severity; the findings mean only that psychotic subtype affective disorder as a category is not associated with elevated risk for death by suicide.

Thus, for example, we found that the isolated symptom of "delusions of thought insertion" was associated with a high risk of death by suicide despite a low base rate in the sample (Fawcett et al., 1987). Of 16 (schizoaffective) patients evidencing delusions of thought insertion in our large sample of patients with affective disorder, 19% died by suicide.

Other investigators have linked delusional depressions with increased suicide risk, but their data do not support their conclusions well. Roose, Glassman, Walsh, Woodring, and Vital-Herne (1983), for example, conducted a follow-up study of patients at one psychiatric hospital who had died while on the hospital census (i.e., as inpatients, on pass, or after eloping) over a 25-year period. They concluded that delusional patients with unipolar major depression were five times more likely to die by suicide than their nondelusional counterparts. Their data had three major shortcomings: (1) they identified only suicides occurring on the hospital census, neglecting to consider suicides in the same cohort occurring outside the hospital during the same period; (2) they counted "probable" cases (i.e., no evidence that the belief was fixed and unamenable to reason) as "definite" cases in their analysis; and (3) their comparison group consisted of a much lower fraction of men than did the suicide group. For these reasons we believe their study was inconclusive.

Schizophrenic patients, who are discussed at greater length later in this chapter, are also associated with an elevated risk for suicide. In particular, young male paranoid schizophrenics whose illness onset took a gradual course may be vulnerable to suicide during early phases of their illness (Westermeyer & Harrow, 1989).

Family History of Suicide

Important insight into the clinical value of family history information for estimating suicide risk is provided by Scheftner and colleagues' analysis of data from the National Institute of Mental Health Collaborative Depression Study, a sample of 954 patients (80% inpatients) entering treatment for affective disorders at five university medical centers. Subjects and relatives were interviewed to characterize all first- and second-degree relatives by means of structured family history interview, yielding information about 5055 relatives, 2046 of whom were interviewed first-hand. The relatives of 27 patients who died by suicide during a 5-year follow-up and relatives of the remainder surviving had comparable lifetime rates of suicide (0.7% and 0.9%, respectively). Thirteen percent of the relatives of patient suicides and eight percent of the relatives of other patients had made a nonlethal suicide attempt, but the two groups of relatives did not differ by intent to die or potential medical lethality of the attempts. Finally, relatives of patient suicides and relatives of other patients did not differ in terms of lifetime rates of unipolar and bipolar depression, schizoaffective disorder, alcoholism, or sociopathy.

For the clinician working with affectively disordered patients, therefore, the study yielded no support for the idea that positive family history of suicide helps identify patients at greater risk for death by suicide during the next 5 years. The lack of any association may be attributable to the fact that all the patients in this study shared a history of major affective disorder (i.e., once a positive family history for affective disorder was common to most subjects, the relationship between patient and familial suicide "disappeared" because it was the diagnosis and not familial suicidal behavior that conferred elevated risk). The results also suggest that the absence of a family history of suicide should not lead the clinician to assume that an affectively disordered patient is at lower risk for suicide during the next 5 years than other patients.

Assessing Suicide Risk in the Context of Alcoholism

The lifetime risk of suicide for alcoholics with a history of inpatient treatment is 3.4%, and that for untreated alcoholics is 1.8% (Murphy & Wetzel, 1990). In their reanalysis of published studies on the lifetime risk of suicide in alcoholism, Murphy and Wetzel demonstrated that alcoholism is implicated in about 25% of all U.S. suicides; that the mean duration of time from the beginning of excessive drinking to completed suicide for alcoholics not using other drugs is 19 years (i.e., most alcoholic suicides occur between the ages of 40 and 59 years); and that only one-third of alcoholic suicide victims have a history of inpatient psychiatric treatment. Risk factors for suicide associated with alcoholism are shown in Table 2.6.

Murphy and Wetzel observed that "suicide in alcoholism is largely dependent on the supervention of depressive episode" in one-half to three-fourths of all cases. When the alcoholic develops a depressive illness complicated by abuse of or dependence on drugs other than alcohol, the combination "shortens the

Table 2.6. Risk Factors Among Persons with Alcoholism

Supervention of a depressive episode
Concomitant drug abuse or dependence
Acute disruption of a major interpersonal relationship
Not currently hospitalized
Imprisonment, especially the first 24 hours

career from onset of substance abuse to suicide, regardless of which [complicating disorder] comes first."

Interpersonal losses appear to play a greater part in precipitating suicidal crises among alcoholics than is true among patients with other diagnoses. Thus Murphy has shown that "disruption of a major interpersonal relationship signals a period of heightened risk of suicide in the alcoholic," but not necessarily in nonalcoholic patients with major depression (Murphy et al., 1979; Murphy & Wetzel, 1990).

Beck, Steer, and Trexler (1989) found that 11% of the patients originally admitted with a diagnosis of alcoholism but only 2% of the remainder died by suicide in a 5- to 10-year follow-up mortality study of 499 patients hospitalized for suicide attempts. The mean amount of time between the original hospitalization and death by suicide for the alcoholic suicide completers was 44 months. The investigators puzzled over a finding that degree of hopelessness assessed at index admission was not related to death by suicide, given the consistency with which other studies have reported this variable correlates with suicidal outcome in a variety of inpatient and outpatient settings (Beck, Brown, Berchick, Stewart, & Steer, 1990; Beck, Steer, Kovaks, & Garrison, 1985).

There is some evidence that alcoholics are less likely to die while hospitalized than patients with other disorders associated with high risk for suicide (e.g., depressive disorders, schizophrenia) (Modestin & Kopp, 1988). This point may be true because the detoxification period of treatment is associated with a relatively low incidence of suicide. The coincidence of drinking and depressive episodes seems to be an important precondition for suicide in alcoholics. In the hospital, drinking is curtailed and depressive states are more amenable to detection and treatment.

Wolk-Wasserman (1987) has argued that psychiatric outpatient services should develop programs specializing in the treatment of suicidal alcohol and drug abusers because they pose different clinical problems from other substance abuse patients and other patients at suicide risk (i.e., the quality of their chaotic lives, their need for help with social problems, and the particularly difficult countertransference reactions they inspire). She took the position that their treatment needs are best served by a team approach defined around shared responsibilities and a colleague support network. Intoxication and a history of substance abuse/dependence at the time of admission to a holding cell, detention cell, jail, or prison, on the other hand, is associated with a greatly elevated suicide risk, particularly during the first 24 hours of imprisonment (Dooley, 1990; Sackett, 1987).

Assessing Suicide Risk in the Context of Schizophrenia

Approximately 10% of schizophrenic persons ultimately die by suicide (Bleuler, 1978; Miles, 1977; Tsuang, 1978). There is also evidence that persons with atypical psychoses (i.e., schizoaffective disorder, schizophreniform disorder,

Table 2.7. Risk Factors Among Persons with Schizophrenia

Young age
Early in the course of the illness
Good premorbid history (i.e., school or work progress)
Good intellectual functioning
Frequent exacerbations and remissions
Painful awareness of the discrepancies between the "normal" future once envisioned and the
  likely degree of chronic disability in the future
Periods of clinical improvement following relapse
Supervention of a depressive episode and increased hopelessness
Suicidal communications
Relative absence of florid psychosis

and other atypical psychoses) are at greatly elevated risk for death by suicide (Buda, Tsuang, & Fleming, 1988). Greater risk among schizophrenic persons (Table 2.7) is associated with male gender, white race, young age, short duration since illness onset, good premorbid history (i.e., good social history, school progress, and employment until the first hospitalization), greater number of exacerbations and remissions, suicidal communications, and painful realistic awareness of the discrepancy between the "normal" future they once envisioned and the severely impaired life style now lying before them (Allebeck, 1989; Allebeck et al., 1987; Allebeck & Wistedt, 1986; Black, 1988; Drake et al., 1985; Harrow, Grinker, Silverstein, & Holzman, 1978; Roy, 1982, 1986; Westermeyer & Harrow, 1989). Suicide among schizophrenics is *more likely* to occur during periods of clinical improvement following relapse (i.e., during the months following hospital discharge) and during periods of depressed mood and hopelessness (Allebeck & Wistedt, 1986). Suicide is *less likely* to occur during exacerbations of florid psychosis (Drake & Cotton, 1986; Drake et al., 1984; Roy, 1982; Westermeyer & Harrow, 1989). The presence of "command hallucinations," or hallucinations instructing the patient to perform specific acts, particularly violent or destructive acts, does not appear to be associated with greater suicide risk (Hellerstein, Frosch, & Loenigsberg, 1987; Roy, 1982; Wilkinson & Bacon, 1984). Several studies have suggested that schizophrenic patients with delusions of persecution (i.e., paranoia) are more likely to take their own lives than other schizophrenic patients (Virkkunen, 1974). Young male paranoid schizophrenics whose illness onset took a gradual course may be particularly vulnerable to suicide during early phases of their illness (Westermeyer & Harrow, 1989).

Help Negation

Many patients at high risk for suicide have reached a state of utter hopelessness concerning treatment and so are prone to soundlessly abandon, politely terminate, or angrily reject treatment as an extension of their pessimism. Sometimes, too, the patient's hopelessness is sufficiently contagious to infect the treating clinician. If the patient is difficult to work with or if the clinician feels that he or she can no longer tolerate the suicide-related anxiety raised by this patient, the clinician may permit or facilitate severance of the treatment relationship. Kullgren (1988), for example, speculated that patients with underlying borderline personality organizations are particularly gifted at provoking rejecting and repressive behavior in close interpersonal relationships, including their relationships with inpatient hospital staff. This speculation led Kullgren to hypothesize that clinicians lacking advanced training and experience in the management of borderline patients were prone to become frustrated and rid themselves of "troublesome" patients by discharging them from hospital prematurely, inadvertently intensifying the patients' suicidal behavior. Maltsberger and Buie (1974) described ways "countertransference aversion" can prematurely interrupt an outpatient therapy relationship with the same kinds of patients, having the same potentially disastrous consequences.

By denying themselves treatment during acutely suicidal phases of their illness, these patients place themselves at even greater risk for death by suicide. By allowing these patients to distance themselves from their therapists and therapy, clinicians court greater risk for losing patients by suicide.

Screening the General Population

We have argued in this chapter that many of the symptoms and features on which clinicians have traditionally relied to estimate suicide risk are *not* effective discriminating tools in specific and narrow psychiatric contexts; that is, professed suicidal ideation and bipolar history do not help pick out those likely to die by suicide within the next year among depressed patients, and command hallucinations do not help pick out those likely to die by suicide among schizophrenic patients. We think that these qualifications are appropriate, given the research evidence, and that they are useful to the clinician trained to evaluate chief complaints, life events, interpersonal problems, and suicide risk in the context of a psychological diagnosis.

It should be realized, however, that many of the symptoms we have "rejected" as risk estimators might well serve as good nonspecific screening items in general population samples. If we were to screen 1500 high school students or 50,000 young men who had enlisted in the army, many of the "rejected" symptom items (e.g., suicidal ideation, history of suicide attempts, diminished appetite and weight loss, delusions, command hallucinations) would help identify subjects who meet criteria for one or more psychiatric disorders and thus are at statistically elevated risk for suicide by virtue of the psychiatric illness. In this vein we want to conclude by emphasizing that to know which specific demographic features and psychological symptoms identify persons more likely to die by suicide, the clinician must read and reevaluate the wealth of retrospective and prospective studies of risk for completed suicide one by one in light of the reference comparison groups posed by each study.

Guidelines for Practice: Chapter Summary

I. A good evaluation for suicide risk necessarily includes assessment of the following.
   A. Ascertainment of sex, age, race, responsibility for children under age 18 years, recent death of spouse.
   B. Structured interview to rule in or rule out the major psychiatric diagnoses, including major mood disorders, alcoholism, drug abuse/dependence, schizophrenia, organic brain syndromes, and panic/anxiety disorders.
   C. Systematic inquiry about thoughts of death, suicidal wishes or ideas, specific methods of suicide contemplated, the degree to which a suicidal plan has been mentally rehearsed, the degree to which a suicidal plan has been behaviorally rehearsed, recent suicide attempts, lifetime history of suicide attempts, and assaultive/homicidal wishes and behaviors.
   D. Systematic inquiry directed to close family members or those living with the patient about suicidal communications and behavior.
   E. Clinical effort to discern the degree to which the patient is preoccupied with and convinced by the internal logic of his or her own suicidal ruminations.
   F. For patients meeting diagnostic criteria for a major depression, systematic inquiry concerning psychic anxiety, loss of pleasure/interest, insomnia, diminished concentration, indecision, acute overuse of alcohol, hopelessness, panic attacks, obsessive-compulsive features, number of prior episodes of major depression, and history of manic/hypomanic disorder.
   G. For patients meeting diagnostic criteria for alcoholism, systematic inquiry concerning all the symptoms that define a major depressive disorder including suicidal ideas and behaviors, all symptoms that define a drug abuse/dependence disorder, duration of the alcoholism episode, and status of current and recent interpersonal relationships.
   H. For patients meeting diagnostic criteria for schizophrenia, systematic inquiry concerning all the symptoms that define a major depressive disorder including hopelessness and suicidal ideas and behaviors, all delusions and hallucinations, premorbid history of functioning, duration since illness onset, and number of prior exacerbations and remissions.
   I. Patient's involvement in evaluation/treatment process and any evidence of help negation.

II. The clinician should tailor risk assessment formulations to the specific principal psychiatric diagnosis (one or several) implicated. The risk factors cited for application within specific diagnostic categories are summarized as Tables 2.5 (major depression), 2.6 (alcoholism), and 2.7 (schizophrenia).

III. The evaluator should keep special considerations in mind.

   A. Persons about to die by suicide tend not to recognize their own psychological symptoms or appreciate the psychological nature of their dilemma, often focusing instead on physical symptoms and physical health concerns.

   B. The presence of a stressor or precipitant for depression does not constitute a reason for judging suicide risk to be low in a particular case.

   C. The "solution" of suicide has seductive, irresistible qualities for the most severely ill patients, who experience nothing but intolerable and unrelieved pain from horizon to horizon. Many persons in *acute suicidal crisis* do not have the capacity to resist the internal logic, emotional tug, and pain relief offered by the idea of suicide. These patients temporarily lose all sense of the impact their death would have on beloved others.

   D. Once suicidal preoccupations have reached a high level of intensity, there is great danger that the patient has become irrefutably convinced of the sense and value of his or her suicidal ideas, so all verbal interventions are rendered totally ineffective.

   E. Many patients at high risk for suicide have reached a state of utter hopelessness concerning treatment and so are prone to soundlessly abandon, politely terminate, or angrily reject treatment as an extension of their pessimism. If the patient is difficult to work with or if the clinician believes that he or she can no longer tolerate the suicide-related anxiety raised by this patient, the clinician may permit or facilitate severance of the treatment relationship, prematurely interrupting the treatment relationship, with potentially disastrous consequences.

REFERENCES

Adelson, L. (1961). Slaughter of the innocents: a study of forty-six homicides in which the victims were children. *New England Journal of Medicine, 264,* 1345–1349.

Allebeck, P. (1989). Schizophrenia: a life-shortening disease. *Schizophrenia Bulletin, 15,* 81–89.

Allebeck, P., Varla, A., Kristjansson, E., Wistedt, B. (1987). Risk factors for suicide among patients with schizophrenia. *Acta Psychiatrica Scandinavica, 76,* 414–419.

Allebeck, P., & Wistedt, B. (1986). Mortality in schizophrenia: a ten-year follow-up based on the Stockholm County Inpatient Register. *Archives of General Psychiatry, 43,* 650–653.

Barraclough, B., Bunch, J., Nelson, B., & Sainsbury, P. (1974). A hundred cases of suicide: clinical aspects. *British Journal of Psychiatry, 125,* 355–373.

Beck, A. T., Brown, G., Berchick, R. J., Stewart, B. L. (1990). Relationship between hopelessness and ultimate suicide: a replication with psychiatric outpatients. *American Journal of Psychiatry, 147*, 190–195.

Beck, A. T., Steer, R. A., Kovacs, M., & Garrison, B. (1985). Hopelessness and eventual suicide: a 10-year prospective study of patients hospitalized with suicidal ideation. *American Journal of Psychiatry, 142*, 559–563.

Beck, A. T., Steer, R. A., & Trexler, L. D. (1989). Alcohol abuse and eventual suicide: A 5- to 10-year prospective study of alcohol-abusing suicide attempters. *Journal of Studies of Alcohol, 50*, 202–209.

Berglund, M. (1984). Suicide in alcoholism: a prospective study of 88 suicides. I. The multidimensional diagnosis at first admission. *Archives of General Psychiatry, 41*, 888–891.

Beskow, J. (1979). Suicide and mental disorder in Swedish men. *Acta Psychiatrica Scandinavica Supplementum, 277*, 1–138.

Black, D. W. (1988). Mortality in schizophrenia—the Iowa record-linkage study: a comparison with general population mortality. *Psychosomatics, 29*, 55–60.

Black, D. W., Warrack, G., & Winokur, C. (1985). The Iowa record-linkage study. I. Suicides and accidental deaths among psychiatric patients. *Archives of General Psychiatry, 42*, 71–75.

Black, D. W., Winokur, G., & Nasrallah, A. (1987). Suicide in subtypes of major affective disorder: a comparison with general population suicide mortality. *Archives of General Psychiatry, 44*, 878–840.

Black, D. W., Winokur, G., & Nasrallah, A. (1988). Effect of psychosis on suicide risk in 1,593 patients with unipolar and bipolar affective disorders. *Archives of General Psychiatry, 145*, 849–852.

Blacker, C. V. R., & Clare, W. (1987). Depressive disorder in primary care. *British Journal of Psychiatry, 150*, 737–751.

Bleuler, M. (1978). *The schizophrenic disorders: long-term and family studies*. New Haven: Yale University Press.

Breier, A., & Astrachan, B. M. (1984). Characterization of schizophrenic patients who commit suicide. *American Journal of Psychiatry, 141*, 206–209.

Brown, J. H., Henteleff, P., Barakat, S., & Rowe, C. J. (1986). Is it normal for terminally ill patients to desire death? *American Journal of Psychiatry, 143*, 208–211.

Buda, M., Tsuang, M. T., & Fleming, J. A. (1988). Causes of death in DSM-III schizophrenics and other psychotics (atypical group). *Archives of General Psychiatry, 45*, 283–285.

Buglass, C. D., & Horton, J. (1974). The repetition of parasuicide: a comparison of three cohorts. *British Journal of Psychiatry, 125*, 168–174.

Chynoweth, R., Tonge, J. I., & Armstrong, J. (1980). Suicide in Brisbane—a retrospective psychosocial study. *Australia and New Zealand Journal of Psychiatry, 14*, 37–45.

Clark, D. C. (1990). Suicide risk assessment and prediction in the 1990s. *Crisis, 11*, 104–112.

Clark, D. C., Gibbons, R. D., Fawcett, J., & Scheftner, W. A. (1989). What is the mechanism by which suicide attempts predispose to later suicide attempts? A mathematical model. *Journal of Abnormal Psychology, 98*, 42–49.

Clark, D. C., Sommerfeldt, L., Schwarz, M., & Watel, L. (1990). Physical recklessness in adolescence and its relationship to suicidal tendencies. *Journal of Nervous and Mental Disease, 178*, 423–433.

Clark, D. C., Young, M. A., Scheftner, W. A., Fawcett, J., & Fogg, L. A. (1987). A field test of Motto's risk estimator for suicide. *American Journal of Psychiatry, 144*, 923–926.

Clayton, P. J. (1985). Suicide. *Psychiatric Clinics of North America, 8*, 203–214.

Coryell, W., Noyes, R., & Clancy, J. (1982). Excess mortality in panic disorder: a comparison with primary unipolar depression. *Archives of General Psychiatry, 39*, 701–703.

Coryell, W., Noyes, R., & House, J. D. (1986). Mortality among outpatients with anxiety disorders. *American Journal of Psychiatry, 143*, 508–510.

Coryell, W., & Tsuang, M. T. (1982). Primary unipolar depression and the prognostic importance of delusions. *Archives of General Psychiatry, 39*, 1181–1184.

Crammer, J. L. (1984). The special characteristics of suicide in hospital in-patients. *British Journal of Psychiatry, 145,* 460–476.

Cullberg, J., Wasserman, D., & Stefansson, C.-G. (1988). Who commits suicide after a suicide attempt? An 8 to 10 year follow-up in a suburban catchment area. *Acta Psychiatrica Scandinavica, 77,* 598–603.

Dooley, E. (1990). Prison suicide in England and Wales, 1972–1987. *British Journal of Psychiatry, 156,* 40–45.

Dorpat, T. L., & Ripley, H. S. (1960). A study of suicide in the Seattle area. *Comprehensive Psychiatry, 1,* 349–359.

Drake, R. E., & Cotton, P. G. (1986). Depression, hopelessness and suicide in chronic schizophrenia. *British Journal of Psychiatry, 148,* 554–559.

Drake, R. E., Gates, C., Cotton, P. C., & Whitaker, A. (1984). Suicide among schizophrenics: who is at risk? *Journal of Nervous and Mental Disease, 172,* 613–617.

Drake, R. E., Gates, C., Whitaker, A., & Cotton, P. G. (1985). Suicide among schizophrenics: a review. *Comprehensive Psychiatry, 26,* 90–100.

Endicott, J., & Spitzer, R. L. (1978). A diagnostic interview: the schedule for affective disorders and schizophrenia. *Archives of General Psychiatry, 35,* 837–844.

Ettlinger, R. W. (1964). Suicides in a group of patients who had previously attempted suicide. *Acta Psychiatrica Scandinavica, 40,* 363–378.

Fawcett, J., Leff, M., & Bunney, W. E. (1969). Suicide: clues from interpersonal communications. *Archives of General Psychiatry, 21,* 129–137.

Fawcett, J. (1972). Suicidal depression and physical illness. *Journal of the American Medical Association, 219,* 1303–1306.

Fawcett, J. (1988). Predictors of early suicide: identification and appropriate intervention. *Journal of Clinical Psychiatry, 49*(suppl.), 7–8.

Fawcett, J., Scheftner, W. A., Clark, D. C., Hedeker, D., Gibbons, R. D., & Coryell, W. (1987). Clinical predictors of suicide in patients with major affective disorders: a controlled prospective study. *American Journal of Psychiatry, 144,* 35–40.

Fawcett, J., Scheftner, W. A., Fogg, L., Clark, D. C., Young, M. A., Hedeker, D., & Gibbons, R. (1990). Time-related predictors of suicide in major affective disorder. *American Journal of Psychiatry, 147,* 1189–1194.

Fowler, R. C., Rich, C. L., & Young, D. (1986). San Diego suicide study. II. Substance abuse in young cases. *Archives of General Psychiatry, 43,* 962–965.

Goodwin, D. W., & Guze, S. B. (1989). *Psychiatric Diagnosis* (4th Ed.). New York: Oxford University Press.

Guze, S. B., & Robins, E. (1970). Suicide and primary affective disorders. *British Journal of Psychiatry, 117,* 437–438.

Hagnell, O., & Rorsman B. (1978). Suicide and endogenous depression with somatic symptoms in the Lundby study. *Neuropsychobiology, 4,* 180–187.

Hagnell, O., & Rorsman, B. (1979). Suicide in the Lundby study: a comparative investigation of clinical aspects. *Neuropsychobiology, 5,* 61–73.

Hagnell, O., & Rorsman, B. (1980). Suicide in the Lundby study: a controlled prospective investigation of stressful life events. *Neuropsychobiology, 6,* 319–332.

Harrow, M., Grinker, R. R., Silverstein, M., & Holzman, P. (1978). Is modern-day schizophrenic outcome still negative? *American Journal of Psychiatry, 135,* 1156–1162.

Hellerstein, D., Frosch, W., & Loenigsberg, H. W. (1987). The clinical significance of command hallucinations. *American Journal of Psychiatry, 144,* 219–221.

Holinger, P. C. (1987). *Violent Deaths in the United States.* New York: Guilford.

Johnson, G. F., & Hunt, G. (1979). Suicidal behavior in bipolar manic-depressive patients and their families. *Comprehensive Psychiatry, 20,* 159–164.

Johnson, J., Weissman, M. M., & Klerman, G. L. (1990). Panic disorder, comorbidity, and suicide attempts. *Archives of General Psychiatry, 47,* 805–808.

Kreitman, N. (1988). Suicide, age, and marital status. *Psychological Medicine, 18,* 121–128.

Kreitman, N., & Casey, P. (1988). Repetition of parasuicide: an epidemiological and clinical study. *British Journal of Psychiatry, 153,* 792–800.

Kullgren, C. (1988). Factors associated with completed suicide in borderline personality disorder. *Journal of Nervous and Mental Disease, 176,* 40–44.

Kuperman, S., Black, D. W., & Burns, T. L. (1988). Excess mortality among formerly hospitalized child psychiatric patients. *Archives of General Psychiatry, 45,* 277–282.

Lindesay, J. (1989). Age, sex and suicide rates within birth cohorts in England and Wales. *Social Psychiatry and Psychiatric Epidemiology, 24,* 249–252.

Linehan, M. M. (1986). Suicidal people: one population or two? In A. Roy (Ed.), *Suicide* (pp. 16–33). Baltimore: Williams & Wilkins.

MacKinnon, D., & Farberow, N. (1975). An assessment of the utility of suicide prediction. *Suicide and Life-Threatening Behavior, 6,* 86–91.

Maltsberger, J. T., & Buie, D. H. (1974). Countertransference hate in the treatment of suicidal patients. *Archives of General Psychiatry, 30,* 625–633.

Marks, J. N., Goldberg, D., & Hillier, V. F. (1979). Determinants of the ability of general practitioners to detect psychiatric illness. *Psychological Medicine, 9,* 337–353.

Martin, R. L., Cloninger, C. R., Guze, S. B., & Clayton, P. J. (1985). Mortality in a follow-up of 500 psychiatric outpatients. I. Total mortality. *Archives of General Psychiatry, 42,* 47–54.

Miles, C. P. (1977). Conditions predisposing to suicide: a review. *Journal of Nervous and Mental Disease, 164,* 231–246.

Modestin, J., & Kopp, W. (1988). A study of clinical suicide. *Journal of Nervous and Mental Disease, 176,* 668–674.

Morgan, H. G., Barton, J., Pottle, S., Pocock, H., & Burns-Cox, C. J. (1976). Deliberate self-harm: a follow-up study of 279 patients. *British Journal of Psychiatry, 128,* 361–368.

Moscicki, E. K., O'Carroll, P., Rae, D. S., Locke, B. Z., Roy, A. & Regier, D. A. (1988). Suicide attempts in the epidemiologic catchment area study. *Yale Journal of Biology and Medicine, 61,* 259–268.

Motto, J. (1965). Suicide attempts: a longitudinal view. *Archives of General Psychiatry, 13,* 516–520.

Motto, J. A. (1980). Suicide risk factors in alcohol abuse. *Suicide and Life-Threatening Behavior, 10,* 230–238.

Motto, J. A., Heilbron, D. C., & Juster, R. P. (1985). Development of a clinical instrument to estimate suicide risk. *American Journal of Psychiatry, 142,* 680–686.

Murphy, G. E. (1972). Clinical identification of suicide risk. *Archives of General Psychiatry, 27,* 356–359.

Murphy, G. E. (1975). The physician's responsibility for suicide. II. Errors of omission. *Annals of Internal Medicine, 82,* 305–309.

Murphy, G. E. (1983). On suicide prediction and prevention. *Archives of General Psychiatry, 40,* 343–344.

Murphy, G. E. (1984). The prediction of suicide: why is it so difficult? *American Journal of Psychotherapy, 38,* 341–349.

Murphy, G. E. (1986a). The physician's role in suicide prevention. In A. Roy (Ed.), *Suicide* (p. 175). Baltimore: Williams & Wilkins.

Murphy, E. G. (1986b). Suicide and attempted suicide. In G. Winokur & P. Clayton (Eds.), *The Medical Basis of Psychiatry.* Philadelphia: Saunders.

Murphy, E. G., Armstrong, J. W., Hermele, S. L. Fischer, J. R., & Clendenin, W. W. (1979). Suicide and alcoholism: interpersonal loss confirmed as a predictor. *Archives of General Psychiatry, 36,* 65–69.

Murphy, G. E., & Wetzel, R. D. (1990). The lifetime risk of suicide in alcoholism. *Archives of General Psychiatry, 47,* 383–392.

Myers, J. K., Weissman, M. M., Tischler, G. L., Holzer, C. E., Leaf, P. J., Orvaschel, H., Anthony, J. C., Boyd, J. H., Burke, J. D., Kramer, M., & Stoltzman, R. (1984). Six-month prevalence of psychiatric disorders in three communities. *Archives of General Psychiatry, 41*, 959–967.

National Center for Health Statistics (1992). Advance report of final mortality statistics, 1989. *NCHS Monthly Vital Statistics Report, 40* (8, suppl.).

Pokorny, A. D. (1966). A follow-up of 618 suicidal patients. *Archives of General Psychiatry, 14*, 1109–1116.

Pokorny, A. D. (1983). Prediction of suicide in psychiatric patients: report of a prospective study. *Archives of General Psychiatry, 40*, 249–257.

Regier, D. A., Goldberg, I. D., & Taube, C. A. (1978). The de facto US mental health services system: a public health perspective. *Archives of General Psychiatry, 38*, 685–693.

Rich, C. L., Fowler, R. C., Fogarty, L. A., & Young, F. (1988). San Diego suicide study. III. Relationships between diagnoses and stressors. *Archives of General Psychiatry, 45*, 589–594.

Rich, C. L., Young, D., & Fowler, R. C. (1986). San Diego suicide study. I. Young vs. old subjects. *Archives of General Psychiatry, 43*, 577–582.

Robins, E. (1981). *The Final Months: A Study of the Lives of 134 Persons Who Committed Suicide.* New York: Oxford University Press.

Robins, E. (1986). Completed suicide. In A. Roy (Ed.), *Suicide* (pp. 123–133). Baltimore: Williams & Wilkins.

Robins, E., Gassner, S., Kaye, J., Wilkinson, R. H., & Murphy, G. E. (1959a). The communication of suicidal intent: a study of 134 consecutive cases of successful (completed) suicide. *American Journal of Psychiatry, 115*, 724–733.

Robins, E., Murphy, G. E., Wilkinson, R. H., Gassner, S., & Kayes, J. (1959b). Some clinical considerations in the prevention of suicide based on a study of 134 successful suicides. *American Journal of Public Health, 49*, 888–899.

Rodenburg, M. (1971). Child murder by depressed parents. *Canadian Psychiatric Association Journal, 16*, 41–48.

Roose, S. P., Glassman, A. H., Walsh, B. T., Woodring, S., & Vital-Herne, J. (1983). Depression, delusions, and suicide. *American Journal of Psychiatry, 140*, 1159-1162.

Rosenbaum, M. (1990). The role of depression in couples involved in murder-suicide and homicide. *American Journal of Psychiatry, 147*, 1036–1039.

Roy, A. (1982). Suicide in chronic schizophrenia. *British Journal of Psychiatry, 141*, 171–177.

Roy, A. (1986). Suicide in schizophrenia. In A. Roy (Ed.), *Suicide* (pp. 97–112). Baltimore: Williams & Wilkins.

Sacket, S. A. (1987). Suicide in Scottish prisons. *British Journal of Psychiatry, 151*, 218–221.

Sawyer, D., & Sobal, I. (1987). Public attitudes toward suicide: demographic and ideological correlates. *Public Opinion Quarterly, 51*, 92–101.

Scheftner, W. A., Young, M. A., Endicott, J., Coryell, W., Fogg, L., Clark, D. C., & Fawcett, J. (1988). Family history and five-year suicide risk. *British Journal of Psychiatry, 153*, 805–809.

Schweizer, E., Dever, A., & Clary, C. (1988). Suicide upon recovery from depression: a clinical note. *Journal of Nervous and Mental Disease, 176*, 633–636.

Shapiro, S., Skinner, E. A., Dramer, M., Steinwachs, S., & Regier, D. A. (1985). Measuring need for mental health services in a general population. *Medical Care, 23*, 1033–1043.

Shneidman, E. (1985). *Definitions of Suicide.* New York: John Wiley & Sons.

Skuse, D., & Williams, P. (1984). Screening for psychiatric disorder in general practice. *Psychological Medicine, 14*, 365–377.

Smith, J. C., Mercy, J. A., & Conn, J. M. (1988). Marital status and the risk of suicide. *American Journal of Public Health, 78*, 78–80.

Tsuang, M. T. (1978). Suicide in schizophrenics, manics, depressives, and surgical controls. *Archives of General Psychiatry, 35*, 153–154.

Tsuang, M. T., Woolson, R. F., & Fleming, J. A. (1980). Premature deaths in schizophrenic and affective disorders. *Archives of General Psychiatry, 37*, 979–983.

Veevers, J. E. (1973). Parenthood and suicide: an examination of a neglected variable. *Social Sciences and Medicine, 7,* 135–144.

Verhaak, P. F. M. (1988). Detection of psychological complaints by general practitioners. *Medical Care, 26,* 1009–1020.

Virkkunen, M. (1974). Suicide in schizophrenia and paranoid psychoses. *Acta Psychiatrica Scandinavica Supplementum, 250,* 1–305.

Walker, M., Moreau, D., & Weissman, M. M. (1990). Parents' awareness of children's suicide attempts. *American Journal of Psychiatry, 147,* 1364–1366.

Whitlock, F. A. (1986). Suicide and physical illness. In A. Roy (Ed.), *Suicide* (pp. 151–170). Baltimore: Williams & Wilkins.

Weiss, J. M. A., & Scott, K. F. (1974). Suicide attempters ten years later. *Comprehensive Psychiatry, 15,* 165–171.

Weissman, M. M., Klerman, G. L., Markowitz, J. S., & Ouellette, R. (1989). Suicidal ideation and suicide attempts in panic disorder and attacks. *New England Journal of Medicine, 321,* 1209–1214.

Westermeyer, J. F., & Harrow, M. (1989). Early phases of schizophrenia and depression: prediction of suicide. In R. Williams & J. T. Dalby (Eds.), *Depression in Schizophrenics* (pp. 153–169). New York: Plenum.

Wilkinson, G., & Bacon, N. A. (1984). A clinical and epidemiological survey of parasuicide and suicide in Edinburgh schizophrenics. *Psychological Medicine, 14,* 899–912.

Wolfgang, M. E. (1958). An analysis of homicide-suicide. *Journal of Clinical and Experimental Psychopathology, 19,* 208–217.

Wolk-Wasserman, D. (1987). Some problems connected with the treatment of suicide attempt patients: transference and countertransference aspects. *Crisis, 8,* 69–82.

Woodruff, R. A., Clayton, P. J., & Guze, S. B. (1972). Suicide attempts and psychiatric diagnosis. *Diseases of the Nervous System, 33,* 617–621.

Younger, S. C., Clark, D. C., Oehmig-Lindroth, R., & Stein, R. J. (1990). Availability of knowledgeable informants for a psychological autopsy study of suicides committed by elderly persons. *Journal of the American Geriatrics Society, 38,* 1169–1175.

3

Clinical Applications of Biological Aspects of Suicide

JEROME A. MOTTO

Information continues to emerge at a steady pace in most aspects of the behavioral sciences, but the rate at which biological data have been reported in recent years has been bewildering. Not only has a flood of hard-science observations been presented to the clinician, but most have been couched in the not-so-familiar languages of psychopharmacology, neurochemistry, neuroanatomy, and psychoendocrinology. Even a superficial understanding of current reports in many leading psychiatric journals requires some familiarity with such concepts as adrenergic, serotonergic, and dopaminergic systems and their metabolites; agonists, antagonists, and ligands; and chlorine and calcium channels and their significance.

The challenge to the clinician has entailed both the effort to remain abreast of this rapidly growing body of technical information and, more important, to determine its usefulness in the everyday tasks of patient evaluation, treatment, and management. The present discussion addresses this question of the practical application of suicide-related biological observations at our present state of knowledge.

Background

The idea of biochemical measures as predictors of behavior has threaded through the history of modern psychiatry. The first with clear implications for suicide was probably the "anger in—anger out" hypothesis of Funkenstein, King, and Drolett (1954), who presented evidence that a high urinary norepinephrine/epinephrine ratio is related to a tendency to turn anger outward (assaultiveness) and a low ratio to directing anger inward (depression and self-destructiveness).

More specific indicators of suicide risk subsequently emerged from the scrutiny of biological compounds during the course of studying the biology of depression. In prospective studies of depressed inpatients, those subjects who were found to have committed suicide after discharge were reviewed to determine if their previously observed levels of various biological measures differed significantly from those of subjects who had survived. This serendipitous sequence was first reported by Bunney and Fawcett (1965), who had observed elevated 17-hydroxycorticosteroids in the urine of the subjects who committed suicide in their sample of depressed patients studied at the National Institute of Mental Health. Increasing interest in this approach was evident through the 1970s, and by the end of the 1980s biological observations related to suicide had increased from a few scattered reports to a relative torrent of information.

Several issues cloud the clinical scene. One is that some observations appear so esoteric as to have a remote, if any, specificity for suicide at this time. A second issue is that most of the available data regarding suicide are of great interest to researchers and thus get to the media but have no present relevance to clinical practice. Third, pertinent studies have focused on correlates of a wide array of suicide-related phenomena, such as a history of suicide attempts or suicidal ideas, a tendency toward violence, or a future suicide attempt or completed suicide by means other than drug ingestion or a single wrist cut. The range of possible implications of various biological markers has led clinicians outside the research arena to take a generally watchful but passive stance as regards the application of these observations to clinical work. The ever present question, of course, is at what point it is appropriate to change that stance to a more active use of biological measures in the service of suicide prevention and optimal clinical care.

Biological Measures with Potential Clinical Use

5-Hydroxyindoleacetic Acid in Cerebrospinal Fluid

Though not the first potential biological marker of suicide risk to be identified, cerebrospinal fluid (CSF) 5-hydroxyindoleacetic acid (5-HIAA) has completely dominated the field during the 1980s in both research reports and popular interest. Its role as the leading candidate during this period stemmed from two considerations. First, it is a major metabolite of the neurotransmitter serotonin (5-hydroxytryptamine), which has played a prominent role in current studies of the biology of depression. It was in the course of such a study that Asberg, Traskman, and Sjostrand (1976) unexpectedly found that depressed subjects with low concentrations of CSF 5-HIAA made more suicide attempts than other depressed subjects. The traditional association of depressive and suicidal states give this observation an aura of clinical relevance, even though the connection between altered serotonin turnover and suicidal behavior was seen as a "formidable question" because of our limited knowledge of serotonergic function.

Second, the previously observed association (in experimental animals) of reduced serotonin synthesis with increased sensitivity to pain and increased aggressive behavior became an immediate focus of attention. Asberg and her coworkers pointed out that the observed low CSF 5-HIAA may not be directly related to depression and suicidal behavior but, rather, an epiphenomenon to other, more important processes, and that its clinical value as an indicator of suicide potential remained to be proved. In a subsequent report by this group

(Traskman, Asberg, Bertilsson, & Sjostraud, 1981), adding further support to the initial observations about low CSF 5-HIAA and suicidal behavior, the investigators reiterated that routine 5-HIAA determinations to assist the evaluation of suicide would be premature.

A steady flow of reports has provided additional findings consistent with the earlier results. They have extended the scope of inquiry beyond adult depressed inpatients to samples of such populations as homicide offenders (Lidberg, Tuck, Asberg, Scalia-Tomba, & Bertilsson, 1985), suicidal schizophrenic patients (Ninan et al., 1984), brain tissue of completed suicides (Arango et al., 1990; Arora & Meltzer, 1989), repeated suicide attempters (Roy et al., 1989), elderly suicide attempters (Jones et al., 1990), alcoholic suicide attempters (Roy, Pickar, DeJong, Karoum, & Linnoila, 1990), and inpatients with mixed diagnoses (Apter et al., 1990). They represent studies of depression from research centers around the world, including Sweden (Traskman et al., 1981), the Netherlands (van Praag, 1982), England (Montgomery & Montgomery, 1982), India (Palaniappan, Ramachdran, & Somasundaram, 1983), Hungary (Banki, Arato, Papp, & Kurcz, 1984), and Spain (Lopez-Ibor, Saiz-Ruiz, & Perez do los Cobos, 1985). The findings (with few exceptions) are consistent with the initial observations of low CSF 5-HIAA in persons who exhibit suicidal, homicidal, assaultive, aggressive, or impulsive behaviors.

Despite the number and agreement of those findings, no recommendation for the clinical application of CSF 5-HIAA levels has been made. Asberg (1989) pointed out that although such determinations might help in the assessment of suicide risk, there are problems in using them in a clinical setting. First, the range of values is such that a large number of "false positives" would be found. Second, there are so many factors that influence CSF concentrations of 5-HIAA (age, sex, height, physical illness, drug treatment, diet, technical aspects of obtaining and analyzing the CSF) that the spinal tap procedure would have to be standardized "to an extent that is rarely practical in a busy clinic." Third, overnight hospitalization is required; and finally, the patient must have been off all antidepressant and neuroleptic drugs or lithium for several weeks before the spinal tap. Despite these points, Asberg allowed that routine taps may nevertheless be a real help in clinical management while acknowledging that there is a need for further prospective studies.

It seems clear that although determination of CSF 5-HIAA remains an intriguing window for the observation of serotonin metabolism in a research setting, it is not yet a usable clinical tool as regards suicide risk assessment. Detailed reviews of this approach have been provided by van Praag (1986), Asberg (1989), Ricci and Wellman (1990), and Brown and Linnoila (1990).

Tritiated Imipramine Binding

The complexities of obtaining spinal fluid to measure serotonin activity in the brain makes it desirable to find a simpler method. Because the receptor sites of blood platelets and of brain tissue possess virtually identical binding characteristics, and because there is evidence that tritiated imipramine binding is associated with the neuronal uptake mechanism for serotonin, it could be predicted that the platelets of depressed persons would show decreased binding sites for this compound. Extending this situation to suicidal states, Stanley, Virgilio, and Gershon (1982) demonstrated that the brain tissue of persons who had committed suicide showed reduced binding compared with brains of persons who died of other causes. The implied marker for suicide risk therefore would be a determination of tritiated imipramine binding in platelets in the peripheral blood as an indicator of the level of serotonin transmission activity in the brain. A low level would imply a high suicide risk. The simplicity of such a blood test would certainly be appealing if a valid cutoff point could be established. Other work however (Gross-Isseroff, Israeli, & Biegon, 1989), suggests that different areas of the brain show different changes in imipramine binding in persons who commit suicide, and that the correspondence between brain and platelets is still inconclusive (DeLeo & Marazzitti, 1988). Clinical application of this biological approach to understanding suicide is not yet at hand.

Urinary 17-Hydroxycorticosteroids

Urinary 17-hydroxycorticosteroids (17-OHCS) provided the first model for the identification of biological markers for suicide (Bunney & Fawcett, 1965). When 3 of 36 depressed research subjects committed suicide during or shortly after a period of inpatient study and treatment, it was recognized that their levels of urinary 17-OHCS had been among the highest in the entire sample. Of special clinical interest was the fact that twice-daily estimates of suicide risk had ranked all three of these subjects below the group mean. The findings were interpreted cautiously with the postulate that 17-OHCS levels might reflect the intensity of psychic pain in a certain subgroup of depressed patients with limited psychological defenses, pointing out that a low 17-OHCS level in other patients (less than 9 mg/24 hours for women, and less than 14 mg/24 hours for men) should offer no reassurance regarding suicide.

A second report 4 years later (Bunney, Fawcett, Davis, & Gifford, 1969) provided an enlarged sample and added the refinement of "long-term" and "short-term" patterns of elevated 17-OHCS. The investigators reiterated that low levels should not be considered an indication of low suicide risk and commented that the clinical significance of their data remained uncertain. Meanwhile, a few scattered contrary reports appeared, based (as was the original report) on small samples and pointing out the confounding effects of body size, psychotropic drugs, and the difficulty of obtaining a complete 24-hour urine sample.

The simplicity and logic of measuring a stress-related compound by a non-invasive technique retains its early appeal, but replication studies have not been forthcoming, and the use of urinary 17-OHCS determinations has not been widely adopted in the clinical setting.

Urinary Norepinephrine/Epinephrine Ratio

The prospect of using urinary compounds for suicide risk assessment attracted renewed interest with an investigation of the hypothesis that a high urinary norepinephrine/epinephrine ratio is indicative of a tendency to turn anger outward and a low ratio to directing it inward (Ostroff et al., 1982). This hypothesis revived an idea explored much earlier in a widely publicized laboratory experiment (Funkenstein et al., 1954). The new report was prompted by finding a low ratio in two suicides and one serious suicide attempter among 22 male research subjects, which was consistent with the anticipation that a low ratio would suggest risk of suicide and a high ratio would indicate risk of assault. Earlier findings consistent with this one had been reported by Woodman (1979), who found a positive relation between a consistently high ratio and assaultiveness in prison inmates. A later effort to demonstrate a low ratio in persons with a history of suicide attempt produced statistically significant differences in mean values, but the overlap was so great that the clinical usefulness of the observation is questionable (Ostroff, Giller, Harkness, & Mason, 1985).

Roy et al. (1989) examined the relation of suicidal behavior and noradrenergic activity, including norepinephrine concentration, as reflected by indices in the CSF and blood plasma as well as in urine. A lower level of noradrenergic function in suicidal subjects would support the concept of a low norepinephrine/epinephrine ratio as a marker for suicide. No consistent pattern was found, which was interpreted as indicating that the noradrenergic system is probably not a major determinant of suicidal behavior in depressed patients.

Despite the lack of confirmation that low norepinephrine or a low norepinephrine/epinephrine ratio are related to suicide, the dynamic appeal of an "anger in" versus "anger out" measurement is still intriguing. Its clinical role, however, is yet to be demonstrated.

Cortisol Levels in Plasma and CSF

Elevated levels of cortisol have long been observed in states of depression and in conditions of increased psychological or physical arousal. Krieger (1974) first focused attention on cortisol as a possible indicator of suicide risk by pointing out that this precursor of 17-OHCS was not only simpler to determine than its urinary metabolite but would more accurately reflect the intensity of emotional distress experienced by the subject. He noted that 13 patients who committed suicide within 2 years of the test had a significantly higher mean serum cortisol level (21.08 μg/dl) than did 39 suicide-risk matched patients who did not commit suicide (16.5 μg/dl). He postulated an elevated "set point" in the hypothalamus of some individuals, resulting from genetic influence or early emotional stress, that leads to a higher than normal basal level of cortisol. This in turn could make persons more vulnerable to suicide in the presence of added stress in their lives. He pointed out further that various clinical states involving similar steroids, such as Cushing's disease or treatment with steroid medications, are known to be associated with depressive and suicidal states.

Krieger (1975) later offered a clear clinical application of the test for serum cortisol concentration by suggesting that if a patient recognized to be at risk for suicide has a level above 20 μg/dl, in the absence of other causes for such a high level, special suicide prevention measures were indicated, including more protective precautions, more frequent therapist contact, and prolonged after-care following discharge. Trials of the usefulness of this risk indicator have not been reported, to our knowledge, though an observation by Meltzer and Lowy (1989) showed that although 18 suicide attempters had a higher mean serum cortisol (14.2 ± 7.3 μg/dl), than 76 nonattempters (11.6 ± 4.8 μg/dl) the 20.0 μg/dl cutoff point (as suggested by Krieger) did not effectively differentiate the two samples.

Most of the attention to serum cortisol has focused on the use of the dexamethasone suppression test (DST) for the diagnosis of depression. Widely used during the 1980s, questions of validity have gradually led to its relatively infrequent use at present. Its potential value as a suicide risk indicator has appeared to be mixed at best, as noted in reviews by Motto (1986) and Meltzer and Lowy (1989). We have probably not heard the last of cortisol as a marker of suicide risk, but it does not yet have a place in our clinical armamentarium.

Homovanillic Acid and 3-Methoxy-4-hydroxyphenolglycol

The examination of neurotransmitters as possible indicators of suicide risk has included, in addition to serotonin, (1) dopamine and its metabolite homovanillic acid (HVA) and (2) norepinephrine and its metabolite 3-methoxy-4-hydroxy-phenolglycol (MHPG). Numerous efforts to relate these compounds to suicidal thinking or behavior have thus far been relatively unproductive—hence their somewhat low visibility in the suicide prevention literature. They are included here for the sake of completeness.

Roy et al. (1986) pointed out that in a sample of depressed patients with a prior suicide attempt three of the four who committed suicide within 1 year of evaluation were melancholic subjects who had shown a CSF HVA level below 75 pmol/ml and were DST nonsuppressors. Similarly, among a sample of elderly depressed patients, those who attempted suicide showed significantly lower CSF 5-HIAA and HVA than those who had not made a suicide attempt (Jones et al., 1990). This finding is consistent with the observation of Asberg (1989) that in studies relating CSF metabolites to personality features of high vitality, self-reported impulsivity, and psychopathy-related features correlations with HVA tend to parallel those with 5-HIAA, though they are weaker.

In an effort to relate suicidal behavior, depression, and noradrenergic function, Roy et al. (1989) found no consistent pattern of norepinephrine or MHPG levels in urine, CSF, or plasma in depressed patients who had attempted suicide. In another attempt to find suicide-related characteristics in a specific clinical subgroup, Roy et al. (1990) examined alcoholic patients who had attempted suicide. They found no significant differences between attempters, nonattempters, or normal controls as measured by CSF 5-HIAA, HVA or MHPG.

Though still of much theoretical interest, it seems clear that at this point practical application of what is known of HVA and MHPG is not yet feasible.

Thyroid-Stimulating Hormone Response to Thyrotropin-Releasing Hormone

The study of the hypothalamic-pituitary-adrenal (HPA) axis in suicidal behavior was recapitulated with the hypothalamic-pituitary-thyroid (HPT) axis in that dysregulation of the latter was initially suggested as a marker for depression. Specifically, evidence was found of a reduced response of thyroid-stimulating hormone (TSH), from the anterior pituitary, to thyrotropin-releasing hormone (TRH), from the hypothalamus, in depressed subjects.

The next step, encountering suicides in a sample of depressed subjects whose TSH response to TRH had previously been determined, was reported from Brussels by Linkowsky, van Wettere, Kerkhofs, Brauman, and Mendlewicz (1983). The observed that all four suicides during a 5-year follow-up of 51 depressed women had shown an absence of TSH response (less than 1 μU/ml) when first evaluated, with the three women with natural deaths all showing greater responses than the suicides. Recognizing that this measurement might provide a biological indicator of high suicide risk in depressed female patients, an expanded sample was subsequently reported to show that over a mean follow-up of 8 years five of seven suicide completers had very low TSH responses compared to the women with nonsuicidal deaths and the controls (Linkowsky, van Wettere, Kerkhofs, Gregoire, & Brauman, 1984). Even among suicide attempters, subjects with a history of "violent" suicide attempts showed significantly lower responses than those with no history of suicidal behavior. In view of a decreased TSH response in normal men over age 40, the test was suggested only for depressed women.

Efforts to replicate these observations have led to inconsistent and contradictory findings, though they have primarily involved suicide attempts (Banki, 1985; Banki et al., 1984; Kjellman, Ljunggren, Beck-Frier, & Wetterberg, 1985; van Praag & Plutchik, 1984) and suicide ideation (Maes et al., 1989) rather than completed suicides. Though the test is relatively simple and inexpensive, it has not been widely reported in either clinical or research settings, so we are still awaiting clarifying information and definitive guidelines for its use.

It seems that although enormous strides have ben made in biochemical knowledge, measurement of biological compounds, and technological resources, the underlying concepts of physical-emotional-behavioral interplay is fairly constant. In this regard, a psychiatric study of seven depressed women that was carried out 60 years ago noted a tendency toward low basal metabolic rates in the sample. It led to the comment that "organotherapy" with thyroid and pituitary compounds "obviously suggests itself" (Bowman & Bender, 1932). It seems that the puzzling role of the HPT axis in depression and its complications is not a new challenge to the researcher.

CSF Corticotropin-Releasing Factor

In a persistent scrutiny of the role of the HPA axis in completed suicide, Arato et al., (1986) in Budapest provided a promising lead. Using "a new research tool in the investigation of complete suicides," specifically post mortem CSF analyses, they found no differences between suicides and controls in terms of CSF cortisol, reflecting adrenal activity, or CSF adrenocorticotropic hormone (ACTH), reflecting pituitary activity. Pursuing this hypothesis one step further, they subsequently found evidence of increased hypothalamic stimulation of the pituitary in the form of increased corticotrophin releasing factor (CRF) levels in the CSF of suicide victims (Arato, Banki, Bissett, & Nemeroff, 1989). This evidence is consistent with findings in depressed patients, who also show elevated CSF CRF concentrations (Banki et a., 1987; Roy, Picker, Linnoila, Crousos, & Gold, 1987). It is also consistent with the finding of decreased CRF receptor density in the frontal cortex of suicide victims, attributed to the downregulation of CRF binding sites induced by CRF hypersecretion by the hypothalamus (Nemeroff, Owens, Bissett, Andorn, & Stanley, 1988).

The hypothesis is advanced that severe, prolonged stress could lead to a breakdown of psychological defenses and eventually to suicide. In that case, the measurement of CRF would be a better indicator of stress than either ACTH or cortisol, as secondary changes at the pituitary and adrenal levels could conceal any hyperactivity of the HPA axis. The investigators assumed, however, that the elevated CRF concentrations found in suicides are due primarily to the patients' underlying depression, as suicidal and nonsuicidal depressed subjects show no differences in CSF CRF (Banki et al., 1987). It remains to be seen whether there are differences between depressed suicides and suicides without depression. In any case, this arena clearly remains in the realm of research. At present, clinicians can only watch with intense interest.

CSF Magnesium

After Banki, Vojnik, Papp, Bolla, and Arato (1985) reported that low CSF magnesium concentrations were highly correlated with suicide attempts in 11 of their 56 female patients, this simple element was added to the growing list of potential indicators of suicide risk. The subjects' diagnoses represented a mixture of depressive states, schizophrenia, and adjustment disorders, and the correlation of suicide attempts with a lower mean CSF magnesium level was found regardless of diagnosis. Magnesium levels correlated significantly with CSF 5-HIAA, especially after correction for height and age, and even showed a stronger correlation with suicide attempts than did 5-HIAA.

Confirming studies have not appeared to our knowledge, so despite its intriguing potential as a simple chemical test with a possible relation to peripheral blood levels, it remains clinically just an interesting observation.

Genetics

Many human characteristics tend to cluster in families, so it is not surprising that suicide does as well. Much evidence exists to support a genetic factor in major psychotic and affective disorders, but the role of genetics in suicide itself remains less clear. Twin studies have been inconclusive, and family and adoption studies have not yet satisfactorily distinguished a genetic risk for suicide from the genetic risk for depression or schizophrenia. A practical application of this situation was suggested by Roy (1989) in the form of giving added attention to a depressed or suicidal adolescent with a family history of suicide. This study would give recognition to the possibility that the young person may be developing a treatable recurrent affective disorder.

The most recent evidence (Roy, Sega, Centerwall, & Robinette, 1991), based on 176 twin pairs, found a clear preponderance of concordance for suicide in the monozygotic (identical) twin pairs (11.3%) compared to the dizygotic (non-identical) pairs (1.8%). When treatment for psychiatric disorders in the twins and their first- or second-degree relatives was considered as well, the conclusion was that "genetic factors related to suicide may largely represent a genetic predisposition for the psychiatric disorders associated with suicide."

Although the precise role of genetics remains to be clarified, we can and should address the issue of suicides in family members with a grasp of the potential psychological and social implications of such a pattern. The biological implications, however, cannot yet be put to clinical use.

Other Biological Considerations

Electroencephalography

The recognition by Struve, Klein, and Saraf (1972) of a significant association of paroxysmal electroencephalograms (EEGs) with suicidal ideation, suicide attempts, and assaultive-destructive behavior raised hopes of a noninvasive biological indicator of risk for these behaviors. Efforts to explore this relation led to "a paucity of positive findings," though suicidal persons of high emotional "reactivity" did exhibit paroxysmal EEGs, whereas those with a more detailed suicide plan had normal records (Struve, Saraf, Arko, Klein, & Becka, 1977).

One approach to this question examined sleep EEG patterns in suicidal and nonsuicidal adolescents with a major depressive disorder (Dahl et al., 1990). The suicidal subgroup showed significantly longer sleep latency than the nonsuicidal group or the controls, though other sleep-related EEG characteristics were similar.

The use of the EEG for suicide risk assessment has the considerable advantages of minimum discomfort for the subject, noninvasiveness, outpatient applicability, and a rich context for interpretation. Its imprecision and lack of specificity are also great; and at present it is not a very active arena of investigation.

Growth Hormone Response to Desipramine

Another biological function explored in depressed adolescents is the response of growth hormone to desipramine. This response was found to be significantly decreased in depressed adolescents with suicidal ideation or attempts compared with those who showed no suicidal manifestations. The difference was attributed to differences in central nervous system (CNS) β-adrenergic or serotonergic function (or both) associated with suicidality (Ryan et al., 1988). These findings were interpreted as supporting the idea that "suicidality per se is the important variable in growth hormone secretion," a far-reaching interpretation that requires much more testing before it is a candidate for clinical application.

Thyroid Dysfunction

Mood disorders are a well-known aspect of thyroid dysfunction, whether hypo- or hyperthyroid states are considered. Though suicide is always an issue with disordered mood, a special clinical consideration was introduced by the identification of early symptomless autoimmune thyroiditis (SAT) (Gold, Pottash, & Extein, 1982). This syndrome, apparently related to atrophic changes in the thyroid gland, reveals no clear thyroid symptoms or laboratory abnormalities but presents in the form of anergia, difficulty concentrating, and depression that can reach suicidal intensity. The patient's despair may be increased if the clinician misinterprets the symptoms as a major depressive disorder that does not respond to treatment.

The investigators pointed out that this condition is characterized by an exaggerated TSH response to TRH (in contrast to a blunted or absent response in depression) combined with a positive test for antithyroid microsomal antibodies. The TSH response is recommended as a screening test for patients who present with anergia or depression and do not respond to traditional treatment. If the TSH response is exaggerated, a complete thyroid evaluation would be indicated, and finding antithyroid antibodies for the thyroid microsome would serve as a marker for SAT. The recognition of SAT and other thyroid disorders is especially important because relatively simple treatment (e.g., thyroid medication) may provide relief of symptoms.

Electroconvulsive Therapy

Electroconvulsive therapy (ECT) deserves mention because of its prominent role in the treatment of mood disorders, which in turn are closely associated with suicide. Though clinical and methodological considerations prevent a controlled study of the potential of ECT to prevent suicide, efforts have been made to explore the question. For example, Avery and Winokur (1978) reported a

1-year follow-up of 519 depressed inpatients, of whom 257 received ECT. None of the subsequent four suicides had received ECT, and the rate of suicide attempts during the first 6 months of treatment was 0.85% in the ECT group compared to 7.0% in the "adequate antidepressant treatment" group.

As with many procedures, the key to effectiveness with ECT is patient selection. The best results have been seen in those with primary endogenous depression, but a number of clinical patterns have been seen to respond, even when the disorder has appeared to be intractable. Patients with personality disorders are less likely to show improvement, but if a superimposed depression is present, the suicide risk may be reduced even though some symptoms persist. An excellent review of the clinical issues involved in the use of ECT with suicidal patients was provided by Frankel (1984).

Suicide-Related Medications and Drugs

The metabolic correlates of suicide discussed above have been considered as possible endogenous biological "markers" or indicators of increased suicide risk when their concentrations are within a specified range. It is also the case that some medications used as therapeutic agents in various physical disorders can induce suicidal states, which may present unexpectedly in persons with no prior evidence of emotional difficulty.

Not surprisingly, in view of the compounds noted above, the best known of the exogenous agents associated with suicide is steroids. An advantage here, of course, is that with the onset of cognitive, mood, thought, or behavioral dysfunction the steroids can be reduced or eliminated until normal nervous system function is restored. What sometimes occurs, however, is that the person does not report early signs of toxicity or may manifest only mild intermittent evidence of confusion, despondency, euphoria, or disordered thinking. If the steroid intake is not reduced, suicide may result (Braunig, Bleistein, & Rao, 1989).

Another source of steroids in the contemporary world is self-administered anabolic compounds by athletes and weight lifters. If not monitored closely, such misuse of steroids may lead to a suicidal episode, especially during the first 3 months following abrupt discontinuation of such use (Pope & Katz, 1988). Brower, Blow, Eliopulos, and Beresford (1989) pointed out that patients may be reluctant to volunteer information about steroid use, and that urine testing may be a useful check in the case of a depressed athlete or body builder.

Interferon alfa-2B, an agent used in the treatment of a number of neoplastic and infectious disorders, has been found to cause depression in a significant proportion of patients. Though a number of medications may induce depressive symptoms (e.g., propranolol), the affective disturbance observed with interferon has in some instances been so intense and unresponsive to any antidepressant measures that the medication had to be stopped to avoid an impending suicide (Fenster, 1990).

An active controversy presently creating concern in the clinical community involves the potential of the widely used antidepressant fluoxetine to induce or intensify both suicidal and homicidal thoughts and behavior. Sparked primarily by a report of six depressed patients who developed intense, violent suicidal preoccupation 2 to 7 weeks after starting treatment with fluoxetine (Teicher, Glod, & Cole, 1990a), the broad issue of "antidepressants" paradoxically triggering suicidal states is coming to the fore.

The observation is not new. Alprazolam has been seen to produce a dramatic increase in aggression, self-mutilation, and suicidal behavior (Gardner & Cowdry, 1985), presumably by a process of disinhibition. Of 28 borderline patients treated with amitriptyline, 15 were reported to show increased suicide threats and assaultiveness (Soloff, George, Nathan, Schulz, & Perel, 1986), and new suicidal ideation and preoccupation have been observed following treatment with desipramine (Dalmuji & Ferguson, 1988). One suicidal patient on fluoxetine recovered on a tricyclic agent, and two suicidal patients on desipramine recovered when switched to fluoxetine. A detailed critique of the Teicher et al. (1990a) report by Tollefson (1990) met an equally detailed rebuttal (Teicher, Glod, & Cole, 1990b); and a large study by Fava and Rosenbaum (1990), which showed no greater suicidality in fluoxetine-treated patients than in those on other antidepressants, was cited by Teicher et al. (1990b) as supporting an opposite view when only those patients without suicidal thoughts prior to treatment are considered.

A clinician might best take note of what most observers agree on: specifically, that essentially all psychotropic drugs can produce atypical and unexpected effects, and that watching closely for that possibility is a routine aspect of pharmacological treatment. Fluoxetine-induced akathisia may be pertinent here.

Discussion

The current flood of biological information as regards suicidal persons shows no signs of abating as researchers continue to pursue an exhaustive search of the body and brain to identify new associations between organic observations and self-destructive behaviors. Although the most frequent paradigm is suicide attempts, correlates of past or present suicidal thoughts, violent or nonviolent suicide attempts, a range of impulsive/aggressive/assaultive characteristics, suicide threats, completed suicides, and various combinations of these behaviors are also reported. Postmortem studies provide an additional dimension by revealing different findings in various areas of the brain. The newest investigative technique, postmortem CSF studies, compares CSF samples from different points in the central nervous system. Still another dimension is represented by specific diagnostic categories (e.g., depression, mania, or schizophrenia) that may account for some biological manifestations independently of suicidal behaviors. Though much of what we have learned about the biology of suicide was derived from studies of depressive states, the value of neuroendocrine tests even in major depression remains controversial (Rubin, 1990).

It is tempting to cite several hundred references to document the many intriguing investigations that have been carried out to explore the questions raised in this area, but at present it would not further our search for practical applications. In view of the frequent methodological problems of small samples, multiple paradigms, and retrospective design, the present body of biological information can best be characterized as reflecting an area of ongoing active research rather than a source of new clinical tools.

The most consistent and promising biological data relate low CSF 5-HIAA levels to a variety of suicidal, impulsive, or disinhibited behavior patterns. Although there are some inconsistent findings, so many confirmatory reports have appeared that they have been said to constitute "one of the most highly replicated set of findings in biological psychiatry" (Brown & Goodwin, 1986). Despite the general agreement among researchers that serotonin is somehow related to aggressive behavior, no clinical guidelines have been suggested to translate that observation into a practical clinical tool. A clinician might ask whether such a tool is needed to recognize vulnerability to aggressive behavior in view of its accessibility by clinical inquiry. Such inquiry makes it possible to assess the person's capacity to control aggressive impulses as well, which may be the determining factor as regards outcome.

We might anticipate that one of the next biological systems to become prominent as a possible suicide marker will involve the pineal gland and melatonin, which in the CSF has been found to correlate with CSF magnesium, and in nocturnal serum to be decreased in association with depression (Claustrat, Chazot, Brun, Jordan, & Sassoias, 1984; Milin, , Previtera, Sovljanski, Previtera, & Milin, 1990). Serotonin is a precursor of melatonin, although little is known of any correlation between the concentrations of the two compounds in humans (Asberg, 1989).

Another compound found unexpectedly to be indirectly associated with suicide and not yet studied regarding its links to suicidal behavior is cholesterol. Examining the cardiovascular effects of this substance in a long-term mortality study, subjects with lowered serum cholesterol levels showed a reduced mortality rate from heart disease but a significantly higher rate of death from accidents, suicide, and violence (Muldoon, Manuck, & Matthews, 1990). That the finding was consistent across all of the six studies analyzed creates inevitable questions as to the mechanisms involved, especially whether there is some link to serotonin metabolism.

The list of still-unexplored potential markers for suicide seems almost endless. Those on the immediate horizon include changes in the mu and delta opioid receptors in the brain, altered skin conductivity, reduced prolactin response to fenfluramine, and structural characteristics of the hippocampus.

A source of confusion that may complicate the practical application of new biological data is the difference between clinical significance and statistical significance. The widespread use of the P-value for judging the importance or merit of a given outcome and adhering to the arbitrary 0.05 levels as the guide to acceptance of new findings can be misleading (Gardner & Altman, 1990).

While the search for a clinically useful role for biological markers continues, we are reminded by Stanley (1986) of serious ethical issues involved in such investigation. These issues are especially pertinent to the danger of prematurely identifying a finding as indicative of high risk. For example, what should a person with a positive finding be told? Could a self-fulfilling prophesy emerge? What effect might it have on health or life insurance, sensitive jobs, or public office? The unsettled state of the field is acknowledged in this thoughtful reminder by granting that these concerns are not immediate, but clinicians have been appropriately cautioned.

Investigators and clinicians alike seem to agree that despite many advances the search for biological markers of suicide risk is still in a tentative stage. Secunda et al. (1986) expressed it succinctly: "It is clear that no unitary clinical or biochemical relationship exists between a clinical diagnosis, e.g., depression, and a behavior—suicide. The most recent work attempting to define the impact of alterations in the biogenic amine system on aggression and impulsivity is interesting and invites much speculation as regards one final common pathway behavior—suicide. However, the work remains too sketchy and it must be for future authors to provide the necessary linkage."

Clinical Guidelines: Chapter Summary

Biological correlates of suicide have emerged over the past 25 years as an incidental finding in the course of investigating the biology of depression. The most consistent candidate for potential clinical use as an indicator of suicide risk has been serotonin and its metabolite 5-hydroxyindoleacetic acid. The relation of these compounds to suicidal acts appears to be their role in a broad range of aggressive, impulsive, and anxiety-related behaviors. Although some tentative ideas have been expressed as regards possible clinical applications, there is general agreement that such use at present would be premature. There have been no published reports to our knowledge of the noninvestigative utilization of any biological correlates of suicide, a reflection of the absence of clear guidelines for doing so.

A new standard of care has not been created by the expansion of knowledge about neurochemical and endocrine functions related to suicidal behavior. Contemporary literature and professional meetings devote increasing attention to the issue, but presentations are still couched more in terms of supporting evidence than confirmation of hypotheses.

Although new information continues to be reported at a remarkable rate, the psychobiology of suicide remains an area of exciting ongoing research rather than a source of practical clinical tools. None of the measurements developed to date have a clear and present role in suicide risk assessment.

REFERENCES

Apter, A., van Praag, H., Plutchik, R., Sevy, S., Korn, M., & Brown, S. (1990). Interrelationships among anxiety, aggression, impulsivity, and mood: a serotonergically linked cluster? *Psychiatry Research, 32,* 191–199.

Arango, V., Ernsberger, P., Marzuk, P., Chen, J., Tierney, H., Stanley, M., Reid, D., & Mann, J. (1990). Autoradiographic demonstration of increased serotonin $5HT_2$ and beta-adrenergic receptor binding sites in the brain of suicide victims. *Archives of General Psychiatry, 47,* 1038–1047.

Arato, M., Banki, C., Tothfalusi, L., Nemeroff, C., Bissett, G., Akil, H., & Watson, S. (1986). Postmortem CSF measurements as a new research tool. *Clinical Neuropharmacology, 9*(suppl. 4), 578–580.

Arato, M., Banki, C., Bissett, G., & Nemeroff, C. (1989). Elevated CSF CRF in suicide victims. *Biological Psychiatry, 35,* 355–359.

Arora, R., & Meltzer, H. (1989). Serotonergic measures in the brains of suicide victims: $5HT_2$ binding sites in the frontal corte of suicide victims and control subjects. *American Journal of Psychiatry, 146,* 730–736.

Asberg, M. (1989). Neurotransmitter monoamine metabolites in the cerebrospinal fluid as risk factors for suicidal behavior. In L. Davidson & M. Linnoila, (Eds.), *Report of the Secretary's task force on youth suicide* (Vol. 2; pp. 193–212). DHHS Publ. No. (ADM) 89-1622, Washington, DC: US Government Printing Office.

Asberg, M., Traskman, L., & Sjostrand, L. (1976). Monoamine metabolites in CSF and suicidal behavior. *Archives of General Psychiatry, 33,* 1193–1197.

Avery, D., & Winokur, G. (1978). Suicide, attempted suicide, and relapse rates in depression: occurrence after ECT and antidepressant therapy. *Archives of General Psychiatry, 35,* 749–753.

Banki, C. (1985). Biochemical markers for suicidal behavior. *American Journal of Psychiatry, 142,* 147–148.

Banki, C., Arato, M., Papp, Z., & Kurcz, M. (1984). Biochemical markers in suicidal patients. *Journal of Affective Disorders, 6,* 341–350.

Benki, C., Vojnik, M., Papp, Z., Balla, K., & Arato, M. (1985). Cerebrospinal fluid magnesium and calcium related to amine metabolites, diagnosis, and suicide attempts. *Biological Psychiatry, 20,* 163–171.

Banki, C., Bissette, J., Arato, M., O'Conner, L., & Nemeroff, C. (1987). CSF corticotropin-releasing factor-like immunoreactivity in depression and schizophrenia. *American Journal of Psychiatry, 144,* 873–877.

Bowman, K., & Bender, L., (1932). The treatment of involution melancholia with ovarian hormone. *American Journal of Psychiatry, 11,* 867–892.

Brower, K., Blow F., Eliopulos, G., & Beresford, T. (1989). Anabolic androgenic steroids and suicide [Letter]. *American Journal of Psychiatry, 146,* 1075.

Brown, G., & Goodwin, F., (1986). Human aggression and suicide. *Suicide and Life-Threatening Behavior, 16,* 223–243.

Brown G., & Linnoila, M. (1990). CSF serotonin metabolite (5-HIAA) studies in depression, impulsivity, and violence. *Journal of Clinical Psychiatry, 51*(suppl.), 31–41.

Braunig, P., Bleistein, J., & Rao, M. (1989). Suicidality and corticosteroid-induced psychosis [Letter]. *Biological Psychiatry, 26,* 209–210.

Bunney, W., & Fawcett, J. (1965). Possibility of a biochemical test for suicide potential. *Archives of General Psychiatry, 13,* 232–239.

Bunney, W., Fawcett, J., Davis, J., & Gifford, S. (1969). Further evaluation of urinary 17-hydroxycorticosteroids in suicidal patients. *Archives of General Psychiatry, 21,* 138–150.

Claustrat, B., Chazot, G., Brun, J., Jordan, D., & Sassoias, G. (1984). A chronobiological study of melatonin and cortisol secretion in depressed subjects: plasma melatonin, a biochemical marker in major depression. *Biological Psychiatry, 19,* 1215–1228.

Dahl, R.,, Puig-Antich, J., Ryan, N., Nelson, B., Dachille, S., Cunningham, S., Trubnick, L., & Klepper, T. (1990). EEG sleep in adolescents with major depression: the role of suicidality and inpatient status. *Journal of Affective Disorders, 19*, 63–75.

Dalmugi, N., & Ferguson, J. (1988). Paradoxical worsening of depressive symptomatology caused by antidepressants. *Journal of Clinical Psychopharmacology, 8*, 347–349.

DeLeo, D., & Marazziti, D. (1988). Biological prediction of suicide: the role of serotonin. *Crisis, 9*, 109–118.

Fava, M., & Rosenbaum, J. (1990). Suicidality and fluoxetine: is there a relationship? In *New Research Program and Abstracts, 143rd Annual Meeting, American Psychiatric Association*, Washington, DC: APA.

Fenster, L. (1990). Gastroenterology update. Lecture at Virginia Mason Medical Center, Seattle.

Frankel, F. (1984). The use of electroconvulsive therapy in suicidal patients. *American Journal of Psychotherapy, 38*, 384–391.

Funkenstein, D., King, S., & Drolette, M. (1954). The direction of anger during a laboratory stress-inducing situation. *Psychosomatic Medicine, 16*, 404–413.

Gardner, M., & Altman D. (1990). Confidence –and clinical importance– in research findings (editorial). *British Journal of Psychiatry, 156*, 472–474.

Gardner, D., & Cowdry, R. (1985). Alprazolam-induced dyscontrol in borderline personality disorder. *American Journal of Psychiatry, 142*, 98–100.

Gold, M., Pottash, A., & Extein, I. (1982). "Symptomless" autoimmune thyroiditis in depression. *Psychiatry Research, 6*, 261–269.

Gross-Isseroff, R., Israeli, M., & Biegon, A. (1989). Autoradiographic analysis of tritiated imipramine binding in the human brain post mortem: effects of suicide. *Archives of General Psychiatry, 46*, 237–241.

Jones, J., Stanley, B., Mann, J., Frances, A., Guido, J., Traskman-Benz, L., Winchel, R., Brown, R., & Stanley, M. (1990). CSF 5-HIAA and HVA concentrations in elderly depressed patients who attempted suicide. *American Journal of Psychiatry, 147*, 1225–1227.

Kjellman, B., Ljunggren, J-G., Beck-Fries, J., & Wetterberg, L. (1985). Effect of TRH on TSH and prolactin levels in affective disorders. *Psychiatry Research, 14*, 353–363.

Krieger, G. (1974). The plasma level of cortisol as a predictor of suicide. *Diseases of the Nervous System, 35*, 237–240.

Krieger, G. (1975). Is there a biochemical predictor of suicide? *Suicide and Life-Threatening Behavior, 5*, 228–231

Lidberg, L., Tuck, J., Asberg, J., Scalia-Tomba, G-P., & Bertilsson, L. (1985). Homicide, suicide and CSF 5-HIAA. *Acta Psychiatrica Scandinavica, 71*, 230–236.

Linkowski, P., van Wettere, J., Kerkhofs, M., Brauman, H., & Mendlewicz, J. (1983). Thyrotrophin response to thyrotrophin response to thyreostimulin in affectively ill women: relationship to suicidal behavior. *British Journal of Psychiatry, 143*, 401–405.

Linkowski, P., van Wettere, J., Kerkhofs, M., Gregoire, F., & Brauman, H. (1984). Violent suicidal behavior and the thyrotropin releasing hormon-thyroid stimulating hormone test: a clinical outcome study. *Neuropsychobiology, 12*, 19–22.

Lopex-Ibor, J., Jr., Saiz-Ruiz, J., & Perex do los Cobos, J. (1985). Biological correlations of suicide and aggressivity in major depressions (with melancholia): 5-hydroxyindoleacetic acid and cortisol in cerebral spinal fluid, dexamethasone suppression test and therapeutic response to 5-hydroxytryptophan. *Neuropsychobiology, 14*, 67–74.

Maes, M., Vandewoude, M., Schotte, C., Martin, M., Block, P., Scharpe, S., & Cosyns, P. (1989). Hypothalamic-pituitary-adrenal and thyroid axis dysfunctions and decrements in the availability of L-tryptophan as biological markers of suicidal ideation in major depressed females. *Acta Psychiatrica Scandinavica, 80*, 13–17.

Meltzer, H., & Lowy, M. (1989). The neuroendocrine system and suicide. In L. Davidson & M. Linnoila (Eds.), *Report of the Secretary's task force on youth suicide*, Vol 2; pp. 235–246). DHHS Publ. No. (ADM) 89-1622, Washington, DC: U.S. Government Printing Office.

Milin, J., Previtera, L., Sovljanski, M., Previtera, M., & Milin, R. (1990). The pineal gland morphofunctional disorder: a risk factor of suicidality? In G. Ferreri, M. Bellini, & P. Crepet (Eds.), *Suicidal behaviors and risk factors* (pp. 693–698). Bologna: Monduzzi Editore.

Montgomery, S., & Montgomery, D. (1982). Pharmacological prevention of suicidal behaviors. *Journal of Affective Disorders, 4*, 291–298.

Motto, J. (1986). Clinical considerations of biological correlates of suicide. *Suicide and Life-Threatening Behavior, 16*, 83–102.

Muldoon, M., Manuck, S., & Matthews, K. (1990). Lowering cholesterol concentrations and mortality: a quantitative review of primary prevention trials. *British Medical Journal, 301*, 309–314.

Nemeroff, C., Owens, M., Bissett, G., Andorn, A., & Stanley, M. (1988). Reduced corticotropin releasing factor binding sites in the frontal cortex of suicide victims. *Archives of General Psychiatry, 45*, 577–579.

Ninan, P., van Kammen, D., Scheinen, M., Linnoila, M., Bunney, W., & Goodwin, F. (1984). CSF 5-hydroxyindoleacetic acid levels in suicidal schizophrenic patients. *American Journal of Psychiatry, 141*, 566–569.

Ostroff, R., Giller, E., Bonese, K., Ebersole, E., Harkness, L., & Mason, J. (1982). Neuroendocrine risk factors of suicidal behavior. *American Journal of Psychiatry, 139*, 1332–1325.

Ostroff, R., Giller, E., Harkness, L., & Mason, J., (1985). The norepinephrine-to-epinephrine ratio in patients with a history of suicide attempts, *American Journal of Psychiatry, 142*, 224–227.

Palaniappan, C., Ramachandran, V., & Somasundaram, O. (1983). Suicidal ideatin and biogenic amines in depression. *Indian Journal of Psychiatry, 25*, 286–292.

Pope, H., & Katz, D. (1988). Affective and psychotic symptoms associated with anabolic steroid use. *American Journal of Psychiatry, 145*, 487–490.

Ricci, L., & Wellman, M. (1990). Monamines: biochemical markers of suicide? *Journal of Clinical Psychology, 46*, 106–116.

Roy, A. (1989). Genetics and suicidal beahvior. In L. Davidson & M. Linnoila (Eds.), *Report of the Secretary's task force on youth suicide*, (Vol. 2; pp. 247–262). DHHS Publ. No. (ADM) 89-1622, Washington, DC: U.S. Government Printing Office.

Roy, A., Agren, H., Pickar, D., Linnoila, M., Doran, A., Cutler, N., & Paul, S. (1986). Reduced CSF concentrations of homovanillic acid and homovanillic acid to 5-hydroxyidoleacetic acid ratios in depressed patients: relationship to suicidal behavior and dexamethasone nonsuppression. *American Journal of Psychiatry, 143*, 1539–1545.

Roy A., Pickar, D., Linnoila, M., Crousos, P., & Gold, P. (1987). Cerebrospinal fluid corticotropin-releasing hormone in depression: relationship to noradrenergic function. *Psychiatry Research, 20*, 229–237.

Roy A., Pickar, D., DeJong, J., Karoum, F., Linnoila, M. (1989). Suicidal behavior in depression: relationship to noradrenergic function. *Biological Psychiatry, 25*, 341–350.

Roy, A., Lamparski, D., DeJong, J., Adinoff, B., Ravits, B., Geoge, D., Nutt, D., & Linnoila, M. (1990). CSF monomine metabolites in alcoholic patients who attempt suicide. *Acta Psychiatrica Scandinavica, 81*, 58–61.

Roy, A., Sega, N., Centerwall, B, & Robinette, D. (1991). Suicide in twins. *Archives of General Psychiatry, 48*, 29–32.

Rubin, R. (1990). Value of neuroendocrine tests in major depression remains controversial. *Psychiatric Times*, August, 42.

Ryan, N., Puig-Antich, J., Rabinovich, H., Ambrosini, P., Robinson, D., Nelson, B., & Navacenko, H. (1988). Growth hormone response to desmethylimipramine in depressed and suicidal adolescents. *Journal of Affective Disorders, 15*, 323–337.

Secunda, S., Cross, C., Koslow, S., Katz, M., Kocsis, J., & Maas, J. (1986). Studies of amine metabolites in depressed patients. *Annals of the New York Academy of Sciences, 487*, 231–241.

Soloff, P., George, A., Nathan, R., Schulz, P., & Perel, J. (1985) Paradoxical effects of ami-triptyline on borderline patients. *American Journal of Psychiatry, 143,* 1603–1605.

Stanley, B. (1986). Ethical considerations in biological research on suicide. *Annals of the New York Academy of Sciences, 487,* 42–46.

Stanley, M., Virgilio, J., & Gershon, S. (1982). Tritiated imipramine binding sites are decreased in the frontal cortex of suicides. *Science, 216,* 1337–1339.

Struve, F., Klein, D., & Saraf, K. (1972). Electroencephalographic correlates of suicide ideation and attempts. *Archives of General Psychiatry, 27,* 363–365.

Struve, F., Saraf, K., Arko, R., Klein, D., & Becka, D. (1977). Relationship between paroxysmal electroencephalographic dysrhythmia and suicide ideation and attempts in psychiatric patients. In C. Shagass, S. Gershon, & A. Friedhoff (Eds.) *Psychopathology and brain dysfunctions,* (pp. 199–221). New York: Raven Press.

Teicher, M., Clod, C., & Cole, J. (1990a). Emergence of intense suicidal preoccupation during fluoxetine treatment. *American Journal of Psychiatry, 147,* 207–210.

Teicher, M., Glod, C., & Cole J. (1990b). Dr. Teicher and associates reply [Letter]. *American Journal of Psychiatry, 47,* 1692–1693.

Tollefson, G. (1990) Fluoxetine and suicidal ideation [Letter]. *American Journal of Psychiatry, 147,* 1691–1692.

Traskman, L., Asberg, M., Bertilsson, L., & Sjostrand, L. (1981). Monoamine metabolites in CSF and suicidal behavior. *Archives of General Psychiatry, 38,* 631–636.

Van Praag, H. (1986). Biological suicide research: outcome and limitations. *Biological Psychiatry, 21,* 1305–1323.

Van Praag, H. (1982). Depression, suicide and the metabolism of serotonin in the brain. *Journal of Affective Disorders, 4,* 275–290.

Van Praag, H., & Plutchik, R. (1984). Depression type and depression severity in relation to risk of violent suicide attempt. *Psychiatric Research, 12,* 333–338.

Woodman, D. (1979). Evidence of a permanent imbalance in catecholamine secretions in violent social deviants. *Journal of Psychosomatic Research, 23,* 155–157.

PART II
Special Populations

4

Suicidal Behavior of Children

MAE S. SOKOL AND CYNTHIA R. PFEFFER

Suicide prevention is an important concern for professionals who work with children. Although completed suicide is rare before puberty, it does exist, as does a range of suicidal behaviors. Suicide is the seventh leading cause of death in 5- to 14-year-olds (National Center for Health Statistics, 1990). About 250 children under the age of 14 die by suicide yearly in the United States, and there has been a gradual increase in the incidence of suicide in this age group since the 1970s. This incidence is in contrast to that among adolescents, for whom suicide is the third leading cause of death; only accidents and homicide are more frequent. In the 15- to 24-year-old age group, 12.9 such deaths per 100,000 population occur yearly in the United States.

Suicidal behavior appears more commonly in boys than girls under 12 years of age, although there is no difference in degree of severity by gender. Children from every background and social status are vulnerable (Joffe & Offord, 1990). In a study of normal children (Pfeffer, Zuckerman, Plutchik, & Mizruchi, 1984), 11.9% had some suicidal ideation or behavior.

There is evidence that children as young as 2.5 years of age can display suicidal proclivities (Rosenthal & Rosenthal, 1984), which is borne out in our clinical practice. One 3-year-old boy, for example, was admitted to a pediatrics service after he took one of his mother's digitalis tablets. There was no concern about suicide risk because of the child's young age; but he soon began to hit himself and jump off furniture, and he talked about his sadness and wish to die.

Many find it difficult to believe that preadolescents might want to take their own lives. Yet there is clear evidence that such propensities exist, even though some of these children may not fully understand the consequences of their acts. Although it is necessary for children to have some concept of death to be considered suicidal, they need not understand the finality of death in adult terms or have achieved Piaget's formal operational stage of cognitive development. The definition of suicide therefore requires some modification for children. "Suicidal behavior in children can be defined as any self-destructive behavior that has an intent to seriously damage oneself or cause death" (Pfeffer, 1986, p. 14).

Certain factors appear to protect preadolescents from suicide (Shaffer, 1974;

69

Shaffer & Fisher, 1981). First is children's lower propensity for depression. Depression is a major risk factor for suicidal behavior; and although it occurs in children, it is much less prevalent than in adults. Second, children usually live in a family or group setting, which can provide emotional support and protect against isolation. Isolation and lack of emotional support are known to be related to suicidal proclivities. Certain governmental welfare agencies exist, in fact, to ensure that no child is emotionally neglected. Third, developmental issues may be involved. A child's cognitive abilities and physical strength may not be advanced enough to carry out a suicidal act. Along these lines, the abstract thinking necessary to succumb to the despair and hopelessness that often accompany suicidal tendencies may not develop until the adolescent period (Piaget, 1936).

Self-Destructive Methods Used by Children

A variety of self-destructive means are employed by children with suicidal tendencies. These methods appear to be similar whether the youngsters contemplate, attempt, or commit suicide (Pfeffer, Conte, Plutchik, & Jerrett, 1979, 1980). Pfeffer and colleagues found jumping from heights to be the most common method in 6- to 12-year-olds. Ingestions, drowning, stabbing, hanging, and getting run over by a vehicle were also frequent in this age group. Hollinger (1978) reported firearms, hanging, and poisoning by ingestion or inhalation to be the most common means used by American 10- to 14-year-olds. In England and Wales these findings were corroborated by Shaffer (1974), who also noted boys to be more likely to hang themselves, whereas girls undertook more drug ingestions.

Method choice shows some interesting contrasts with adolescents. Preadolescents are much less likely to use firearms or substance ingestion than their adolescent counterparts (Pfeffer, Newcorn, Kaplan, Mizruchi, & Plutchik, 1988). In a record review of completed suicides, Hoberman and Garfinkel (1989) found that children under 14 years of age were more likely to hang themselves than were adolescents ages 15 to 19. Adolescents were three times as likely to poison themselves with carbon monoxide. The most frequent cause of death in the adolescent group was the use of firearms (44%), which was much less common in the younger group (24%).

A combination of factors determine method choice. Access to dangerous objects and skill in their use are important. Ultimately, one must know the individual child to understand the reasons for a particular choice. It is important to explore the psychodynamic aspects of the behavior, as unconscious factors often determine method choice. when aggressive motives are present, for example, violent means such as cutting and stabbing are more likely to be employed (Bender & Schilder, 1937). More systematic research in this area would be helpful. One 11-year-old boy who attempted to choke himself described longing to be held by his cold and distant parents, yet feeling stifled by their rules.

Risk Factors for Suicide in Children

No one can predict which child will take his or her life. Certain factors, however, have been demonstrated to be associated with suicidal tendencies in children (Pfeffer, 1988; Shaffer & Fisher, 1981; Shafii, Carrigan, Whittinghill, & Derrick, 1985). Freud (1917) stated that the most significant influence on suicidal behavior is conflict with loved persons. He pointed out that actual or perceived lack of love is present in almost all cases.

Depression is known to be associated with suicidal propensities at all ages. Studies have supported this relation in children (Mattsson, Seese, & Hawkins, 1969; Pfeffer et al., 1984; Shaffer, 1974). It is sometimes difficult, however, to discern depression in youngsters. Mood in depressed children may rapidly alternate between sadness and euthymia (Ryan et al., 1987), or the children may be irritable rather than depressed. DSM-III-R (1987) includes irritable mood as one of the possible symptoms of a major depressive episode only in this age group. Also, children frequently do not verbalize their depressed feelings. One must surmise it from their play or demeanor.

These children may be apathetic or anhedonic, lose interest in their usual activities, and have a depressed appearance. Somatic signs of depression may be encountered, including changes in appetite, sleeping pattern, activity level, and ability to concentrate. Helplessness, hopelessness, worthlessness, or excessive guilt may be expressed. They may be withdrawn, making it difficult to communicate with them and assess their suicidal tendencies. Academic performance, which may be compromised in depressed children, is an important factor to consider.

Other risk factors for suicidal behavior in youngsters include antisocial behavior, conduct disorder, and impulsivity (Alessi, McManus, Brickman, & Grapentine, 1984; Shaffer, 1974). Males with concomitant affective disorder and antisocial behavior are considered to be at particular risk. Drug and alcohol abuse correlate with suicidal propensities in adolescents (Pfeffer et al., 1988; Rich, Young, & Fowler, 1986), but preadolescents are much less likely to use these substances. Previous attempts also increase risk (Pfeffer et al., 1988). The "contagion effect," in which suicidal behavior increases among youngsters who know or hear about others exhibiting suicidal behaviors, has been studied in adolescents and is known to be a salient factor (Gould & Shaffer, 1986). Less is known about this effect in the preadolescent age group. Especially in light of children's exposure to such material in the media, further investigation would be fruitful.

The family constellation must be carefully assessed for risk factors of suicidal behavior. As previously stated, youngsters may be relatively protected from suicidal behavior by the structure and support provided by the family. When there are problems in the family, there is often an increase in suicidal tendencies among the children. Divorce, parental death, and absence from the home are risk factors. Parental psychopathology, especially suicidal behavior, depression, and alcohol or drug abuse, correlates highly with suicidal proclivities in the offspring. Children are at higher risk for suicide in homes in which there is abuse, neglect, or a chaotic environment.

There may be a biological basis for the predisposition to suicide in certain families. Genetic studies are ongoing. There is also speculation that suicidal behavior in individuals and families may be correlated with dysregulation of the neurotransmitter serotonin (Mann, Stanley, McBride, & McEwen, 1986).

Only rudimentary research has been done thus far on risk factors for suicide in children. Further studies will hopefully elucidate the issues needed to determine strategies for assessment and intervention of potentially suicidal youngsters. Clinicians must be cognizant of the factors that predict risk for suicidal behavior. This knowledge can aid in the assessment of suicide potential and help to individualize treatment.

Assessing Suicide Potential in Children

Children with any psychological problems should be thoroughly and repeatedly assessed for suicide potential, especially if any of the above-mentioned risk factors are present. Frankly suicidal patients should be appropriately protected and assessed for acute medical danger. The possibility of overdose, blood loss, and so on must be addressed immediately by a physician.

Once the child's immediate safety has been established, he or she should be asked about suicidal impulses in a manner as calm, direct, compassionate, and objective as possible. One should determine if the child has any thoughts of hurting or killing himself or herself. Any expression of suicidal tendencies should be taken seriously and warrants full exploration. Some professionals fear "putting ideas in the child's mind" by bringing up the subject. Not discussing it, however, leaves the child unprotected and creates the impression that the subject is "taboo." Semistructured interviews dealing specifically with this area are available for those clinicians who are more comfortable with such a format. For example, the Child Suicide Potential Scales (Pfeffer, 1986; Pfeffer et al., 1984) comprise a comprehensive battery useful for methodically eliciting information from the child, parents, and caretakers.

It is essential to remember that suicidal impulses fluctuate over time, along a spectrum ranging from ideas to threats, attempts, and suicide itself. The nature and acuteness of suicidal impulses is therefore essential to assess. Does the child have a specific plan? Can this plan lead to death or serious harm? How intent is the child to carry out the plan? The child's potential for suicide should not necessarily be discounted if the plan does not appear medically dangerous. A child who threatens to take two aspirins to kill himself is making an important statement about his state of mind. He may also go on to employ more dangerous methods as he develops cognitively and physically and gains access to more resources. For example, a 9-year-old boy made a serious suicide attempt by taking an overdose of his prescribed imipramine. He had not been thought suicidal prior to this episode. Upon careful history-taking, however, it was learned that he had made a series of ingestions, starting at age 3 when he drank a cup of laundry detergent, ostensibly mistaking it for lemonade.

A history of suicidal behavior should be assessed. If an attempt has already been made, it must be discussed in detail. Did the child actually wish to die? How well did the child understand the consequences of the act? What environmental stressors were present at the time? Was there an attention-seeking or vengeful element to the behavior? Even if the latter is the case, the attempt must be taken seriously, as children can harm or kill themselves in an attempt to manipulate others. One 10-year-old patient wanted to kill himself after his mother accused him of stealing from her purse. He said he hoped that when the stealing continued after his death his mother would realize he was not the thief.

A full mental status examination is essential. This child's affective state is of particular importance. A depressed or irritable child is at risk for suicidal behavior. Reality testing must be noted. A child may deny suicidal ideation yet admit to auditory hallucinations and feel compelled to obey voices telling him to kill himself. How much insight does the child have into the nature of the problem? How good is his judgment? What is his cognitive level? How capable is he of carrying out a suicidal plan? Can he comprehend the consequences of his actions and the finality of death? These abstract concepts are acquired as part of a maturational process (Sahler & Friedman, 1981). Although children may not understand the nature of death as do adults, they should still be considered at risk if they wish to die and are capable of an attempt to hurt or kill themselves.

Even young children are capable of suicidal behaviors. Some find it easier to express their feelings in play than in words. One 5-year-old boy, for example, spoke about feeling sad that he had been taken away from his mother and put into foster care. When asked if he ever felt so sad that he wanted to die, he said nothing, but made motions of stabbing and choking himself.

Play observation should be part of the assessment process for any suicidal child. A detailed history of the child's play should be obtained as well. The play of suicidal children has certain characteristic features (Pfeffer, 1979, 1986). These children repetitively play out themes of self-destruction and aggression, often metaphorically. They typically become more out of touch with their surroundings while engaging in play, making it more difficult to interrupt them. Dangerous play is frequent, such as jumping off furniture, running through traffic, or riding on top of elevator cars. Breaking or misusing toys is also typical.

A complete medical evaluation should take place for any child with suicidal proclivities. Hypothyroidism and other endocrine disorders, for example, may cause depression and suicidal behavior. Certain medications may also affect mood. An ill child may express suicidal tendencies by altering compliance with a medical regimen. For example, diabetic children may misuse insulin or engage in dietary indiscretions to harm or even kill themselves. Older children need to be frankly and carefully questioned about substance abuse. Cognitive abilities and judgment can be altered by drug or alcohol use. Overdosing is also a frequently used method of suicide.

Physical examination is essential. New and old injuries can be indications of past suicide attempts, self-injurious behavior, accidents, or abuse. Drug abuse should be suspected if a damaged nasal septum or needle tracks are present. Blood tests can be used to determine the presence of endocrine disorders or drug use.

Psychological evaluation can clarify various aspects of suicidal behavior. Cognitive tests can help determine the developmental level of the child's understanding of the consequences of his behavior. Learning difficulties, which undiagnosed can lead to feelings of depression and inferiority, can be elucidated.

To understand any suicidal child, one must evaluate the home. A child's suicidal behavior is often an expression of family dynamics. For example, a 10-year-old suicidal boy's mother was depressed and suicidal herself. It was believed within the extended family that the mother was to blame for the boy's problems. He felt guilty for "getting her in trouble," which made him more confused, depressed, and suicidal.

Stress within the family, such as divorce, family turmoil, or change in household composition, may increase a child's suicidal propensity (Cohen-Sandler, Berman, & King, 1982; Mattsson et al., 1969; Paulson, Stone, & Sposto, 1978). One must explore the meaning of such factors to a particular child. Young children, especially, may see themselves as responsible for changes in their environment. For example, a 5-year-old suicidal girl repeatedly played out the story of a mother and father dog who fought constantly about their puppy. At the puppy's funeral, they always kissed and made up.

The problems and strengths of individual family members warrant evaluation. As mentioned previously, parental psychopathology, especially suicidal tendencies, depression, and drug abuse, increases suicide risk in the offspring (Mattsson et al., 1969; Paulson et al., 1978; Pfeffer et al., 1984). These factors have important ramifications for treatment planning. Concerned parents who work in alliance with the therapist are more likely capable of maintaining such a child at risk for suicide at home. Hospital admission is more often necessitated when there is a chaotic family that does not follow recommendations.

It is essential to establish a working alliance with all family members and work closely with them throughout the assessment and treatment process. Family members can provide salient information that cannot be obtained from the youngster. A detailed history must be obtained from the parents, siblings, and all other significant caretakers. Ideally, one should meet with the family as a unit, as well as with each individual separately. One should ask about the child's suicidal behavior, especially the emotional and environmental factors that may have been influential.

Psychosocial functioning is important to consider. The ability to maintain interpersonal relationships and engage in age-appropriate activities may be interrelated with a child's propensity for suicidal behavior. Although systematic study has not yet been done in this area, it has long been acknowledged that social and adaptive skills, as distinct from diagnosis, have an important influence on suicide risk.

Countertransference issues must be given special attention. Working closely with children and parents in turmoil is overwhelming at times. The need for frequent assessment of the patient's changing suicide potential can be stressful. Deciding to hospitalize a patient can be perceived as a personal, not just a treatment, failure. Listening to a patient's veiled threats or expressed hopelessness and helplessness can produce anxiety in the therapist. This situation is compounded when one must deal with the parents' concerns, especially if they are not working in alliance with the therapist.

Countertransference is also a valuable tool when working with these patients. Analyzing one's countertransference helps one understand the patient's unconscious. It also gives one a sense of what the patient wants the therapist to experience. A child who threatens suicide just before the therapist's vacation may make the therapist feel helpless and frustrated because the child is feeling this way. A therapist in such a situation can imagine how the parents feel, dealing with the child on a constant basis.

Anger and sadness are also among the emotions that can be elicited by these patients and their families. It is important for the therapist to acknowledge these feelings and the power that patients can have over us. Talking to a supervisor can be helpful. Each therapist must also know his or her limit and not treat too many suicidal patients at a time.

Treatment of Suicidal Children

Keeping patients alive is the primary goal when treating suicidal children. To this end, difficult decisions sometimes must be made. They may include invoking hospitalization, medication trials, or removal from the home. Multimodal treatment, coordinating the efforts of therapist, child psychiatrist, pediatrician, psychologist, social worker, teachers, and family, is indicated for optimal outcome. Compartmentalized, achievable goals are essential to treatment planning for suicidal children. It is necessary for a variety of reasons, not the least of which are countertransference issues involved in dealing with such difficult patients.

The family must always be encouraged to stay involved in the treatment, no matter which modality is chosen. Professionals can recommend treatment, but parents must ultimately decide what is best for their child. If a family's decision would jeopardize the child's safety, however, it may constitute negligence. For example, parents' refusal to hospitalize an acutely suicidal child may put that child in danger. A report is then made to the local child welfare authorities, who must determine the appropriate course of action. Not to report would constitute negligence on the therapist's part. It is best, of course, to inform the parents of your intent before calling the authorities.

Outpatient psychotherapy may be viable when the child's safety can be maintained at home. Not only must these children not be a danger to themselves, but the parents must be cooperative, supportive, and capable of providing continuous supervision and protection. In individual therapy, the goal is to help the child develop more adaptive means of dealing with the feelings that lead to suicidal behaviors. Crisis intervention and long-term planning are necessary. The therapist should regularly assess suicidal propensity. The appropriateness of a particular treatment approach should be determined on an individual basis. Some children benefit from exploratory treatment, through which they gain an understanding of their thoughts and feelings. Play therapy can be a useful type of exploratory work, especially for younger or less verbal children. Even for older children, it is often easier to "play out" self-destructive impulses than to admit them verbally to an adult therapist. Supportive work may be more beneficial to others, especially if they are in the midst of overwhelming life circumstances or are acutely suicidal. Behavior modification can also be effective.

Cognitive techniques work well with certain children, especially older ones (Beck, Rush, Shaw, & Emery, 1979; Cayton and Russo, 1985). Cognitive therapy is based on the theory that an individual's unique perception of experiences determines habitual ways of feeling and behaving. Psychological disturbances are believed to be due to specific habitual, distorted conceptualizations and dysfunctional beliefs (cognitive distortions). For instance, children may incorrectly interpret life situations, believing themselves responsible for the parents' divorce, or that they must stop the parents' fighting. Children also may judge themselves too harshly, feeling that they do not "deserve to live" because their grades at school did not meet the parents' expectations. The strategy of this technique is to help the patient identify, test the reality of, and then correct these habitual cognitive distortions. The patient learns to think and act more adaptively in specific situations, which leads to a decrease of symptoms. Originally designed for adults, modifications can be made to accommodate the adolescent's (Wilkes & Rush, 1988) or child's cognitive level of development. After several weeks of cognitive therapy, a 12-year-old girl, who had made repeated suicide threats said, "I learned that I really didn't hate myself or want to kill myself. I just didn't like everything about myself, but I can change that. The world is not as black and white as I thought it was. There's a lot more gray."

Combinations of techniques are often the most effective. For example, a 10-year-old boy hit himself and said, "I'm going to kill myself," several times daily in his parents' presence. The parents responded by giving the child added, though sometimes "negative," attention during the episodes. At first they comforted him and spent more time with him; but eventually they became impatient, telling him this behavior was "bad" and "crazy." The father hit him several times, hoping it would "snap him out of it." This action led to arguments between the parents, often in the boy's presence, about how the problem should be managed. This family was taught to administer a behavioral management plan in which rewards were given for self-protective behaviors, and negative behaviors were ignored. A star chart was constructed, on which the boy received a star for each 15 minutes during which he did not hit himself. At the end of each day, he could trade in the stars for special reading time with each parent. In this way, he received more attention for positive behaviors, and the negative behaviors diminished. The parents were relieved and more in control of the situation, which augmented the working alliance, making the parents more amenable to supportive work and obtaining marriage counseling with a different therapist. The boy was more available for insight-oriented work, in which he was able to explore his thoughts and feelings about his parents and himself, which had led to the suicidal behavior.

Therapists must be cognizant of their rationale for choosing a particular treatment approach, including their own psychological reasons. For example, the intense transference and countertransference often elicited in exploratory work with suicidal youngsters may lead the therapist to choose other, more structured techniques.

Group psychotherapy can be particularly beneficial for these children. Youngsters are often more willing to discuss suicidal impulses in a group of peers than alone with an adult therapist. This method serves to improve peer interaction and decreases the isolation and loneliness often characteristic of suicidal children. Group leaders facilitate the interchange of ideas and help youngsters choose more adaptive, age-appropriate methods of dealing with problems.

Working with the family is an integral part of the treatment of any suicidal youngster. Parents need to learn more constructive ways to deal with their child's suicidal behavior and to understand the seriousness of the problem. Parental counseling can be done in sessions with the parents separately or by observing the parents and child perform activities together. Direct observation probably provides the most valuable information about family dynamics. A father, for instance, who cannot tolerate being outdone at sports by his child or cannot set a limit when the child's play gets too rough has particular trouble when the child becomes suicidal. Marital therapy may be necessary if conflicts detract from parents' availability to the child. Family therapy is often indicated to reduce conflict, improve communication, and create a more supportive and empathic home environment. Members of the extended family, such as grandparents, can provide the therapist with a different point of view about the nature of the problem and be an invaluable resource for the child.

Choice of treatment setting is a crucial factor to consider with a suicidal child. The goal is to keep the child in the mainstream as much as possible while maintaining safety. Parents must be involved in the choice of treatment setting. Often they are not prepared to accept the level of care recommended, especially when it involves the child's removal from the home. A less structured setting may have to be accepted, at least temporarily, provided the child's safety can be maintained.

For a child who can remain at home, an appropriate school must be chosen. If the child can attend a mainstream school, at least one staff member should be chosen to serve as an ombudsman. This person should know about the child's problems, be available to the child throughout the day, and be able to contact the therapist or family quickly if problems arise. If a child can remain at home only with highly structured and supervised treatment, a psychiatric day hospital may be appropriate. In the day hospital, schooling, therapy, medication, and milieu treatment are provided in one setting.

A variety of treatment settings are available for a child who cannot remain at home, either because of individual psychopathology or because of the home environment. One choice is a residential treatment center, a highly structured therapeutic community where round-the-clock supervision and multimodal interventions are provided. Foster care placement and group homes are other possible placements.

Psychiatric hospitalization may be necessary for an acutely suicidal youngster. The logistics of hospitalization depend on the particular community's resources. Referral can be made through a child psychiatrist or directly to the hospital. If placement at the hospital is unavailable, the child can be sent to a hospital emergency room for evaluation. Many communities also provide crisis intervention teams.

Psychotropic medication may enhance other forms of treatment for suicidal youngsters. The pharmacological agent chosen depends on the child's underlying psychopathology. The pharmacological armamentarium of child psychiatry is limited, and medication choice is based on target symptoms, rather than diagnoses. Benefits and risks of medication must be carefully weighed throughout the treatment process. Sometimes several drug trials are necessary before the optimal regimen can be determined. Developmental issues are of particular concern; medication is administered to children not only to decrease symptoms but to allow appropriate development.

The parents and child need to be intimately involved in the process, not just passive recipients of a substance. How a family views medication can be pivotal in the treatment process. This issue can be difficult if, for example, the child has taken a medication overdose in the past, or there is a familial history of drug abuse. The indications, goals, untoward effects, and approximate time frame of the medication trial must be clearly spelled out to the family. It is essential to determine who will be administering the medication and where it will be kept. Although an adult must always be responsible, the child should know what the pill looks like and what to do if a problem arises.

School personnel often need to be involved in pharmacological treatment. If a noontime dose is required, for example, the school's ability to administer it, how the medicine will be delivered to the school, and the child's concerns about being singled out are among the issues that warrant attention. Children often respond differently in the school environment than at home, so reports from school can be crucial for monitoring results.

Depressed suicidal youngsters most often are treated with tricyclic antidepressants. Imipramine is the antidepressant most commonly used for depressed, suicidal children (Ambrosini & Puig-Antich, 1985; Campbell, Green, & Deutsch, 1985). Double-blind controlled studies, however, do not demonstrate antidepressants' superiority over placebo in children (Puig-Antich et al., 1987) or adolescents (Kramer & Feiguine, 1981). Furthermore, extreme caution must be exercised when administering these substances to individuals because of cardiac and behavioral toxicity (to which youngsters are more susceptible) as well as potential for overdose (Gilman, Goodman, Rale, & Murad, 1985).

A psychotic child may benefit from neuroleptics, such as haloperidol or thioridazine. Such children may jump out a window believing they can fly or hear voices telling them to kill themselves. Neuroleptics can decrease psychotic symptoms, increasing the children's cognizance of their actions' consequences and their availability to other therapeutic interventions. An aggressive, conduct-disordered suicidal child may respond to lithium or haloperidol (Campbell et al., 1984). A variety of medications may be included in the treatment of a drug-abusing youngster, such as disulfiram for alcoholism. It must be emphasized that medication is only a useful adjunct to other forms of treatment and cannot be expected to independently eliminate the problem (Ambrosini & Puig-Antich, 1985; Campbell et al., 1985).

Summary

Suicidal behavior is a complex symptom that is not uncommon in youngsters. It is one of the most difficult problems facing professionals who work with children. Helping young patients and their families deal with such a difficult problem can be particularly stressful. Countertransference issues may affect a therapist's ability to be effective unless appropriately addressed. To maximize preventive efforts, it is vital to take seriously any expression of suicidal tendencies. Risk factors must be taken into account and appropriate interventions promptly instituted. The coordinated expertise of several professional disciplines is necessary to keep these children safe and for optimal treatment planning.

Guidelines for Practice

I. Epidemiology
    A. Suicide is the seventh leading cause of death in 5- to 14-year-olds.
    B. Approximately 250 children aged 5 to 14 years die by suicide yearly in the United States.

    C. There has been an increase in suicidal behaviors in this age group during the last several decades.

    D. Completed suicide is rare before puberty, but suicidal ideation, threats, plans, and attempts are not.

II. Risk factors

    A. Individual factors

       1. Depression

          a. Mood—depressed or irritable

          b. Loss of interest or pleasure in usual activities

          c. Appetite—decreased or increased

          d. Sleep pattern—insomnia, hypersomnia, or daytime sleepiness

          e. Activity level—decreased or increased

          f. Loss of energy

          g. Difficulty concentrating

          h. Distractibility

          i. Feelings of helplessness, hopelessness, worthlessness, or excessive guilt

       2. Antisocial behavior or conduct disorder

       3. Impulsive behavior

       4. Alcohol and drug abuse

       5. Academic and behavior problems at school

       6. Previous attempts or self-destructive behavior

       7. History of frequent accidents

       8. "Contagion effect"—hearing in the mass media or knowing about someone who exhibited suicidal behavior

    B. Family factors

       1. Chaotic home environment

       2. Divorce

       3. Death of family member

       4. Absence of parent from home

       5. Physical or sexual abuse

       6. Parental psychopathology, especially:

          a. Suicidal behavior

          b. Depression

          c. Alcohol and drug abuse

III. Assessment

    A. Acute medical management

       1. Evaluate for possible overdose, blood loss, etc.

       2. Remove access to pills, sharp objects, etc.

       3. Keep the child safe and under constant observation

    B. Interview the child, family, and other caretakers to obtain a complete history of present episode. Certain questions need to be answered:

       1. Does he/she have suicidal ideation?

       2. Is there a suicide plan?

      3. Does the child have access to dangerous implements?

      4. Is suicidal intent present?

      5. What environmental stressors are at play?

  C. Obtain a complete psychiatric history.

      1. Past suicidal behaviors

      2. Past psychiatric treatment and medication trials

      3. Psychosocial history

      4. Family history

  D. Mental status examination

  E. Physical examination

  F. Educational history

  G. Psychological testing

    Note:

      1. Don't treat suicide as a taboo subject by refusing to discuss it.

      2. Questions or confidentiality: Although it is not necessary to repeat details of what the child says, this subject is too serious not to share it with parents or responsible others. It is best if the child shares information or at least knows in advance what must be shared.

IV. Interventions

Determine acuity of the problem.

  A. Acute treatment is warranted when a child is in danger of hurting himself or herself.

      1. Inform parents or caretakers.

      2. Perform complete psychiatric evaluation

      3. Hospitalize if necessary (invoke local crisis intervention team and/ or child welfare agency, if necessary).

  B. Long-term treatment

      1. Individual psychotherapy

      2. Parent counseling

      3. Group therapy

      4. Family therapy

      5. Psychopharmacological intervention if indicated

REFERENCES

Alessi, N. E., McManus, M., Brickman, A., & Grapentine, L. (1984). Suicidal behavior among serious juvenile offenders. *American Journal of Psychiatry, 141,* 286–287.

Ambrosini, P. J., & Puig-Antich, J. (1985). Major depression in children and adolescents. In D. Shaffer, A. A. Ehrhardt, & L. L. Greenhill (Eds.), *The clinical guide to child psychiatry* (pp. 182–191). New York: Free Press.

Beck, A. T., Rush, A. J., Shaw, B. F., & Emery, G. (1979). *Cognitive therapy for depression.* New York: Guilford Press.

Bender, L., & Schilder, P. (1937). Suicidal preoccupation and attempts in children. *American Journal of Orthopsychiatry, 7,* 225–235.

Campbell, M., Small, A. M., Green, W. H., Jennings, S. J., Perry, R., Bennett, W. G., &

Anderson, L. (1984). Behavioral efficacy of haloperidol and lithium carbonate: a comparison in hospitalized aggressive children with conduct disorder. *Archives of General Psychiatry, 41*, 650–656.

Campbell, M., Green, W. A., Deutsch, S. I. (1985). *Child & Adolescent Psychopharmacology.* Beverly Hills, CA: Sage Publications.

Cayton, T. G., & Russo, D. C. (1985). The behavior therapies. In D. Shaffer, A. A. Ehrhardt, & L. L. Greenhill (Eds.), *The clinical guide to child psychiatry* (pp. 519–538). New York: Free Press (Macmillan).

Cohen-Sandler, R., Berman, A. L., & King, R. A. (1982). Life stresses and symptomatology: determinants of suicidal behavior in children. *Journal of the American Academy of Child Psychiatry, 21*, 178–186.

Diagnostic & statistical manual of mental disorders (3rd Ed., revised) (1987). Washington, DC: American Psychiatric Association.

Freud, S. (1917). Mourning and melancholia. In: *The standard edition of the complete psychological works of Sigmund Freud, 20.* London: Hogarth Press

Gilman, A. G., Goodman, L. S., Rale, T. W., & Murad, F. (Eds.) (1985). *Goodman & Gilman's the pharmacological basis of therapeutics.* New York: Macmillan.

Gould, M. S., & Shaffer, D. (1986). The impact of suicide in television movies: evidence of imitation. *New England Journal of Medicine, 315*, 690–694.

Hoberman, H. M., & Garfinkel, B. D. (1989). Completed suicide in youth. In C. R. Pfeffer (Ed.), *Suicide among youth: perspectives on risk & prevention* (pp. 21–40). Washington, DC: American Psychiatric Press.

Holinger, P. C. (1978). Epidemiologic issues in youth suicide. In C. R. Pfeffer (Ed.), *Suicide among youth: perspectives on risk & prevention* (pp. 41–62). Washington DC American Psychiatric Press.

Joffe, R. T., & Offord, D. R. (1990). Epidemiology. G. MacLean (Ed.), *Suicide in children & adolescents* (pp. 1–14). Toronto: Hogrefe & Huber.

Kramer, A. D., & Feiguine, R. J. (1981). Clinical effects of amitriptyline in adolescent depression: a pilot study. *Journal of the American Academy of Child Psychiatry, 20*, 636–644.

Mann, J. J., Stanley, M., McBride, P. A., & McEwen, B. S. (1986). Increased serotonin and beta-adrenergic receptor binding in the frontal cortices of suicide victims. *Archives of General Psychiatry, 43*, 954–959.

Mattsson, A., Seese, L. R., & Hawkins, J. W. (1969). Suicidal behavior as a child psychiatric emergency. *Archives of General Psychiatry, 20*, 100–109.

National Center for Health Statistics (1991). Births, marriages, divorces & deaths for October 1990. *Monthly vital statistics report* (Vol. 39, No. 10). Hyattsville, MD: U.S. Public Health Service.

Pfeffer, C. R. (1979). Clinical observations of play of hospitalized suicidal children. *Suicide & Life-Threatening Behavior, 9*(4), 235–244.

Pfeffer, C. R. (1986). *The suicidal child.* New York: Guilford Press.

Pfeffer, C. R. (1988). Suicidal behavior among children and adolescents: risk identification and intervention. In A. J. Frances & R. E. Hales (Eds.), *American Psychiatric Press Review of Psychiatry, 7*, 386–402.

Pfeffer, C. R., Conte, H. R., Plutchik, R., & Jerrett, I. (1979). Suicidal behavior in latency-age children: an empirical study. *Journal of the American Academy of Child Psychiatry, 18*, 679–692.

Pfeffer, C. R., Conte, H. R., Plutchik, R., & Jerrett, I. (1980). Suicidal behavior in latency-age children: an empirical study. *Journal of the American Academy of Child Psychiatry, 19*, 703–710.

Pfeffer, C. R., Newcorn, J., Kaplan, B., Mizruchi, M. S., & Plutchik, R. (1988). Suicidal behavior in adolescent psychiatric inpatients. *Journal of the American Academy of Child Psychiatry, 27*(3), 357–361.

Pfeffer, C. R., Zuckerman, S., Plutchik, R., & Mizruchi, M. S. (1984). Suicidal behavior in normal school children: a comparison with child psychiatric inpatients. *Journal of the American Academy of Child Psychiatry, 23*, 416–423.

Phillips, D. P., & Carstensen, L. L. (1986). Clustering of teenage suicides after television news stories about suicide. *New England Journal of Medicine, 315*, 685–689.

Piaget, J. (1936). *The origins of intelligence in children.* New York: International Universities Press.

Puig-Antich, J., Perel, J. M., Lupatkin, W., Chambers, W. J., Tabrizi, M. A., King, J., Goetz, R., Davies, M., & Stiller, R. L. (1987). Imipramine in prepubertal major depressive disorders. *Archives of General Psychiatry, 44*, 81–89.

Rich, C. L., Young, D., & Fowler, R. C. (1986). San Diego suicide study. I. Young versus old subjects. *Archives of General Psychiatry, 43*, 577–582.

Rosenthal, P. A., & Rosenthal, S. (1984). Suicidal behavior by preschool children. *American Journal of Psychiatry, 141*, 520–525.

Ryan, N. D., Puig-Antich, J., Ambrosini, P., Rabinovich, H., Robinson, D., Nelson, B., Iyengar, S., & Twomey, J. (1987). The clinical picture of major depression in children and adolescents. *Archives of General Psychiatry, 44*, 854–861.

Sahler, O. J., & Freidman, S. B. (1981). The dying child. *Pediatrics in Review, 3*(5), 159–165.

Shaffer, D. (1974). Suicide in childhood and early adolescence. *Journal of Child Psychology & Psychiatry, 15*, 275–291.

Shaffer, D., & Fisher, P. (1981). The epidemiology of suicide in children and young adolescents. *Journal of the American Academy of Child Psychiatry, 20*, 545–565.

Shafii, M., Carrigan, S., Whittinghill, J. R., & Derrick, A. (1985). Psychological autopsy of completed suicide in children and adolescents. *American Journal of Psychiatry, 142* 1061–1064.

Wilkes, T. C. R. (C.), & Rush, A. J. (1988). Adaptations of cognitive therapy for depressed adolescents. *Journal of the American Academy of Child and Adolescent Psychiatry, 27*(3), 381–386.

5

*Suicidal Behavior of Adolescents**

ALAN L. BERMAN AND DAVID A. JOBES

The similarity between Jerry Minnick and Paul Sloan ends with the fact that both were 18-year-old graduating high school seniors whose suicides will be etched permanently into the memories of the survivors, especially their peers. Jerry Minnick was a deeply disturbed young man who felt chronically abandoned. "Everyone whom he ever got close to dumped on him, rejected him, abandoned him," remarked his therapist. "His suicide was the ultimate act of revenge. . . . It was a statement to the world [to] 'Go to Hell!' "

Jerry's mother killed herself when he was 4 years old. His father abandoned him soon thereafter. Jerry was then adopted into a home ruled by harsh discipline and fundamentalist religion. This home never felt comfortable to Jerry, and he ran away often. With the court's intervention Jerry was placed into foster care. Bitter toward parents and their authority, Jerry spent his adolescence drunk and stoned. He fathered and abandoned an illegitimate child, and he attempted to kill himself by overdosing on drugs.

Therapy did not help. He was depressed and noncompliant, refusing recommended medications. Jerry dropped out completely 5 months before his death. His suicide was most public and received considerable media attention. After writing a suicide note and putting it inside his motorcycle helmet, Jerry deliberately drove his cycle at full speed into a wall of his high school, instantly identifying with all the aggressors in his life by rejecting both life and others. In his note he expressed his ultimate identification, asking to be buried next to his mother.

Paul Sloan was the youngest of three sons in an intact, highly successful family. One brother was following in his father's footsteps having achieved Law Review at a prestigious East Coast law school. Paul's other brother was completing his junior year in college with a straight A average. Paul was soon to give the valedictory address to his graduating high school class. He had recently returned from an extraordinary 5 weeks in Moscow as one of a handful of

*The material presented in this chapter has been reprinted, abstracted, and adapted from A. L. Berman & D. A. Jobes (1991). *Adolescent suicide: Assessment and intervention.* Washington, DC: American Psychological Association. Permission to reprint has been granted by the American Psychological Association and is gratefully acknowledged.

84

students from his class participating in an exchange program. Awaiting his return were letters of acceptance from each of the elite universities to which he had applied.

Paul was well liked by his peers, noted to be only slightly uncomfortable around girls but popular within a group of friends. His parents appeared to have a stable marriage with no overt conflict. His life appeared charmed.

Only his suicide belied that appearance. On a Saturday night 1 week before his graduation he overdosed on more than 100 over-the-counter medications. He left no note. The family remarkably was able to keep the media off the story and Paul's name off the front page.

For those of us not even remotely involved in their lives, Jerry's and Paul's suicides quickly translate into mortality statistics. Jerry and Paul were only two of the approximately 1900 suicides of adolescents 19 years old and younger that occurred in the United States last year. Their deaths were not typical of all youth suicides. There simply is no such thing as a typical youth suicide. Each of these unwanted deaths makes its own unique contribution to the gestalt of youth suicide. These aggregated mortality statistics, however, summarized by modal characteristics, both demographic and situational, help alert our attention to factors of risk common to each of these youthful suicides. In turn, these risk factors serve as the basis for clinicians to better detect the risk of suicide by others like Jerry and Paul. These assessments also further our ability to intervene early in order to prevent such premature and tragic consequences. It is through this process of understanding the nomothetic net of youth suicide that the idiographic case is best understood and treated.

Epidemiology of Adolescent Suicide

Within the constraints and limitations of the epidemiological study of suicide we reviewed extensively elsewhere (Berman & Jobes, 1991), it must be noted that our concern for adolescents who commit suicide has grown commensurate with and as a direct result of epidemiological study. Adolescent suicide is not a modern phenomenon. However, marked changes in its incidence and prevalence in recent decades have alerted the attention of clinicians and public health specialists, educators, and the public at large to a problem now demanding response. Correspondingly, that concern is warranted ought not be equated with the need for hyperbole and distortion. Increased rates do not an "epidemic" make. The data are sufficiently instructive.

In 1987 there were 4924 officially recorded suicides for young people between the ages of 15 and 24 (National Center for Health Statistics, 1989), translating into a rate of 12.9 per 100,000. This rate was only slightly lower than the peak rate this century for this age group, 13.3 per 100,000 in 1977. When separately examined, however, rates for 15 to 19-year-olds have continued to rise since the late 1970s, whereas those of the 20 to 24-year-old group have been on a downward trend.

These most recent rates are markedly higher than those derived 30 years ago. Between 1957 and 1987 completed suicide among 15 to 24-year-olds more than tripled (from 4.0/100,000 to 12.9/100,000—an increase of 222%). For 15 to 19-year-olds, rates have quadrupled during this time period (from 2.5/100,000 to 10.3/100,000—an increase of 312%). Youth suicide rates in the United States have always been higher among whites than among blacks. The modal race/ sex group of completers is that of white males. However, rates among black males have increased 206% between 1960 and 1987. Although rates for females are only about one-fifth those for males (the 1987 ratio was 5.1:1.0), both white and black female rates have also increased, more than doubling since 1960.

Holinger, Offer, and Zola (1988) have examined period effects (temporal trends) among youth suicide rates, showing that increases and decreases in U.S. youth suicide rates have corresponded with increases and decreases in the proportion of adolescents in the population as a whole. Their consequent projection anticipates a short-term decline in rates followed by an upturn from the mid-1990s on. In contrast, Easterlin (1978) examined cohort effects. He hypothesized that as post-World War II "baby boomers" move toward old age they may be expected to bring with them problems (e.g., higher suicide rates) inherent in such larger birth cohorts. Newman and Dyck (1988), for example, were able to document such an effect but noted that it is difficult to distinguish cohort from period effects.

Increases in youth suicide rates have not been limited to the United States. Lester (1988) investigated rates among 15 to 24-year-olds in countries with a minimum of one million population (and a minimum of 100 suicides annually) and found increases in 23 of 29 countries. Some increases, however, were limited to only one gender. Diekstra (1989) reported increases (1970–1985) in 9 of 13 countries examined.

Methods of Adolescent Suicide

The choice of method or means used to effect a self-inflicted death depends on a number of factors, most importantly one's psychological intention (aim, goal). Intention and lethality are linked significantly (Brent, 1987), with greater lethality of method being related to an intention of death. In addition to intentionality, factors such as (1) the availability and accessibility of a particular means; (2) one's knowledge, experience, and familiarity with a method; (3) the personal meaning or symbolic or cultural significance of a method; and (4) the actor's state of mind play a role in method choice.

In 1987 the modal method of choice for completed suicide among youth in the United States (57%) was that of firearms and explosives. More than three-fourths of these self-inflicted gunshot wound victims were white males. Hanging accounted for 19% of self-inflicted deaths among youths. Since 1967 there has been a slight increase in these more active methods of suicide relative to slight decreases in all other methods. It has been argued (Boyd, 1983) and documented (Sloan, Rivara, Reay, Ferris, & Kellerman, 1990) that the greater availability and accessibility of firearms has had a direct, measurable, and explanatory effect on observed increases in youth suicide in the United States.

Suicide-Attempting Behaviors Among Adolescents

Nonfatal suicide behavior among adolescents is difficult to measure epidemiologically. Most youth suicide attempts (about seven of every eight) are of such low lethality as to not require medical intervention (Smith & Crawford, 1986). Thus many "attempts" are never reported but for surveyed self-identifications of adolescents.

Two surveys of high school students (Harkavy-Friedman, Asnis, Boeck, & DiFiore, 1987; Smith & Crawford, 1986) estimated the suicide attempt rate at 8 to 9%. The estimated ratio of attempts to completions in this population is about 100:1 (Jacobziner, 1965), although one report (Garfinkel, 1989) of a rural population of self-referred attempters places the ratio at 350:1. This discrepancy is no doubt even greater among female adolescents, who represent the modal low lethality attempter and the rare completer. Most of these attempts involve drug ingestions taken at home often in front of others. Although clearly not of life-threatening consequence, the significance of these behaviors should not be discounted by clinicians. Once an attempt has been made, irrespective of lethality, the risk for future and more serious attempts and completions increases significantly.

Risk Factors

Clearly the modal attempter and the modal completer during adolescence are different. Inferences drawn from the study of one and applied to the other population must be done so with great caution. For the most part, what is known about adolescent suicide comes from studies of relatively nonlethal attempters (Berman & Cohen-Sandler, 1982; Maris, 1981), groups composed largely of female drug ingestors (80-plus %) who comply with recommended treatment regimens. It is therefore helpful to distinguish attributes associated with different suicidal behaviors in order to arrive at useful commonalities for the assessment of risk (refer to Shneidman, 1985).

To date, little research has been published assessing the psychological characteristics of adolescent completers. Two studies—those by Shaffi, Carrigan, Whittinghill, and Derrick (1985) and by Shaffer and Gould (1987)—together yield important data contrasting these youths with both nonsuicidal controls and nonfatal attempters. For example, each of these studies highlights greater frequencies of both conduct disorders and substance use among largely male groups of adolescent completers. In contrast, nonfatal attempters present with a range of associated pathologies, perhaps the most frequent of which are affective disorders. These three disorders predominate in reports of completed and attempted suicide of adolescents and often present co-morbidly, with the frequency and lethality of attempts increasing with the degree of co-morbidity (Frances & Blumenthal, 1989).

Depression, both unipolar and bipolar, has been frequently reported among both completers (Brent et al., 1988; Rich, Young, & Fowler, 1986; Shaffer & Gould, 1987; Shafii et al., 1985) and nonfatal attempters (Carlson, 1983; Friedman, Corn, Aranoff, Hurt, & Clarkin, 1984). The relation between depression and suicidality, however, is complex (Carlson & Cantwell, 1982). It should be emphasized that most depressed youths are not suicidal. The estimated ratio of depressed to depressed suicidal adolescents is 660:1 (Shaffer & Bacon, 1989).

Substance use and abuse have been found with great frequency among both completers (Rich et al., 1986; Shaffer & Gould, 1987; Shafii et al., 1985) and nonfatal attempters (Garfinkel, Froese, & Hood, 1982; Riggs, Alvario, McHorney, DeChristopher, & Crombie, 1986). Among adolescent substance users, suicide attempts have been found to occur three times as frequently as among controls, with the wish to die increasing dramatically *after* the onset of substance use (Berman & Schwartz, 1990). Most important, substance use at the time of suicidal behavior has been found to be related to the lethality of the method used (Brent, Perper, & Allman, 1987).

Rosenstock (1985) has reported concurrent increases in inpatient admission diagnoses of depression (350%), substance abuse (200%), and suicidal ideation (300%) in a 9-year longitudinal study. Common to these symptom presentations are a number of intervening variables significant to weaving together a pattern of risk: cognitive distortions, impulsiveness, frequent and serious interpersonal loss, and family pathology.

Few suicides appear free of psychiatric symptoms prior to death (Shaffer, Garland, Gould, Fisher & Trautman, 1988). Even more profoundly, Rich and colleagues (1986) asserted that "psychiatric illness is a necessary . . . condition for suicide." Exemplifying this assertion, Runeson (1989) reported on an older sample (ages 15–29) of Swedish completors, finding *only one* of 58 cases not having an Axis I or II diagnosis.

The third commonly found pathology—conduct disorder—describes behaviors of dyscontrol and difficulties with authorities and systems of external control. Apter, Bleich, Plutchik, Mendelsohn, and Tyano (1988) found higher scaled scores for suicidality on the K-SADS for hospitalized conduct-disordered adolescents than those with major depressive disorder. Central to the diagnosis may be difficulties in the control of aggression. Plutchik, van Praag, and Conte (1989) proposed that suicide risk is heightened when aggressive impulses are triggered and not attenuated by opposing forces. Such a relation, then, may underlie reported frequencies of diagnosed borderline personality disorder among adolescents making more serious suicide attempts (Friedman, Clarkin, & Corn, 1982).

Common to these diagnoses are symptoms of affect dysregulation, intense rage, and impulsive behavior, personality traits commonly reported in studies of adolescent parasuicides (Berman & Jobes, 1991). Related, as well, to these conditions is hopelessness, a cognitive-emotional variable often found significantly more often among adult and adolescent parasuicides (Spirito, Williams, Stark, & Hart, 1988). However, during adolescence the relation between suicidality and hopelessness has not been upheld universally (Rotheram-Borus & Trautman, 1988).

As noted at the beginning of this chapter, there is no typical suicidal adolescent. Subgroups of completers and attempters have been found showing evidence of anxiety, perfectionism, and distress, particularly at times of change and dislocation (e.g., Paul Sloan) (Shaffer et al., 1988). Shaffer et al. (1988) also reported learning disabilities and schizophrenia as characterizing adolescent completers. The most common descriptors applied to Hoberman and Garfinkel's (1988) sample of adolescent completers were withdrawn, lonely, and supersensitive, labels remarkably similar to those Shafii et al. (1985) found in two-thirds of their subjects.

The behavioral withdrawal noted above may be a sign of depression or a coping strategy (Spirito, Brown, Overholser, & Fritz, 1989). Social isolation and alienation have consistently identified the suicidal adolescent (Faberow, 1989). Adolescent substance users who have attempted suicide significantly more often than controls (nonsuicidal users and normals) characterize their early childhood as lonely (Berman & Schwartz, 1990), a finding consistent with that of other studies reporting suicidal adolescents to be lonely and socially withdrawn (Petzel & Cline, 1978; Rubenstein, Heeren, Houseman, Rubin & Stechler, 1989).

Social alienation may be predictive of emotional disturbance generally and of depression in particular. These problems present symptomatically through a variety of school problems (e.g., poor performance, acting-out) or through runaway behavior, reflective of signs of family conflict. Similarly, adolescents involved in santanism have been reported to have frequent histories of suicide attempt (Bourget, Gagnon, & Bradford, 1988), but our understanding of these youths awaits more control group study.

Among the most studied of variables is the influence of the family and, in particular, the parental system. Reviews of these studies are reported elsewhere (Berman & Jobes, 1991; Pfeffer, 1989) but may be summarized as follows. Compared to normal adolescents, suicidal adolescents suffer greater family stress, particularly due to changes and threatened changes in the parental system, such as loss, death, separation, and divorce and a consequent lack of support. In addition, their families are characterized by greater parental dysfunction, suicidality, and psychopathology, the latter ranging from generalized psychiatric problems to depression and substance abuse and include consequent aggression and violence, abuse, and neglect of children.

Exposure to the suicidal behavior of another person in the social network or family is itself a significant factor of risk for ideators, attempters, and completers (Garfinkel et al., 1982; Harkavy-Friedman et al., 1987; Shafii et al., 1985; Smith & Crawford, 1986). Murphy and Wetzel (1982), after reviewing the literature, estimated that 6 to 8% of adolescent attempters had a family history of suicide.

Such exposure to another's suicidal behavior may be an accelerating factor (versus a causative factor) for those already predisposed to be at risk. With regard to media as a source of this exposure, early and widely publicized reports of aggregate increases in youth suicide after television presentations on suicide (e.g., Gould & Shaffer, 1986) have been contradicted by later studies (Berman, 1987; Kessler, Downey, Milavsky, & Stipp, 1988). However, media exposure may still exert an effect on an idiopathic level, influencing some adolescents toward loss of hope and, in contrast, exemplifying an alternative (i.e., suicide) to a life of perceived pain.

Imitation or modeling appears to be the primary mechanism by which exposure effects follow-up suicides in clusters as well (Davidson, Rosenberg, Mercy, Franklin, & Simmons, 1989). Adolescents, in general, are highly susceptible to suggestion and imitative behavior. To a disturbed youngster, particularly one with preexisting suicidal impulses and diffuse ego boundaries, the perceived attention given to a suicidal event might easily stimulate irrational cognition: Attention and notoriety may be both gained and appreciated through suicidal behavior, even in death.

Psychopathology and disordered personality characteristics observed in suicidal adolescents may be explained through any of a variety of mechanisms from social learning or modeling to genetics and biochemistry. There is both research support and contradiction for each of these modes of influence, none of which we have space to elucidate here. One clear artifact of the disturbed familial and social contexts is notably increased levels of stress in the presuicidal lives of these adolescents (Cohen-Sandler, Berman, & King, 1982; Rubenstein et al., 1989); however, Paykel (1989) concluded that there is a dearth of studies employing careful methodology and controlled comparisons of recent life events of suicidal versus nonsuicidal youths. In addition, other variables (e.g., psychopathology) may mediate between stress and suicidality rather than result from stressful life events.

The very occurrence of suicidal behavior in a context of stressful life events strongly implicates the absence or ineffectiveness of coping strategies and alternative cognitive problem-solving. Deficits in problem-solving skills in suicidal children and adolescents have been well documented (Berman & Jobes, 1991). These deficits likely distort perceptions, narrow the range of alternatives, and greatly increase a sense of hopelessness and the risk of impulsive behavior. Under such conditions the risk of suicidal behavior greatly increases. Moreover, once a suicidal behavior occurs, the risk for more lethal consequence increases as well.

In studies of completed youth suicide, a history of prior attempt consistently stands out as a risk factor. Summarizing the literature, Farberow (1989) reported prior attempts, threats, ideation, or a combination of these factors to be "one of the strongest indicators" of risk, with reported frequencies varying from 22% to 71% of those making subsequent attempts. Similarly, Shafii et al. (1985) concluded that a prior attempt was the single most important risk factor for ultimate completion.

Conversely, follow-up studies of teenage attempters find the rate of completed suicide to be as high as 9% among males and 4% among females (Motto, 1984; Shaffer, 1988). Conversely, Shaffer and Gould (1987) reported that 21% of male and 33% of female completers in their metropolitan New York City sample of completers had a history of attempt. As a related issue, the frequency of repeated attempts has been estimated to be between 14 and 26% within 1 year of the first attempt (Reynolds & Eaton, 1986), with increased risk for serious attempt and completion among repeaters. For example, Kotila and Lonnquist (1987) reported on a 5-year follow-up study of adolescent repeaters and found a rate of completion five times that of "first-timers."

Protective Factors

Garmezy (1985) has written extensively on protective factors, those personality characteristics that increase resilience and adaptive capacity. Protective factors oppose factors of risk, making some youths more adaptive than others in similar situations of stress or pain. Such characteristics as self-esteem, feelings of autonomy and control, the presence of external supports and resources, family cohesion and warmth, and the absence of familial discord and neglect serve to counteract suicidality. Rubenstein et al. (1989) reported on a study of 300 high school students and were able to document decreased suicidality among those adolescents who perceived their families to be more cohesive and adaptable and saw themselves as a valued member of a peer group.

Common Themes: Risk Factors

Embedded in the foregoing summary of risk factors are a number of common themes, operating synergistically. None, by itself, is either necessary or sufficient to produce a suicidal adolescent. However, in as yet ill-defined combinations they may prove deadly. The clinician's attention to these "between the lines" issues may make the art of risk assessment rest appropriately on its scientific base.

1. *Negative personal history*: Life stress, skill deficit, negative models, and family history of suicidality and psychopathology increase vulnerability and wound narcissistically.

2. *Psychopathology and/or significant negative personality attributes*: Axis I and Axis II diagnoses describe personalities less likely to attach in healthy ways to systems or develop good object relations. When exacerbated by alcohol or drugs, personality attributes of aggression, poor frustration tolerance, loneliness, and impulsivity may be significantly potentiated.

3. *Stress*: Stimuli that threaten the adolescent's ability to maintain a developing yet fragile sense of self and self-esteem or that potentially overwhelm available coping skills create a context for vulnerability. Anticipated losses,

threatened humiliations, guilt, rejections, or feared punishments can over-whelm as well.

4. *Defensive breakdown; affective and behavior dysregulation*: Evidence of cog-nitive rigidity, irrationality, thought disturbance, or loss of reality testing, as well as acute behavioral changes, panic, and disorientation, signal a breakdown in adaptive coping skills and ego defenses.

5. *Social/interpersonal isolation and alienation*: When adolescents move away from typical attachments, do not form alliances with systems, and are noncom-pliant with help-givers—or, alternatively, isolate themselves or identify with marginal and fringe groups—there is less social regulation and, correspond-ingly, an increased tolerance for deviance, substance use, and so on.

6. *Self-deprecatory ideation, dysphoria, hopelessness*: Statements of unhappiness or pessimism; expressed feelings of worthlessness, uselessness, and other neg-ative views of self; or expressed death and suicidal fantasies may place the adolescent on the brink of life-or-death decisions.

7. *Method availability/accessibility/knowledgeability*: With sufficient impulse to-ward self-destructive action or with intent to die or use self-harm for instru-mental goals, the easy accessibility of a weapon (and knowledge about its use) may be an irresistible call to action.

Multifocal Assessment of Risk

Clinicians have little rational basis on which to predict outcomes of expected actions and, indeed, do so only with minimal precision (Kaplan, Kotter, & Frances, 1982). Despite this fact, clinical assessments of risk are necessary to determine decisions to act (e.g., to hospitalize) in order to protect patients from potential self-harm and lethal consequence. With attention to multiple frames of assessment and a stepwise screening procedure (Berman & Jobes, 1991; Rotherman, 1987), we believe that clinicians can make reasoned judgments of risk leading to reliable early detection and appropriate treatment planning.

The assessment of short-term versus long-term risk for suicidal behavior of adolescents begins with observations of alerting signs for risk. The presence of one or more research-based risk factors, as outlined above, should signal the clinician that there is a context for the consideration of suicide by the adolescent. Also, as noted above, the presence of one or more diagnosable psychopath-ologies should "red flag" the adolescent as having potential for suicidal behavior.

1. *Imminent risk*: Signals for imminent risk first and most often come in the form of verbal or behavioral messages from the adolescent. Indeed, Brent et al. (1988) found that 83% of their sample of adolescent completers had made suicidal threats to others during the week prior to their death. Although less frequently writers of suicide notes than their adult counterparts (Posener, LaHaye, & Cheifetz, 1989), adolescents communicate their suicidality in words, school essays, poems, diaries, journals, and art work, often with themes suggesting a preoccupation with death, dying, an afterlife, or suicide. Questions asked out

of context with more than ordinary interest, as well as communicated dreams with death themes, should alert the observer to follow up with appropriate concern. All potential and direct suicidal communications require further inquiry as the first step to a possible interventive response.

The possibility of suicidal behavior appears greatest, and is most difficult to predict temporally, when there are behavioral communications such as poor impulse control and rage. In particular, a history of violent, assaultive behavior implicates strongly the potential that such behavior may occur without directionality (i.e., as free-floating rage). For example, Cairns, Peterson, and Neckerman (1988) compared matched samples of emotionally disturbed, assaultive youngsters (ages 4–18) with or without histories of suicide attempts and found no difference between the groups regarding the frequency and severity of other-directed violence. Impulsivity and expressed rage are pathognomonic of poor control. This loosening of ego regulatory mechanisms may be due to pathology or substance abuse. Irrespective, when there exists a context to believe that suicide intent is present, poor control in concert with an available weapon spells great danger for imminent self-harm. Other behavioral markers should be noted as alerting signs: prior suicidal behavior, termination behaviors (making specific preparations for death), and acute behavioral changes, particularly those out of character and persistent or otherwise unexplained.

2. *Lethality and intent*: The potential for life-threatening impact from any self-harm behavior (lethality) must be evaluated in order to determine the proper preventive response to that evaluation. An excellent resource for understanding and assessing lethality has been presented by Smith, Conroy, and Ehler (1984).

Closely related to the concept of lethality is that of psychological intent: the purpose or goal of self-destructive behavior (refer to discussion by Shneidman, 1980). When the goal is to escape a painful condition or context by seeking a condition of nonpain (i.e., by wanting death), intent is high. Similarly, when the aim is to reunite with a deceased loved one, to be reborn, or to transcend to a new life, intent is defined as that of ending one's known life; and the possibility of hastening that end through some lethal behavior is great.

During adolescence, most often the stated intent of a suicidal motive is both instrumental and interpersonal, that is, to alter life circumstances, to communicate an unheard message, to mobilize others to change. The functional analysis of such motives is most important to therapeutic planning. The clinician must attend well to the stated intent of the adolescent and be wary of too quick an interpretation of the "true meaning" of the adolescent's behavior. Adolescents, particularly suicidal ones, cry out to be heard as *they* think. Themes of escape and relief, control, and power frequently describe their suicidal goals, which are clear replacements for the experience of pain, helplessness, hopelessness or powerlessness, affects and cognitions common to the suicidal state (Shneidman, this volume).

3. *Predisposing conditions and precipitating events*: Most suicidal adolescents have suffered serious blows to their self-esteem, their sense of self, and their ability

to cope often because of years of conflict and stress. Familial-genetic, biophysical, sociocultural, psychopathological, and related factors may determine a suicidal vulnerability to react to specific precipitating events with suicidal action.

Precipitants common to suicidal actions among adolescents are interpersonal conflict (a disciplinary crisis with parents or arguments with peers) and events that threaten loss (e.g., rejection), failure, guilt, humiliation, or punishment. As these events are common to the human experience, their significance for risk assessment lies only in the context of predisposing vulnerability.

4. *Psychopathology*: Predisposing vulnerability is perhaps best described by evidence of current mental disorder. As noted above, overwhelmingly most adolescent suicidal behaviors are characterized by diagnosable mental disorders. Whereas certain diagnoses, as noted, are more commonly observed (depression, conduct disorder, and substance abuse), the clinician should be alerted that a number of others are also noted in the literature (Berman & Jobes, 1991). Furthermore, different studies report markedly different prevalences of the same disorder. For example, Brent et al. (1988) found the prevalence of bipolar disorder to be 11 times greater than that reported by Hoberman and Garfinkel (1988). The proper evaluation of symptoms or behaviors suggestive of an overt or an underlying pathology allows an immediate interventive response (e.g., through medication) that may make a life-saving difference in the regulation of irrational cognitions or behavioral dyscontrol.

5. *Coping skills and resources*: Suicidal behavior may be construed as an end result of despair and hopelessness, self-contempt, rage, and the unavailability of sustaining resources (Maltsberger, 1986). Alternatively stated, the risk of suicidal behavior is attenuated by available internal (e.g., cognitive) and external (e.g., interpersonal) resources for coping with stress and conflict. To the extent that interpersonal supports are accessible and used, the adolescent is not left alone during short-term crisis. In the long term, intolerable feelings of aloneness can be reduced and problem-solving aided. Cognitive strategies (e.g., problem-solving skills) are essential to arrive at rationally derived alternatives, to engage cognitive rehearsal, to think hierarchically versus dichotomously, to assess self, and to tolerate ambivalence.

Among the more significant contraindications of risk are certain protective factors that increase stress resistance (Garmezy, 1985). Rubenstein et al. (1989) have validated the importance, for example, of family cohesion and adaptability to offset the effects of increased stress in the lives of at-risk youngsters.

6. *Compliance*: Crucial to the decision not to hospitalize a youth in suicidal crisis is the assessment of his or her compliance with recommended outpatient treatment regimens and the family's support in effecting that compliance. The history obviously teaches us what we may expect in the future. Perhaps one of the better measures available is the adolescent's cooperation with the assessment interview and alliance with the interviewer as help-giver. Motto (1984) reported that future adolescent male completers, significantly more often than controls, were rated as having a negative or ambivalent attitude at the time of intake on

average 2 years prior to their suicides. Adolescents with a history of not being able to align with authorities or to form behavioral contracts and those having a thought disorder or a pattern of impulsive, emotionally reactive behavior have difficulty complying with recommended procedures and agreements.

Screening for Risk

With attention to the foregoing observations, the clinician now may make reasoned judgments regarding the adolescent's potential for self-harm and the development of treatment plan. The screening of risk involves a hierarchy of significant questions regarding possible lethal outcome.

Level A: Does this adolescent have the potential for self-harm?
Level B: If yes, might this adolescent possibly harm himself or herself?
Level C: If yes, what is the probability of self-harm behavior, under what circumstances, to what degree of life-threat, and how imminently?

Level A is the red-flagging of an adolescent by virtue of shared risk factor(s) of psychopathology. Level B reflects evidence of communications, poor impulse control, rage, cognitive constriction, or social isolation/alienation. Level C introduces the notion of situation specificity to risk assessment, of the availability/ obtainability of a character-syntonic method of self-harm, and the planfulness or impulsivity of that character. Along these lines, Shneidman (1988) has offered a compelling conceptualization in his assertion that all completed suicides are fundamentally driven by a synergy of three primary and converging forces: acute psychic pain, extreme pressures (Murray's notion of presses), and a high level of perturbation (i.e., agitation/emotional upsetness).

Treatment

Decision to Hospitalize

Assessment is the first step in the process of treatment. If done well, the transition into successful treatment is significantly eased. The assessment interview is both an opportunity to observe and gather necessary information and a structuring vehicle to stabilize the adolescent in suicidal crisis. Should risk remain high despite this initial intervention, hospitalization should be considered to provide sufficient sanctuary and time to achieve that stabilization. Other considerations, such as high levels of unresolvable stress, the presence of a thought disorder or ego decompensation, symptoms of dyscontrol, high levels of rage, panic, or violent behavior, and the absence of interpersonal supports during the period of crisis, also point to the need for careful consideration of the potential costs and benefits of short-term hospitalization. Among the costs of hospitalizing an adolescent are the isolation of the youth from available supports, the reinforcement of dependency, and the perceived stigma attached to removal from the everyday environment.

Macrotherapeutic Issues

HELP-SEEKING AND COMPLIANCE. The stigma is both real (i.e., peers do label, and one's dignity must be maintained) and a corollary of the adolescent's projected introspection that results in a belief that others are paying attention. As a consequence, adolescents are notoriously reluctant to seek help, and they view mental health professionals as helpers of last resort (Aaronson, Underwood, Gaffney, & Rotheram-Borus, 1989). Also, this resistance to help-seeking is rooted in the fact that those who provide treatment are both adults (members of the same group from whom the teenager has been striving to gain autonomy), and transference objects (parent figures akin to those real parents who may have inflicted trauma and narcissistic injury). Even when a parasuicidal event brings the adolescent to the emergency room, family systems with a pathological history often collude to deny the seriousness of the attempt for fear of exposing the parent(s) as bad, to avoid feared punishment, to maintain control and so on (Rinsley, 1980).

One significant consequence of parental noncompliance with recommended treatment is the probability of repeated attempts by the adolescent in an increasingly lethal cycle of cause–effect behavior in the family system. This cycle, then, defines a major task of therapy, namely, that of helping the family interpret the suicidal behavior as a family event and, if a therapeutic alliance can be secured, helping the family to change.

THERAPEUTIC ALLIANCE AND COUNTERTRANSFERENCE. The therapeutic alliance has been described as the most critical factor determining whether the process of psychotherapy is successful (Goldfried, 1980). However, suicidal adolescents often have histories of disturbed early relationships prototypical of difficulties in later, and therapeutic, relationships. They enter treatment with negative and often off-putting affects of hopelessness, helplessness, despair, ambivalence, regressive dependency, anger, or passivity. As therapists, we often expect patients to be motivated to use offered help, to follow therapeutic advice, and to actively engage. Suicidal adolescents rarely meet that expectation.

In consequence, therapists often respond with similarly negative attitudes and feelings, ranging from irritation to hostility. Such negative attitudes may have iatrogenic effects when countertransference reactions are not effectively controlled. Less help may be offered, emotional distancing may increase, and pejorative labeling is common. For example, suicidal gestures are referred to as "pseudocides," or "manipulative," rather than in terms more descriptive of the adolescent's need for treatment.

Elsewhere (Berman & Jobes, 1991; Jobes & Berman, 1991) we have made a number of suggestions regarding therapeutic strategies for dealing with these macrotherapeutic issues, increasing the probability that treatment may proceed beyond "band-aid" interventions. In addition, we have outlined at length our perspective on treating the suicidal adolescent, which space constraints prohibit us from describing in greater detail here. A brief summary and overview must suffice.

Treatment Plan

There are few if any research-based guidelines for effective treatment of the suicidal adolescent. Nevertheless, the therapist needs a plan that addresses at a minimum the site of treatment (inpatient versus outpatient), the modalities of treatment (individual, group, family therapy, pharmacotherapy), and the related strategies for obtaining desired therapeutic goals.

If hospitalization is not necessary, outpatient therapy is preferred. Individual psychotherapy is the mainstay of the suicidal adolescent. However, family therapy may be considered the therapy of choice, particularly where family conflict or pathology is central to the adolescent's suicidality or where family enmeshment inhibits the adolescent's struggle toward autonomy. Pharmacotherapy plays a small, but not inconsiderable role, in the treatment of the suicidal adolescent. When underlying pathology may be successfully treated by medication and symptom reduction would allow for greater accessibility to verbal forms of therapy, drugs should be considered. It should be noted, however, that only a few well-controlled drug studies with adolescent patient-subjects have appeared to date in the literature, with little support for effectiveness (Trautman, 1989).

Protecting the Adolescent from Self-Harm: Crisis Intervention

The initial target goal of any therapeutic plan must be that of preventing self-harm behavior. Thus decreasing lethality and perturbation (Shneidman, 1985) must precede therapeutic work on predisposing conditions. Sometimes, as noted, it requires removal from a stressful to a nonstressful environment (e.g., hospitalization). Also involved is a mobilization of available resources, both environmental and internal (e.g., the adolescent's strengths). Most often it involves an active, collaborative, mutual problem-solving strategy within the context of interpersonal support, empathy, and warmth.

A series of strategic steps must be considered and implemented to ensure the adolescent's immediate safety, to diffuse lethality, and to prepare the patient for treatment of underlying problems.

1. Restricting access to available means for self-harm
2. Offering a time-limited, renewable, structuring and linking "agreement" between the adolescent and therapist (a "no-suicide contract")
3. Decreasing isolation by placing the adolescent on "suicide watch" in the natural environment

4. Decreasing symptom-caused perturbation, e.g., by carefully prescribed and monitored medication

5. Structuring treatment by outlining goals, roles, factual information, and so on regarding the process of therapy

6. Problem-solving through collaborative exploration, definition, generation of alternative solutions, and hypothesis testing (central to this process is a functional assessment of the motive of the adolescent's suicidal behavior)

7. Providing maximum contact availability and telephone accessibility

Beyond Crisis Intervention

With the goals of developing a working relationship, establishing a treatment plan, and protecting the adolescent from self-harm, sufficient time should have been bought to allow the work of psychotherapy to begin. The focus now may shift to broadening the adolescent's linkages to the future and to interpersonal resources and reducing the effect of psychopathology and maladaptive cognitions, increasing self-esteem, enhancing instrumental skills, and improving object relations.

These goals may be achieved through any of a variety of therapeutic strategies, the choice of which is more dependent on therapist orientation, temperament, and training than anything else. A number of these models are available to the reader in well-developed manuals (e.g., Beck, Rush, Shaw, & Emery, 1978; Linehan, 1984) or are delineated in great detail elsewhere (Berman & Jobes, 1991).

Note on Termination

As problems of attachment often are central to the adolescent's suicidality, separation/individuation issues deserve crucial therapeutic attention. Therapists experienced with working with suicidal patients know well to be aware of increasing separation anxiety and the adolescent's denial of its existence as termination approaches. We believe that termination should not occur with adolescents. Rather, it is replaced by a weaning process from formal, frequently scheduled sessions to infrequently scheduled follow-ups, to long-term contactability by phone or postcard (or both). This strategy allows continued monitoring of therapeutic gains, a continued link (therapist constancy) should the adolescent regress, a referral source to the adolescent away at a distance, and booster shots (of therapy sessions) as needed.

From Good Practice to Malpractice

The potential for a completed suicide by a suicidal adolescent in the care of a treating psychotherapist is a constant in the process of that treatment. Were that potential to translate into a reality, survivor rage may be externalized and blame denied and projected to the psychotherapist in the form of a malpractice claim. As noted elsewhere in this volume, these claims are increasing in frequency and costs to the mental health practitioner.

The court expects the therapist to attempt reasonably to prevent the suicide of the patient. The standard of care is established essentially through the opinions of experts as treatment not significantly and indefensibly deviant from that of reasonable professionals of similar training and experience. Essentially, there are two factors that determine liability: foreseeability and reasonable care. Foreseeability is based on a risk assessment and a consequent and documented judgment of risk for suicide. Reasonable care is reflected in the appropriateness of interventions that follow the assessment of risk. Inherent in reasonable care is the notion of dependability, i.e., responsible follow-through on such factors as the treatment plan and suicide precautions ordered.

The topic of malpractice is too important for this simple a statement to suffice. The reader is directed to other sources for reviews of case law (VandeCreek & Young, 1989), recommended guidelines for good practice (Berman, 1990; Berman & Cohen-Sandler, 1982; Pope, 1989), and a case presentation including views of experts (Berman, 1990).

Prevention

Clinicians increasingly are being called on to remove themselves from the confines of the office and direct clinical practice in order to extend the reach of clinical knowledge and skills to larger groups (the community) and to preventive interventions. With regard to youth suicide, the impact of clusters of suicide among the young and the reverberations of even one suicide within the adolescent community (e.g., the school) has demanded a shift of focus to prevention.

The federal government has taken some initiative to spur the development of preventive interventions. For example, community-wide models for the prevention and containment of suicide clusters have been proposed by the Centers for Disease Control (CDC) (1988). The CDC has also articulated priority goals for the nation's public health including a clearly stated objective to reverse the rising trend in youth suicidal deaths. The Department of Health and Human Services (DHHS) has now published its four-volume *Report of the Secretary's Task Force on Youth Suicide* (Alcohol, Drug Abuse, and Mental Health Administration, 1989), with 33 recommendations and accompanying action plans for their implementation.

The American Association of Suicidology (AAS) (1989) has proposed specific guidelines for the media in their reporting of suicidal death in order to minimize the potential for imitative suicides. In addition, the AAS has actively networked those involved in school suicide prevention programs in order to maximize program development toward school prevention models.

No specific model of intervention is believed to be sufficient, nor is any viewed as particularly effective in reducing large numbers of youth suicides (Eddy, Wolpert, & Rosenberg, 1989). However, a number of approaches to the problem of youth suicide prevention have been proposed and theoretically are worth consideration. Most common among these proposals are restricting access to lethal means, improving the effectiveness of community crisis clinics and telephone hotlines, and developing school-based prevention programs. Arguments in favor and critical of these models are reviewed elsewhere (Berman & Jobes, 1991). As outcome studies evaluating the effectiveness of these models increasingly are reported, it is probable that second-generation models will have a more profound impact on rates of youth suicide.

Obviously, taking the long view, the most effective and probably the most cost-effective strategy is that of primary prevention. Programs designed to teach health-enhancing behavior through teaching behavioral skills, begun in the primary grades and reinforced through booster sessions annually, have the most promise of reducing the pool of adolescents at risk. Adaptive skill/ competence models of relevance to this goal already exist in the literature, for example, in the areas of problem-solving training (Shure & Spivack, 1978), depression management (Lewinsohn, Antonuccio, Steinmetz & Teri, 1984), and anger management (Feindler & Ecton, 1986; Lochman & Curry, 1986; Novaco, 1979).

However, the most time-honored model of prevention is that of early identification and treatment of a youth at risk. To the extent that the community (schools, parents, adolescents) can be educated about the signs of risk and referral-making skills, it remains for the clinician to do what the clinician is best trained to do: provide direct care for the suicidal youth. Should only one adolescent at risk be turned away from a focus on death and toward a more meaningful life, the value of that effort is immeasurable.

Guidelines for Practice: Chapter Summary

A. To ensure early detection and appropriate treatment of suicidal youth, we recommend that clinicians employ a multifocal assessment of suicide risk. Multiple frames of assessment and a stepwise screening procedure can be used to assess the following.
   1. Imminent risk
   2. Lethality and intent
   3. Predisposing conditions and precipitating events
   4. Psychopathology
   5. Coping skills and resources
   6. Compliance
B. Assessment of the preceding variables may lead to reasoned judgments regarding an adolescent's potential for self-harm and subsequent treatment planning.

C. Screening for risk must address important questions related to potentially lethal outcomes and the possible need for hospitalization in cases of clear and imminent danger.

D. The initial (crisis) treatment plan must ensure the immediate safety of the patient through the following strategic interventions.
1. Restricting access to available means for self-harm
2. Time-limited contracting between the patient and therapist
3. Decreasing interpersonal isolation
4. Decreasing perturbation/agitation
5. Structuring treatment goals
6. Collaborative problem-solving
7. Maximum availability and telephone accessibility

E. Beyond crisis intervention, treatment may proceed along traditional lines of psychotherapy with great care and consideration given to the process of termination.

F. Good practice requires that clinicians foresee and assess a potential suicide risk and provide reasonable care in the form of a treatment plan and follow-through.

G. Consultation and careful documentation are essential.

REFERENCES

Aaronson, S. L., Underwood, M., Gaffney, D., & Rotheram-Borus, M. J. (1989). *Reluctance to help-seeking by adolescents.* Presented at the annual meeting of the American Association of Suicidology, San Diego.

Alcohol, Drug Abuse, and Mental Health Administration (1989). *Report on the Secretary's task force on youth suicide. Vol. 1. Overview and recommendations.* DHHS Publ. No. (ADM)89-1621. Washington, DC: U.S. Government Printing Office.

American Association of Suicidology (1989). Suicide and media response, *Newslink, 15*(4), 1.

Apter, A., Bleich, A., Plutchick, R., Mendelsohn, & Tyano, S. (1988). Suicidal behavior, depression, and conduct disorder in hospitalized adolescents. *Journal of the American Academy of Child and Adolescent Psychiatry, 27,* 696–699.

Beck, A. T., Rush, A. J., Shaw, B. F., & Emery, G. (1978). *Cognitive therapy of depression: A treatment manual.* New York: Guilford Press.

Berman, A. L. (1987). Fictional suicide and imitation effects. Presented at the annual meeting of the American Association of Suicidology, San Francisco.

Berman, A. L. (Ed.) (1990). *Suicide prevention: case consultations.* New York: Springer Publishing Co.

Berman, A. L. (1990). Standard of care in the assessment of suicide potential. *Psychotherapy in Private Practice, 8,* 35–41.

Berman, A. L., & Cohen-Sandler, R. (1982). Suicide and the standard of care: optimal vs. acceptable. *Suicide and Life-Threatening Behavior, 12,* 114–122.

Berman, A. L., & Jobes, D. A. (1991). *Adolescent suicide: assessment and intervention.* Washington, DC: American Psychological Association.

Berman, A. L., & Schwartz, R. (1990). Suicide attempts among adolescent drug users. *American Journal of Diseases of Children, 144,* 310–314.

Bourget, D., Gagnon, A., & Bradford, J. M. (1988). Satanism in a psychiatric adolescent population. *Canadian Journal of Psychiatry, 33,* 197–202.

Boyd, J. H. (1983). The increasing rate of suicide by firearms. *New England Journal of Medicine*, *308*, 872–874.

Brent, D. A. (1987). Correlates of medical lethality and suicide attempts among children and adolescents. *Journal of the American Academy of Child and Adolescent Psychiatry, 26*, 87–91.

Brent, D. A., Perper, J. A., & Allman, C. J. (1987). Alcohol, firearms and suicide among youth: temporal trends in Allegheny County, Pennsylvania, 1960–1983. *Journal of the American Medical Medical Association, 257*, 3369–3372.

Brent, D. A., Perper, J. A., Goldstein, C., Kolko, D. J., Allan, M., Allman, C., & Zelenak, J. (1988). Risk factors for adolescent suicide: a comparison of adolescent suicide victims with suicidal inpatients. *Archives of General Psychiatry, 45*, 581–588.

Cairns, R. B., Peterson, G., & Neckerman, H. J. (1988). Suicidal behavior in aggressive adolescents. *Journal of Clinical Child Psychology, 17*, 298–309.

Carlson, G. A. (1983). Depression and suicidal behavior in children and adolescents. In D. P. Cantwell & G. A. Carlson (Eds.), *Affective disorders in childhood and adolescence*. New York: Medical & Scientific.

Carlson, G. A., & Cantwell, D. P. (1982). Suicidal behavior and depression in children and adolescents. *Journal of the American Academy of Child Psychiatry, 21*, 361–368.

Centers for Disease Control (1988). Centers for Disease Control recommendations for a community plan for the prevention and containment of suicide clusters. *Morbidity and Mortality Weekly Report, 37*(suppl. S-6), 1–12.

Cohen-Sandler, R., Berman, A. L., & King, R. (1982). Life stress and symptomatology: determinants of suicidal behavior in children. *Journal of the American Academy of Child Psychiatry, 21*, 178–186.

Davidson, L. E., Rosenberg, M. L., Mercy, J. A., Franklin, J., & Simmons, J. T. (1989). An epidemiologic study of risk factors in two teenage suicide clusters. *Journal of the American Medical Association, 262*, 2687–2692.

Diekstra, R. F. W. (1989). Suicidal behavior in adolescents and young adults: an international picture. *Crisis, 10*, 16–35.

Easterlin, R. (1978). What will 1984 be like? Socioeconomic implications of recent twists in age structure. *Demography, 15*, 397–432.

Eddy, D. M., Wolpert, R. L., & Rosenberg, M. L. (1989). Estimating the effectiveness of interventions to prevent youth suicides: a report to the Secretary's Task Force on Youth Suicide. In *Report of the Secretary's task force on youth suicide. Vol. 4. Strategies for the prevention of youth suicide* (pp. 37–81.). DHHS Publ. No. (ADM)89-1624. Washington, DC: U.S. Government Printing Office.

Farberow, N. L. (1989). Preparatory and prior suicidal behavior factors. In *Report to the Secretary's task force on youth suicide. Vol. 2. Risk factors for youth suicide* (pp. 34–55). DHHS Publ. No. (ADM)89-1622. Washington, DC: U.S. Government Printing office.

Feindler, E. L., & Ecton, R. B. (1986). *Adolescent anger control: cognitive-behavioral techniques*. New York: Pergamon Press.

Frances, A., & Blumenthal, S. J. (1989). Personality as a predictor of youth suicide. In *Report of the Secretary's Task Force on Youth Suicide. Vol. 2. Risk Factors for Youth Suicide* (pp. 160–171). DHHS Publ. No. (ADM)89-1622. Washington, DC: U.S. Government Printing Office.

Friedman R. C., Clarkin, J. F., & Corn, R. (1982). DSM III and affective pathology in hospitalized adolescents. *Journal of Nervous and Mental Disorders, 170*, 511–521.

Friedman, R. C., Corn, R., Aaronoff, M. S., Hurt, S., & Clarkin, J. F. (1984). The seriously suicidal adolescent: affective and character pathology. In H. S. Sudak, A. B. Ford, & N. B. Rushford (Eds.), *Suicide in the young* (pp. 209–226). Boston: John Wright PSG.

Garfinkel, B. D. (1989). Depression and suicide among adolescents. Presented at Treatment of Adolescents with Alcohol, Drug Abuse, and Mental Health Problems Conference, Alcohol, Drug Abuse, and Mental Health Administration, Arlington, VA.

Garfinkel, B. D., Froese, A., & Hood, J. (1982). Suicide attempts in children and adolescents. *American Journal of Psychiatry, 139*, 1257–1261.

Garmezy, N. (1985). Stress-resistant children: the search for protective factors. In J. E. Stevenson (Ed.), *Recent research in developmental psychopathology. Journal of Child Psychology and Psychiatry Book Supplement 4* (pp. 213–233). Oxford: Pergamon Press.

Goldfried, M. R. (Ed.) (1980). Some views on effective principles of psychotherapy. *Cognitive Therapy and Research, 4*, 269–306.

Gould, M. S., & Shaffer, D. (1986). The impact of suicide in television movies: evidence of imitation. *New England Journal of Medicine, 315*, 690–694.

Harkavy-Friedman, J., Asnis, G., Boeck, M., & DiFiore, J. (1987). Prevalence of specific suicidal behaviors in a high school sample. *American Journal of Psychiatry, 144*, 1203–1206.

Hoberman, H. M., & Garfinkel, B. D. (1988). Completed suicide in children and adolescents. *Journal of the American Academy of Child and Adolescent Psychiatry, 27*, 689–695.

Holinger, P. C., Offer, D., & Zola, M. A. (1988). A prediction model of suicide among youth. *Journal of Nervous and Mental Disease, 176*, 275–279.

Jacobziner, H. (1965). Attempted suicide in adolescence. *Journal of the American Medical Association, 10*, 22–36.

Jobes, D. A. & Berman, A. L. (1991). Crisis intervention and brief treatment for suicidal youth. In A. Roberts (Ed.), *Contemporary perspectives on crisis intervention and prevention* (pp. 53–69). Englewood Cliffs, NJ: Prentice Hall.

Kaplan, R. D., Kottler, D. B., & Frances, A. J. (1982). Reliability and rationality in the prediction of suicide. *Hospital and Community Psychiatry, 33*, 212–215.

Kessler, R. C., Downey, G. D., Milavsky, J. R., & Stipp, H. (1988). Clustering of teenage suicides after television news stories about suicides: a reconsideration. *American Journal of Psychiatry, 145*, 1379–1383.

Kotila, L., & Lonnquist, J. (1987). Adolescents who make suicide attempts repeatedly. *Acta Psychiatrica Scandinavica, 76*, 386–393.

Lester, D. (1988). Youth suicide: a cross-cultural perspective. *Adolescence, 23*, 955–958.

Lewinsohn, P. M., Antonuccio, D., Stenmetz, J., & Teri, L. (1984). *The coping with depression course: a psychoeducational intervention for unipolar depression*, Eugene, OR: Castalia.

Linehan, M. (1984). *Dialectial behavior therapy for treatment of parasuicidal women treatment manual*. Unpublished manuscript, University of Washington, Seattle.

Lochman, J. E., & Curry, J. F. (1986). Effects of social problem-solving training and self-instruction training with aggressive boys. *Journal of Clinical Psychology, 15*, 159–164.

Maltsberger, J. T. (1986). *Suicide risk: the formulation of clinical judgment*. New York: New York University Press.

Maris, R. (1981). *Pathways to suicide*. Baltimore: The Johns Hopkins University Press.

Motto, J. A. (1984). Suicide in male adolescents. In H. S. Sudak, A. B. Ford, & N. B. Rushford (Eds.), *Suicide in the young* (pp. 227–243). Boston: John Wright PSG.

Murphy, G., & Wetzel, R. (1982). Family history of suicidal behaviour among suicide attempters. *Journal of Nervous and Mental Diseases, 170*, 86–90.

National Center for Health Statistics (1989). *Monthly vital statistics report* (Vol. 37, no. 13). Hyattsville, MD: U.S. Public Health Service.

Newman, S. C., & Dyck, R. J. (1988). Age, period, cohort analyses of suicide rates. Presented at the annual meeting of the American Association of Suicidology.

Novaco, R. (1979). The cognitive-behavioral regulation of anger. In P. C. Kendall and S. D. Hollon (Eds.), *Cognitive-behavioral interventions: theory, research, and procedures* (pp. 241–286). Orlando FL: Academic Press.

Paykel, E. S. (1989). Stress and life events. In *Report of the Secretary's task force on youth suicide. Vol. 2. Risk factors for youth suicide* (pp. 110–130). DHHS Publ. No. (ADM)89-1622. Washington, DC: U.S. Government Printing Office.

Petzel, S., & Cline, D. (1978). Adolescent suicide: epidemiology and biological aspects. In S. Feinstein & P. Giovacchini (Eds.), *Adolescent psychiatry* (Vol. 6; pp. 239–266). Chicago: University of Chicago Press.

Pfeffer, C. R. (1989). Family characteristics and support systems as risk factors for youth suicide. In *Report of the Secretary's task force on youth suicide. Vol. 2. Risk factors for youth suicide* (pp. 71–81). DHHS Publ. No. (ADM)89-1622. Washington, DC: U.S. Government Printing Office.

Plutchik, R., van Praag, H. M., & Conte, H. R. (1989). Correlates of suicide and violence risk. III. A two stage model of countervailing forces. *Psychiatric Research, 28,* 215–225.

Pope, K. S. (1989). Malpractice suits, licensing disciplinary actions, and ethics cases: frequencies, causes, and costs. *Independent Practitioner, 9,* 22–26.

Posener, J. A., LaHaye, A., & Cheifetz, P. N. (1989). Suicide notes in adolescence. *Canadian Journal of Psychiatry, 34,* 171–176.

Reynolds, P., & Eaton, P. (1986). Multiple attempters of suicide presenting at an emergency department. *Canadian Journal of Psychiatry, 31,* 328–330.

Rich, C. L., Young, D., & Fowler, R. C. (1986). San Diego suicide study. I. Young vs. old subjects. *Archives of General Psychiatry, 43,* 577–582.

Riggs, S., Alario, A., McHorney, C., DeChristopher, J., & Crombie, P. (1986). Abuse and health related risk-taking behaviors in high school students who have attempted suicide. *Journal of Developmental-Behavioral Pediatrics, 7,* 205–206.

Rinsley, D. B. (1980). *Treatment of the severely disturbed adolescent.* New York: Jason Aronson.

Rosenstock, H. A. (1985). The first 900: a 9 year longitudinal analysis of consecutive adolescent inpatients. *Adolescence, 20,* 959–973.

Rotheram, M. J. (1987). Evaluation of imminent danger for suicide among youth. *American Journal of Orthopsychiatry, 57,* 102–110.

Rotherman-Borus, M. J., & Trautman, P. (1988). Hopelessness, depression, and suicide intent among adolescent suicide attempters. *Journal of the American Academy of Child and Adolescent Psychiatry, 27,* 700–704.

Rubenstein, J. L., Heeren, T., Houseman, D., Rubin, C., & Stechler, G. (1989). Suicidal behavior in "normal" adolescents: risk and protective factors. *American Journal of Orthopsychiatry, 59,* 59–71.

Runeson, B. (1989). Mental disorder in youth suicide. *Acta Psychiatrica Scandinavica, 79,* 490–497.

Shaffer, D. (1988). The epidemiology of teen suicide: an examination of risk factors. *Journal of Clinical Psychiatry, 49,* 36–39.

Shaffer, D., & Gould, M. (1987). *A study of completed and attempted suicide in adolescents* (grant no. MH 38198): progress report. Rockville, MD: NIMH.

Shaffer, D., Garland, A., Gould, M., Fisher, P., & Trautman, P. (1988). Preventing teenage suicide: a critical review. *Journal of the American Academy of Child and Adolescent Psychiatry, 27,* 675–687.

Shaffer, D., & Bacon, K. (1989). A critical review of preventive efforts in suicide, with particular reference to youth suicide. In *Report of the Secretary's task force on youth suicide. Vol. 3. Prevention and interventions in youth suicide* (pp. 31–61). DHHS Publ. No. (ADM)89-1622. Washington, DC: U.S. Government Printing Office.

Shafii, M., Carrigan, S., Whittinghill, J. R., & Derrick, A. (1985). Psychological autopsy of completed suicide in children and adolescents. *American Journal of Psychiatry, 142,* 1061–1064.

Shneidman, E. S. (1980). Suicide. In E. S. Shneidman (Ed.), *Death: current perspectives* (pp. 416–434). Palo Alto, CA: Mayfield Publishing.

Shneidman, E. S. (1985). *Definition of suicide.* New York: John Wiley & Sons.

Shneidman, E. S. (1988). Some reflections of a founder. *Suicide and Life-Threatening Behavior, 18,* 1–12.

Shure, M. B., & Spivack, G. (1978). *Problem-solving techniques in childrearing.* San Francisco: Jossey-Bass.

Sloan, J. H., Rivara, F. P., Reay, D. T., Ferris, J. A., & Kellerman, A. L. (1990). Firearm regulations and rates of suicide: a comparison of two metropolitan areas. *New England Journal of Medicine, 322,* 369–373.

Smith, K., Conroy, R. W., & Ehler, B. D. (1984). Lethality of suicide attempt rating scale. *Suicide and Life-Threatening Behavior, 14*, 215–242.

Smith, K., & Crawford, S. (1986). Suicidal behaviors among normal high school students. *Suicide and Life-Threatening Behavior, 16*, 313–325.

Spirito, A., Brown, L., Overholser, J., & Fritz, G. (1989). Attempted suicide in adolescence: a review and critique of the literature. *Clinical Psychology Review, 9*, 335–363.

Spirito, A., Williams, C., Stark, L. J., & Hart, K. (1988). The Hopelessness Scale for Children: psychometric properties and clinical utility with normal and emotionally disturbed adolescents. *Journal of Abnormal Child Psychology, 16*, 445–458.

Trautman, P. D. (1989). Specific treatment modalities for adolescent suicide attempters. *Report on the Secretary's task force on youth suicide. Vol. 3. Prevention and interventions in youth suicide* (pp. 253–263). DHHS Publ. No. (ADM)89-1623. Washington, DC: U.S. Government Printing Office.

VandeCreek, L., & Young, J. (1989). Malpractice risks with suicidal patients. *Psychotherapy Bulletin, 24*, 18–21.

6

Suicide of the Elderly

JOHN L. MCINTOSH

Despite attention to youth suicide and reports of large increases in their suicide rates, the risk of suicide is highest among the elderly in the United States and most other nations (e.g., Shulman, 1978). Certainly it is important not to downplay the significance of the increasing incidence of youth suicide. However, the elderly are an age group where the incidence and prevention of suicidal behavior have been relatively ignored.

Demographic Aspects

Completed Suicide

Suicide rates, both currently and in the past, can be best described as increasing with age such that the highest suicide risks occur among the oldest age groupings. Indeed, as can be seen in Figure 6.1 for 1987 (National Center for Health Statistics [NCHS], 1989) suicide rates peak after the age of 65 (in the age group 75–84). In addition to the highest rates, suicide is overrepresented among the aged. That is, the elderly commit a higher proportion of the suicides annually than their representation in the population. For example, in 1987 the elderly comprised 12.3% of the U.S. population (U.S. Bureau of the Census, 1988) but committed 21.0% of the suicides (NCHS, 1989). By comparison, the young comprised 16.0% of the population and contributed 15.7% of the suicides.

An examination of major demographic characteristics (sex, race, age subgroupings) reveals that men are at markedly higher risk than are women during old age. Male rates (65 years of age and above) were 43.4 per 100,000 population in 1987 compared to 6.8 for elderly women. In fact, old age is the life period during which the greatest sex differential in rates is observed (Fig. 6.2; for the nation as a whole, the 1987 rates were 20.5 and 5.2 for males and females, respectively).

Rates of suicides for whites are higher than those for nonwhites in all age groups (the national rate for whites in 1987 was 13.7 compared to 6.9 for nonwhites), particular among older adults. Elderly whites had a suicide rate of 23.0 for 1987, whereas the rate for elderly nonwhites was 9.4 (there are substantial ethnic differences) (McIntosh & Santos, 1981). Furthermore, when

106

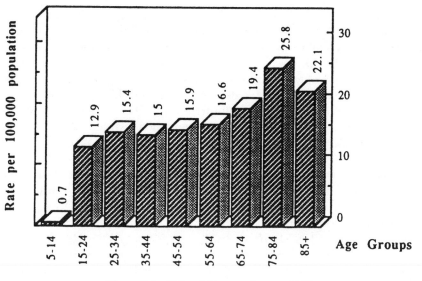

Figure 6.1 U.S. suicide rates by age, 1987.

combining race with sex it is found that white males (46.1 per 100,000 population) are the highest risk for suicide, and nonwhite males (19.0) at lower risk and females of both races at markedly low risk by comparison (white females 7.3; nonwhite females 2.8).

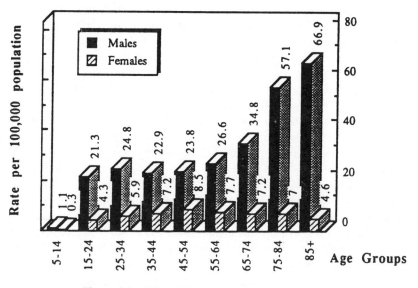

Figure 6.2 U.S. suicide rates by age and sex, 1987.

As shown in Figure 6.1, the old-old (those over the age of 75) are at highest risk of suicide, with the young-old (65–74 years of age) displaying slightly lower rates. It is also typically observed that the rate for those over age 85 is slightly lower than for the 75–84 age group, where peaks occur. However, the rate for the age 85 and older group is higher than for any age group other than 75–84; and the 65–74 rate exceeds those for all nonelderly age groups (i.e., under age 65). When considering all three factors together, the highest rates of suicide by age, sex, and race are observed among white males 85 years of age and older (1987 rate was 71.9; for white females of the same age the rate was 4.8).

Marital status is another major demographic variable that has been shown to influence suicide risk among the elderly and other age groups. It has been theorized that being married and integrated into a family, with its regulation and intimate involvements, increases meaning and lessens the likelihood of suicide (Durkheim, 1897/1951; Gibbs, J. P. & Martin, 1964; Henry & Short, 1954). These sociological explanations have been largely supported by research findings (Danigelis & Pope, 1979; Gibbs, J. P., 1969; Gibbs, J. P., & Martin, 1981; Stack, 1982; an exception was Rico-Velasco & Mynko, 1973; but see Heer & MacKinnon, 1974). Some researchers (Kozak & Gibbs, J. O., 1979; Veevers, 1973; Wenz, 1981) have focused on the presence or number of dependent children, finding it an important factor in married individuals' lower likelihood of suicide. DeCatanzaro (1981) suggested a sociobiological theory in which suicide more often occurs where it has the least effect on the gene pool of the population—among those not married and particularly the old.

Males are at markedly higher risk for suicide than females for all marital statuses and at all ages (McIntosh, 1987b). Among other sex differences noted, it has been observed that widowhood has an especially greater toll among widowers (i.e., men), including and particularly with respect to a higher suicide risk (for a review see Stroebe & Stroebe, 1983). The widowed and divorced usually are observed to have higher suicide rates than the single (and married). In 1987, for example, widowed men aged 65 and above had a rate of 87.5 per 100,000 compared to only 8.1 per 100,000 for elderly women. Divorced elderly men had a similarly high rate of suicide (88.5), and divorced elderly women had higher rates (16.0) than widowed women. Single elderly men were at lower risk, although the rate was still high (67.1), whereas single elderly women had low rates (4.9). The married among elderly men had the lowest rate of suicide for men (33.7), and the rates were once again low for women (5.9).

Bereavement (Kaprio, Koskenvuo, & Rita, 1987; MacMahon & Pugh, 1965), disruption, and isolation caused by the loss of the important relationship of marriage through either death or divorce are most often cited as important factors in these patterns of suicide at all ages (Stack, 1980, 1981; Trovato, 1986; Wasserman, 1984; Wenz, 1977). Isolation at any age has been found to be an important variable with respect to suicide (Gove & Hughes, 1980; Trout, 1980) and has been cited prominently as a factor in suicide among the widowed elderly (Bock, 1972; Bock & Webber, 1972).

LONG-TERM TRENDS. Two long-term trends by age have taken place that over time have somewhat altered the characterization of U.S. suicide rates by age. The first trend, as noted above, is a significant increase in suicide rates for the young (Holinger & Offer, 1982; Maris, 1985). The group most often mentioned here are those 15 to 24 years of age (Fig. 6.3). The increase in rates has been approximately 200% since the 1950s. For example, the 1965 rate for the 15–24 age group was 4.0 per 100,000 population, whereas in 1987 it was 12.9 per 100,000, an increase of 223%. On the other hand, the second trend, among the elderly, has involved a decline of equally dramatic proportions as the increase for the young (McIntosh, 1985). Suicide rates for the old (Fig. 6.3) have decreased markedly and consistently since their peak during the Great Depression years of the early 1930s. For example, elderly (age 65 +) suicide rates were 45.3 per 100,000 in 1933, declining to 26.2 for 1953, and to 21.7 in 1987. The percent decreases were therefore 42%, 17%, and 52%, from 1933 to 1953, 1953 to 1987, and 1933 to 1987, respectively.

When the data are displayed on the same graph, as in Figure 6.3, the differences in suicide rates for these two high-risk age groups are obvious. However, despite the increases for the young and declines for the aged, the suicide rate for the elderly in 1987 was more than 50% (actually 68%) higher than that for the 15–24 age group (21.7 versus 12.9, respectively) (NCHS, 1989). Clearly, then, the high suicide risks and those among whom the greatest change are taking place are the young and the old, with the old consistently at higher risk.

Reasons for the long-term decline in elderly suicide rates are not apparent, and there is inadequate information to fully explain these changes (McIntosh, 1988–89), as few investigators or publications have focused on this issue. Busse

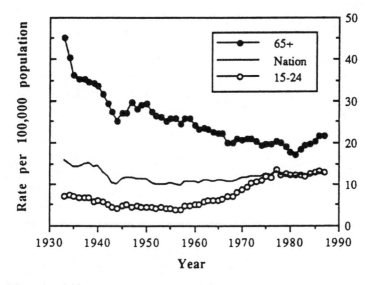

Figure 6.3 U.S. suicide rates among the young (15–24 years of age) and older adults (65 + years of age), 1933–1987.

(1974) suggested that antidepressants and increases in economic security and health care are possible factors in the decline.

Research has provided at least some suggestions regarding possible contributing factors. Although increases in the elderly population as a whole do not appear to be closely associated with their decreased suicide rates (Holinger & Offer, 1982), the increasing representation of women compared to men in the older age categories has been identified as being related to the decline (McIntosh, Hubbard, & Santos, 1980). As was shown above, women are a low-risk group for suicide at all ages (and especially at old age), whereas men are generally a high-risk group overall and particulary during old age. Thus as the total older population has come to include larger proportions of women, it has contributed to the decline in suicide rates among the elderly. In addition, most of the declines in geriatric suicide have been contributed by men, which has accelerated the influence that might be attributable to an increasing number of women alone.

Another factor that has been identified is the economy. Marshall (1978) found a significant correlation between decreased suicide rates for white men 65–74 years of age and improved economic status among the group. The impact of this factor on the most important elderly group with respect to suicide rates, namely, old white men, has had a major influence on the decline among the elderly as a whole. McCall (1991) observed similar relations in an investigation between the suicide rates of elderly white men and measures of well-being and societal affluence (i.e., percentage of white elderly living below the poverty level, average monthly Social Security benefit payments to men, elderly men enrolled in Medicare, and the rate of medical doctors per 1000 population).

From a demographic standpoint, the declines in suicide rates have been noted for both men and women, but they have been much greater for men. The declines for the aged as a whole can be attributed primarily to decreases among elderly men, and the small declines among aged women have not greatly affected the overall rates. Also, the declines have occurred almost exclusively among whites, as nonwhite elderly suicide rates are low and have changed little over time. With respect to the various subgroups that comprise the elderly population, declines have occurred for all aged groups but have been slightly larger among the young-old (65–74 or 55–74) than among the old-old (75 +). Within the old-old population, greater declines in rates have been observed for the 75–84 population than for those 85 and above. Once again, the declines for both the young-old and old-old have been predominantly a white male phenomenon (McIntosh, 1984). Whatever the ultimate explanation for the decline, successive cohorts of elderly have become increasingly lower risks for suicide than were their predecessors. Although the current generation of elderly are at lower risk than elderly in the past, it should be kept in mind that they remain the highest-risk age group for suicide.

SHORT-TERM TRENDS. In 1981 the U.S. elderly suicide rate was the lowest ever recorded, at 17.1 per 100,000 aged individuals. Rates have risen during each successive year from 1982 to 1987 to reach a rate of 21.7 in 1987. That figure represents a rate increase of 26.9% over a 6-year period and a 44.4% increase in the *number* of aged suicides during the same period (from 4478 in 1981 to 6464 in 1987). Six data years are too few to constitute a definitive change in trends, but the increases would be consistent with predicted trends (see below). By comparison, data for the 15–24 age grouping during the same time period show rates that changed only 4.9%, from 12.3 to 12.9 for 1981 and 1987, respectively (comparable national figures are +5.8% and from 12.0 to 12.7). The number of suicides actually declined for this age group by 237 deaths (4.6%), from 5161 to 4924 (comparable national figures are +11.6% and from 27,596 to 30,796). Rates for those 15–24 peaked at the highest recorded level (13.6) in 1977 and have remained either stable or declined slightly since then.

PREDICTIONS FOR FUTURE TRENDS. Although from a long-term perspective suicide rates have declined over time for the old, Pollinger-Haas and Hendin (1983) and Manton, Blazer, and Woodbury (1987) have predicted that the increasing numbers of older adults in the future will produce more stresses for them and lead to increases in their suicide rates. In other words, although the current cohort of elderly are at generally lower risk than generations of older adults before them, these authors did not expect this trend to continue among the current younger cohorts. When these cohorts reach older ages, they may be at higher risk for suicide than their predecessors (the present elderly). The reasoning is that the "baby boomers" in particular have been at higher risk for suicide than earlier cohorts at all ages thus far, and they are expected to carry their high risk with them into their older adulthood, the period of the life-span with traditionally highest risk.

McIntosh (1990) has presented arguments, on the other hand, that higher rates among baby boomers are not a certainty when they reach age 65 and older between the years 2010 and 2030. For instance, the larger size of the baby boom cohort along with its potential political influence may produce over time greater attention and planning for older adulthood and its important economic, social, physiological, and psychological issues. Additionally, if baby boomers kill themselves at high rates at younger ages it is possible that the most suicide-prone among their cohort will have succumbed to suicide prior to reaching older adulthood, leaving a cohort with lower than expected suicide/risk susceptibility. McIntosh provided estimates of suicide levels for the years 2010 and 2030 that demonstrate that even if suicide rates were to remain stable over time the large number of older adults during those decades of the next century will produce a large number of suicides. That is, even without increases in elderly suicide rates in the future, suicide among the elderly is likely to be an important issue during the next century.

As a final demographic note, although their rates are lower than those among the old, suicide among the young should not be ignored. It is particularly important not to ignore a group (1) in which suicide represents an increasing phenomenon; (2) for which suicide ranks third as a cause of death (1987 statistics, NCHS, 1989) behind accidents and homicides (compared to eighth for the nation as a whole and 13th for those 65 and above); and (3) in which approximately 5000 official deaths are self-inflicted annually (more than 10% of all deaths for that group compared to only slightly more than 1% for the nation as a whole and less than 0.5% for the aged). Therefore it is argued here that both the old *and* the young should be given special attention with respect to suicidal behavior.

Attempted Suicide/Parasuicide

Another factor in the significance of elderly suicide relates to suicide attempts (i.e., nonfatal actions, often referred to as parasuicide). Rates of attempted suicide have consistently been shown to peak at high levels among the young and decline with increasing age to very low levels among the elderly of both sexes (e.g., Weissman, 1974; Wexler, Weissman, & Kasl, 1978). However, *when older adults attempt suicide*, compared to younger age groups, the old are more often "successful," so their act ends fatally. Estimates of the ratio of attempted to completed suicides further demonstrate this point. For the population as a whole there are somewhere between 8 and 20 attempts for every completion (i.e., a ratio of 8:1 to 20:1) (Shneidman, 1969; Wolff, 1970). Ratios for the young are as high as 200:1 (McIntire & Angle, 1981) or 300:1 (Curran, 1987), whereas for the old they are generally estimated to be around 4:1 or lower (Stenback, 1980). An interesting aspect of this markedly higher risk of death among elderly attempters is that older adults who attempt suicide and die may be more similar to older adults who die by suicide than are their young counterparts and elderly suicide attempters may be different from younger attempters in a number of ways (Frierson, 1991).

Several explanations exist for this increased risk of death among older adults who attempt suicide (see McIntosh, 1992, McIntosh, Hubbard, & Santos, 1981, McIntosh & Santos, 1985–86). It has been suggested that older adults have a greater intent to die than do younger individuals, and thus their motivation differs. The old use lethal methods in their suicides in higher proportions than do younger persons. Physiological changes associated with age increase the likelihood of succumbing to the physical damage resulting from the suicide attempt of an older adult compared to a younger individual. Social isolation is greater among older adults and is associated with high suicide risk in general. Additionally, greater levels of social isolation among older adults who attempt suicide may mean that it is less likely that someone would be present to effect rescue or to intervene in a suicidal situation. Individually or in combination, these aspects probably increase the likelihood of death among older adults.

Motivations to Suicide Among the Elderly

Suicidal motivations for the elderly (for reviews see McIntosh et al., 1981; Miller, 1979; Osgood, 1985; Osgood & McIntosh, 1986) are more often intrapersonal or social in nature compared to those of the young. For the aged, suicidal etiology is likely to involve several factors interacting rather than a single one. As Barter (1969, p. 9) stated, suicide by an older adult "may appear to be a reaction to total life situation." This multifactor nature of suicide is present for suicides of all ages but is particularly relevant among the elderly. It is unlikely that most of the single factors listed below would produce suicide in an older adult if they occurred alone. The combination of a number of circumstances produces a situation with which the older adult feels unable to cope.

Motivations to suicide in the elderly derive from psychological, physiological, and social realms. Among psychological aspects, depression is an important factor contributing to suicides in the elderly (and other age groups as well). There is a strong relation between suicide and depression, and it is particularly important with respect to older persons where depression underlies two-thirds or more of the suicides (Gurland & Cross, 1983). Psychologists and gerontologists have also included negative evaluations of one's life in old age (especially as a part of "life review") (Butler, 1963), issues of controlling the time of death, and negative attitudes toward aging and old age as possible contributors to suicidal ideation. Physiological factors experienced during old age that may lead to suicide include physical decline and illnesses as well as incurable illness accompanied by pain (Conwell, Rotenberg, & Caine, 1990).

Social explanations of elderly suicide have emphasized the lessened contact and integration of older adults in society and in the social realm more generally (Durkheim, 1897/1951). The loss of loved ones and interpersonal restriction in general contribute greatly to lower levels of integration (Sainsbury, 1962, 1963) as does the disengagement that often accompanies older adulthood, as with retirement (with its concomitant changes in income, status, roles, and independence). These losses and changes, of course, do no invariably mean that suicide will occur. Durkheim (1897/1951) also emphasized the lack of clear norms (i.e., anomie) associated with suicide, and this factor has often been utilized to understand suicide in the context of changing roles by older adults. There is still much to be learned about predictive variables with respect to geriatric suicides (and among other age groups as well). Certainly the individual's personality and established coping style, experience in dealing with traumas and stresses, availability of a supportive network, and previous losses and problems are important factors.

Prevention

In most cases, suicidal persons, including older adults, communicate their intention to commit suicide in a number of ways (see Shneidman, 1965, for a presentation of these clues) to those around them. The presence of these verbal, behavioral, situational, and syndromatic warning signs provide the opportunity to recognize the actively suicidal individuals among our elderly population and provide appropriate mental health interventions. Furthermore, Shneidman

(1985) suggested that there is a consistency and predictability of suicidal actions based on lifelong patterns of coping; similarly, Maris (1981) referred to "suicidal careers." This predictability, even early in life (before age 30) is particularly important and holds great promise for detecting the suicidal person and preventing suicide early *and later* in life.

General prevention (predominantly "primary prevention") efforts that would likely lessen suicides among the elderly follow almost directly from the motivations presented above. As noted elsewhere by McIntosh (1987a) these efforts include

combating physical illness by regular medical examinations and the early detection and treatment of even minor health problems; lessening of loneliness and isolation by recruiting significant others, maintaining social ties, and increasing involvement in social activities and groups; easing of economic problems and retirement-related stress by economic, psychological, and social planning for retirement as well as the cultivation of roles and activities to replace those frequently lost in association with retirement and advancing age; and enabling the work role to continue as long as possible. [p. 134]

Added to this list of suggested measures would be more aggressive diagnosis and treatment of depression in older adults.

Suicide prevention centers have likely had little effect on the declining rates of older adult suicide noted above because the elderly rarely utilize these programs (Atkinson, 1971). Present suicide prevention centers and other crisis intervention services, as well as mental health services generally (Santos & VandenBos, 1982, p. 264), are greatly underutilized by the elderly, so the elderly are underrepresented in the clientele. For example, the old typically represent only 1 to 2% of the caseloads of suicide prevention centers (McIntosh et al., 1981). It means, unfortunately, that the highest-risk group for suicide (the aged) is the *least* served by current suicide prevention schemes. These utilization patterns may be related to the experiences of this cohort of older adults with institutions and agencies as well as earlier beliefs, stigma, and attitudes surrounding mental health programs and therapy. One result is that older adults often present with physical rather than mental health symptoms directly and therefore see physicians and not mental health professionals (Miller, 1979). Physicians and other medical personnel are often not well prepared to recognize mental health problems.

It is not implied that the old are unable to benefit from current suicide prevention and intervention techniques. Indeed, Shneidman (1985) has suggested commonalities among the suicidal that apparently transcend characteristics such as age. (Some empirical evidence for predominant commonalities among the young and old was provided, for example, by McIntosh and Santos [1990].) There is every reason to believe that therapy techniques employed with the suicidal person in general (e.g., psychotherapy, antidepressants, electroconvulsive therapy), adapted when necessary for use with older adults (see below), can be effective with suicidal older adults as well. Getting older adults into therapy settings is where the major gap in service delivery currently exists.

Therefore, because older adults do not seek out or self-refer to mental health services, new techniques and approaches are necessary to identify the suicidal elderly and put them in contact with therapy settings. Among the methods that might be employed to identify and provide treatment to older adults are vigorous and aggressive outreach programs that go to the elderly because they do not come to mental heath service outlets. Relatedly, new or innovative programs and approaches are needed to identify the at-risk and suicidal elderly. Such measures might include establishing separate suicide prevention centers or crisis intervention lines for older adults, including elderly volunteers among those working in crisis intervention with whom the aged can identify and who may better understand and empathize with those who contact the center, training nonprofessionals and other groups who come into frequent contact with older adults to recognize warning signs of suicide and make referrals when warranted, and making efforts to increase the awareness and availability of crisis intervention and general mental health services among the older adult population.

On an individual level, threats of suicide and direct verbal statements about suicide must be taken seriously among all age groups but particularly among the old who are more likely to die than other groups if they make an attempt (i.e., greater risk of death). These clues should be taken as the clearest indication we have that active suicide ideation is present rather than dismissed as a "normal" part of old age. Morgan (1980) suggested that clues be taken seriously because the older adult may have a more completely formulated plan for their suicide than is found among younger individuals (this planning is an aspect in the assessment of suicide risk), and the elderly adult may have an "enhanced degree of comfort with the idea of suicide" that also increases risk.

The multicausal nature of suicide during old age and the many problematic life circumstances that may be present for the older adult often lead individuals to believe that their situation is hopeless and they are personally helpless to effect any improvement. These subjective feelings are a major factor in suicidal ideation. Indeed, the many problems that are present (i.e., physical, social, and psychological and their interactions) make treatment of suicidal thoughts in the elderly especially difficult. A team/consultative approach, rather than an individual therapist acting in isolation, has been suggested as "optimal" for dealing with suicidal individuals (Shneidman, 1985, pp. 225–226), and this ideal would likely be particularly appropriate with older adults. The approach would be a case-solving one from multiple disciplines to deal with the many sources of difficulty that lead an older adult to suicide crisis. Shneidman made another point that is appropriate with respect to suicidal elderly. He suggested that one need not solve or eliminate the source of the problem (or problems) that lead the person to consider suicide. What is necessary to prevent suicide is lessening of the intolerable pain the person feels—to bring that pain to a bearable level from its currently unbearable one. Older adults have undergone a lifetime of coping and adjusting to stressful circumstances. The complexity, sheer number of sources of problems, and changes in their lives that may have altered previous coping ability have likely exhausted the current resources of the older adult and produced feelings of hopelessness and helplessness. The diminution of subjective pain to tolerable levels may also reduce feelings of hopelessness and helplessness.

Another aspect of individual suicidal crises in older adults has to do with shrinking interpersonal networks. Older adults lose others in a number of ways, especially through death and lessened contact. The elderly are more isolated and more often live alone, both factors associated with high suicide risk. In addition to general measures outlined above, another way to confront this interpersonal shrinkage is to consult with and involve family members in therapy whenever possible.

Several special issues arise with respect to the elderly individual as the client in a suicidal crisis. An excellent overview source for information regarding mental health issues and the elderly as therapy clients may be found in the work of Butler, Lewis, and Sunderland (1991). They asserted that, provided there is no severe brain damage, the elderly are good candidates for individual and group psychotherapy techniques. Among the first issues the therapist must keep in mind with elderly clients is the therapist's own fears, attitudes, and beliefs regarding the aged, aging, and old age. The therapist must confront issues of ageism and accompanying aspects of transference (Butler et al., 1991; Morgan, 1989). Kahn (1990) further cautioned the therapist to be aware of countertransference issues associated with the therapist's and the elderly client's philosophy, personal values, and so on in the context of an incurably ill individual and suicide. Related to possible ageism and countertransference, Cohen (1990) suggested that changes in memory, cognitive processes, sleep patterns, and sexual capacity should not be summarily dismissed as a concomitant of the aging process and should instead be investigated as possible signals of mental disorder.

Another special feature associated with older adults is the atypical presentation related to depression. Depression may be masked by vague physical decline or multiple somatic complaints. An additional complication relates to the misdiagnosis of depression as dementia (i.e., pseudodementia). This point makes it important to carefully assess potential symptoms of dementia to ensure that the symptoms are not in reality the result of a depressive and not an organic disorder (Cohen, 1990; Morgan 1989).

When assessing suicide potential in older adults the therapist must be mindful of such issues as quicker fatigability among older adults and various sensory and other physical changes associated with the aging process (e.g., hearing deficits, visual changes) (Butler et al., 1991). It is recommended (Butler et al., 1991; Morgan, 1989; Osgood, 1985) that clinicians take special care to establish trust and a sense of safety among elderly clients in the therapy setting. Addressing elderly clients respectfully (i.e., as "Mr." or "Mrs." and not first names alone) may be a sensitive issue in therapy. Therapists might also consider the importance and significance of touch in therapeutic settings as well as the need for patience with a possibly slower pace. Butler et al. (1991) suggested active inclusion of the older adult to the largest degree possible in decisions about his or her own mental health care.

A final set of special issues arising when an elderly individual is the client relates to biological interventions. The use of electroconvulsive therapy (ECT) in cases of severe depression with suicidal ideation may be effective for older adults, as it is for those of younger ages. That is, age per se does not seem to contraindicate the use of ECT (Butler et al., 1991; Morgan, 1989; Osgood, 1985). Similarly, the use of antidepressant and other psychotropic medications is both appropriate and potentially effective for older adults. At the same time, however, the side effects of medications and the generally lower dosages indicated for the elderly are important issues when prescribing (Butler et al., 1991; Morgan, 1989). The related factors of probable polypharmacy in older adults and possible drug interactions, as well as drug compliance, must be kept in mind (Butler et al., 1991).

Indirect Self-Destructive Behavior

To this point the discussion has been confined to overt suicidal actions of older adults. The elderly, however, have many potential methods available to them that are less clearly and quickly suicidal but nonetheless result in premature, avoidable death. These more subtle, covert, indirect, and perhaps unconscious methods of suicide exist at all ages but are probably prevalent during old age (for an expanded discussion of this topic see McIntosh & Hubbard [1988]). Among the general behaviors that result in death more slowly and indirectly than what we usually term a suicide are overeating and obesity, smoking, alcoholism and other drug addition, risk-taking behaviors, and accident proneness (see Farberow [1980a] for a consideration of these behaviors). The omission of behaviors that would sustain life and health is an especially relevant additional category for the old. Such behaviors include neglect of prescribed medical treatment for illnesses, neglect of routine medical examinations, ignoring or delaying medical aid when needed, refusal of medications or nourishments, smoking or drinking against medical advice, and placing oneself in a hazardous situation or environment (Kastenbaum & Mishara, 1971; Patterson, Abrahams, & Baker, 1974). Rarely would such a death be classified as a suicide, although such individuals are playing a role in hastening their death.

The concept of indirect suicide has been discussed in the literature for some time (Durkheim, 1897/1951; Kastenbaum & Mishara, 1971; Matarazzo, 1980; Meerloo, 1968; Menninger, 1938; Shneidman, 1973) but most recently and systematically by Farberow and Nelson (Farberow, 1980a,b: Nelson & Farberow, 1980, 1982) who utilized the concept of indirect self-destructive behavior (ISDB). Despite the rather lengthy history of the concept of ISDB, the extent and incidence of such behavior are unknown. However, the sentiment of writers on this topic is that direct self-destructive behavior (i.e., what is generally meant by use of the term "suicide") is not nearly as common as ISDB, particularly among the aged. Kastenbaum and Mishara (1971, p. 74) summarized their belief that "the number of self-aided deaths in old age that cannot be classified as suicide equals or exceeds those which properly would be classified as suicidal were all facts known."

In a study of institutionalized elderly, Kastenbaum and Mishara (1971) noted that 44% of the men and 22% of the women exhibited self-injurious behavior during the 1-week period they were observed. Patterson et al. (1974) found lower percentages but various forms of self-injurious behavior among a representative sample of community-dwelling elderly as well. Subsequent research has followed initial attempts by Nelson and Farberow (1977, 1980, 1982) to develop a scale to measure ISDB. A sample of chronically ill elderly male patients in an intermediate care unit of a Veterans Administration medical center were studied by Nelson and Farberow (1982). They found that 88% engaged in some form of ISDB during the week-long observation period. These scant research investigations, then, suggest that ISDB among the old is likely to occur at an incidence significant enough to warrant the attention of gerontologists and those who work with the elderly.

The issue of ISDB is presented here to raise clinician's awareness of this phenomenon and its manifestations. The inclusion of such actions on a continuum of self-destructive behaviors with more overt suicidal actions increases the level of suicide among the old, an already high-risk group. Among older adults who may believe they have lost control over their lives and who feel helpless to change what they perceive as a hopeless situation, ISDB behaviors, though negative, may be an attempt to reassert some control (Nelson & Farberow, 1980). These feelings and perceptions must be dealt with and, as much as possible, relieved in these elderly if the likelihood of both ISDB and direct-self destructive behavior is to be lessened.

The identification of ISDB may require clinicians to alter their history taking and assessment procedures in order to acquire the broad clinical perspective necessary to determine the presence of such a complex set of behaviors. At a minimum, direct and indirect suicidal acts should be assessed in order to determine an effective approach to intervention and treatment. Neither direct nor indirect self-destructive behavior should be taken lightly or ignored. Just as in the case of overt suicidal ideation and behavior, ISDB may be perceived as a cry for help or as an individual in an unbearable situation from which no escape or solution is seen. Greater awareness on the part of clinicians should increase early intervention before the situation becomes almost impossible to change and overt self-destructive behavior occurs.

Institutionalized Elderly Adults

Elderly adults who reside in nursing homes and other institutions have largely gone unstudied with respect to suicidal behaviors. One study (Osgood, Brant, & Lipman, 1991), however, provided tentative data about this special population of older adults. In a national random sample of nursing and related care homes, Osgood et al. noted high levels of both overt and indirect self-destructive behaviors among institutionalized elderly. As among the elderly in the nation as a whole, men and whites were at higher risk for such behaviors than were women and nonwhites. Osgood and colleagues noted that the old in nursing homes are likely to have several high-risk factors associated with suicide (discussed earlier), including greater physical health problems, less contact with family or no living family members, and more advanced age than among the elderly in the nation as a whole. As our population ages and larger numbers of individuals reside in institutional settings, awareness of self-destructive behaviors of all kinds must be raised and measures taken to lessen their occurrence.

Suicide Prevention

A final issue regarding suicide among the elderly that demands greater attention is postvention ("tertiary prevention"). Postvention refers to treatment efforts intended to deal with the impact of suicidal death on the surviving loved ones. Little attention has been directed toward elderly suicide survivors (an exception is the work of Farberow, Gallagher, Gilewski, & Thompson, 1987) and how their grief is altered by the circumstances of suicide compared to other modes of death. The stigma and shame as well as other emotional reactions following a suicide death and how they influence the lives of the survivors must be studied further to better understand and aid this additional group that is affected by suicide during old age (Dunne, McIntosh, & Dunne-Maxim, 1987).

Guidelines for Practice: Chapter Summary

Suicide and suicidal behavior among older adults can often be prevented. The problem of elderly suicide will remain serious and is likely to escalate with the aging of our population. Current discussions of suicide among the old with respect to ethical and philosophical viewpoints (Battin, 1987; Moody, 1984; Portwood, 1978; Prado, 1990) must certainly increase as well. Permissive attitudes toward suicide among the old must be confronted as they affect the devaluation of the life of an individual over age 65 simply because of his or her age. Improvement of training, education, and services for the suicidal elderly should be an immediate priority. Death by suicide is premature and represents personal and social loss of a valuable resource. It is no less true among the longest lived of our populations.

There are certain points to remember regarding suicide among the elderly.

1. The elderly are the highest risk group by age for suicide.
2. Suicide risk is particularly high among men, whites, the old-old (75 years of age and above), and the widowed or divorced. Therefore at especially high risk are white elderly men who have experienced marital disruption.

These high-risk demographic factors are one component when assessing suicidal risk.

3. If older adults make an attempt on their lives, the probability of death is generally much greater than for younger individuals, particularly among men.

4. Motivations for suicide among the elderly, as for other ages, are multiple. Several factors in combination (and not any single factor alone) have almost certainly produced the feelings of hopelessness and most often depression that accompany the decision to kill oneself.

5. There are certain factors that seem to most often be involved in the decision to commit suicide among the old.
   a. Depression
   b. Physical health problems
   c. Losses in any biopsychosocial sphere, including the losses of significant others and social support, roles, and economic security.

6. The multiple-etiological aspect of suicide implies that effective prevention and the reduction of future risk involve intervention with respect to several factors.

7. As with other suicidal individuals, there are nearly always many warning signs and clues that the older adult is contemplating suicide. It is especially crucial to take these signs seriously in the high-risk context of an older adult.

8. Assessment of suicide risk is largely identical among older adults and younger populations. The demographic factor of age, however, places the individual in a higher-risk category. Otherwise, assessment of depression, hopelessness and helplessness, stress, suicidal ideation, suicidal plans, and availability of methods should be made as with other groups.

9. Similarly, older adults, with few alterations in technique, benefit from and respond to therapeutic methods employed for suicidal and depressed individuals of other ages. Both drug and psychotherapeutic approaches, most often in combination, are justified and useful.

10. Total or nearly total "solutions" to the many problems of older adults are not necessary for suicide risk to be lessened. Diminution of the level of subjective psychological pain associated with these factors usually leads to a decision to live rather than die.

11. "Survivors" of the elderly individual's suicide (i.e., family members and close friends) may experience altered grief and bereavement as a result of the mode of death. These bereaved individuals may require support and possibly therapy to deal with their loss and its resolution.

REFERENCES

Atkinson, M. (1971). The Samaritans and the elderly: some problems in communication between a suicide prevention scheme and a group with a high suicide risk. *Social Sciences and Medicine, 5,* 483–490.

Barter, J. T. (1969). Self-destructive behavior in adolescents and adults: similarities and differences. In *Suicide among the American Indians: two workshops* (pp. 7–10). USPHS Publication No. 1903. Washington, DC: U.S. Government Printing Office.

Battin, M. P. (1987). Choosing the time to die: the ethics and economics of suicide in old age. In S. F. Spicker, S. R. Ingman, & I. R. Lawson (Eds.), *Ethical dimensions of geriatric care: value conflicts for the 21st century* (pp. 161–189). Dordrecht, Holland: D. Reidel.

Bock, E. W. (1972). Aging and suicide: the significance of marital, kinship, and alternative relations. *Family Coordinator, 21,* 71–79.

Bock, E. W., & Webber, I. L. (1972). Suicide and the elderly: isolating widowhood and mitigating alternatives. *Journal of Marriage and the Family, 34,* 24–31.

Busse, E. W. (1974). Geopsychiatry: social dimensions. In G. J. Maletta (Ed.), *Survey report on the aging nervous system* (pp. 195–225). DHEW Publication No. (NIH) 74-296. Washington, DC: U.S. Government Printing Office.

Butler, R. N. (1963). The life review: an interpretation of reminiscence in the aged. *Psychiatry, 26,* 65–76.

Butler, R. N., Lewis, M., & Sunderland, T. (1991). *Aging and mental health: positive psychosocial and biomedical approaches* (3rd Ed.). New York: Merrill.

Cohen, G. D. (1990). Psychopathology and mental health in the mature and elderly adult. In J. E. Birren & K. W. Schaie (Eds.), *Handbook of the psychology of aging* (3rd Ed.; pp. 359–371). San Diego: Academic Press.

Conwell, Y., Rotenberg, M., & Caine, E. D. (1990). Completed suicide at age 50 and over. *Journal of the American Geriatrics Society, 38,* 640–644.

Curran, D. K. (1987). *Adolescent suicidal behavior.* New York: Hemisphere.

Danigelis, N., & Pope, W. (1979). Durkheim's theory of suicide as applied to the family: an empirical test. *Social Forces, 57,* 1081–1106.

DeCatanzaro, D. (1981). *Suicide and self-damaging behavior: a sociobiological perspective.* Orlando, FL: Academic Press.

Dunne, E. J., McIntosh, J. L., & Dunne-Maxim, K. (Eds.) (1987). *Suicide and its aftermath: understanding and counseling the survivors.* New York: Norton.

Durkheim, E. (1897/1951). *Suicide: a study in sociology* (J. A. Spaulding & G. Simpson, trans.). New York: Free Press.

Farberow, N. L. (Ed.) (1980a). *The many faces of suicide: indirect self-destructive behavior.* New York: McGraw-Hill.

Farberow, N. L. (1980b). Indirect self-destructive behavior: classification and characteristics. In N. L. Farberow (Ed.), *The many faces of suicide: indirect self-destructive behavior* (pp. 15–27). New York: McGraw-Hill.

Farberow, N. L., Gallagher, D. E., Gilewski, M. J., & Thompson, L. W. (1987). An examination of the early impact of bereavement on psychological distress in survivors of suicide. *Gerontologist, 27,* 592–598.

Frierson, R. L. (1991). Suicide attempts by the old and the very old. *Archives of Internal Medicine, 151,* 141–144.

Gibbs, J. P. (1969). Marital status and suicide in the United States: a special test of the status integration theory. *American Journal of Sociology, 74,* 521–533.

Gibbs, J. P., & Martin, W. T. (1964). *Status integration and suicide: a sociological study.* Eugene: University of Oregon Books.

Gibbs, J. P., & Martin, W. T. (1981). Still another look at status integration and suicide. *Social Forces, 59,* 815–823.

Gove, W. R., & Hughes, M. (1980). Reexamining the ecological fallacy: a study in which aggregate data are critical in investigating the pathological effects of living alone. *Social forces, 58,* 1157–1177.

Gurland, B. J., & Cross, P. S. (1983). Suicide among the elderly. In M. K. Aronson, R. Bennett, & B. J. Gurland (Eds.), *The acting-out elderly* (pp. 55–65). New York: Haworth.

Heer, D., & MacKinnon, D. (1974). Suicide and marital status: a rejoinder to Rico-Velasco and Mynko [Letter]. *Journal of Marriage and the Family, 36,* 6–10.

Henry, A. F., & Short, J. F., Jr. (1954). *Suicide and homicide: some economic, sociological, and psychological aspects of aggression.* New York: Free Press.

Holinger, P. C., & Offer, D. (1982). Prediction of adolescent suicide: a population model. *American Journal of Psychiatry, 139* 302–307.

Kahn, A. (1990). Principles of psychotherapy with suicidal patients. In S. J. Blumenthal & D. J. Kupfer (Eds.), *Suicide over the life cycle: risk factors, assessment, and treatment of suicidal patients* (pp. 441–467). Washington, DC: American Psychiatric Press.

Kaprio, J., Koskenvuo, M., & Rita, H. (1987). Mortality after bereavement: a prospective study of 95,647 widowed persons. *American Journal of Public Health, 77,* 283–287.

Kastenbaum, R., & Mishara, B. L. (1971, July). Premature death and self-injurious behavior in old age. *Geriatrics, 26,* 71–81.

Kozak, C. M., & Gibbs, J. O. (1979). Dependent children and suicide of married parents. *Suicide and Life-Threatening Behavior, 9,* 67–75.

MacMahon, B., & Pugh, T. F. (1965). Suicide in the widowed. *American Journal of Epidemiology, 81,* 23–31.

Manton, K. G., Blazer, D. G., & Woodbury, M. A. (1987). Suicide in middle age and later life: sex and race specific life table and cohort analyses. *Journal of Gerontology, 42,* 219–227.

Maris, R. W. (1981). *Pathways to suicide: a survey of self-destructive behaviors.* Baltimore: Johns Hopkins University Press.

Maris, R. W. (1985). The adolescent suicide problem. *Suicide and Life-Threatening Behavior, 15,* 91–109.

Marshall, J. R. (1978). Changes in aged white male suicide: 1948–1972. *Journal of Gerontology, 33,* 763–768.

Matarazzo, J. D. (1980). Behavioral health and behavioral medicine: frontiers for a new health psychology. *American Psychologist, 35,* 807–817.

McCall, P. L. (1991). Adolescent and elderly white male suicide trends: evidence of changing well-being? *Journal of Gerontology: Social Sciences, 46,* S43–S51.

McIntire, M. S., & Angle, C. R. (1981). The taxonomy of suicide and self-poisoning—a pediatric perspective. In C. F. Wells & I. R. Stuart (Eds.), *Self-destructive behavior in children and adolescents* (pp. 224–249). New York: Van Nostrand Reinhold.

McIntosh, J. L. (1984). Components of the decline in elderly suicide: suicide among the young-old and old-old by race and sex. *Death Education, 8*(suppl.), 113–124.

McIntosh, J. L. (1985). Suicide among the elderly: levels and trends. *American Journal of Orthopsychiatry, 55,* 288–293.

McIntosh, J. L. (1987a). Suicide: training and education needs with an emphasis on the elderly. *Gerontology and Geriatrics Education, 7,* 125–139.

McIntosh, J. L. (1987b). Current levels of suicide by marital status in the United States. Presented at the joint meeting of the American Association of Suicidology and the International Association for Suicide Prevention, San Francisco.

McIntosh, J. L. (1988–89). Official U.S. elderly suicide data bases: levels, availability omissions. *Omega, 19,* 337–350.

McIntosh, J. L. (1990). Older adults: the next suicide epidemic? Presented at the meeting of the American Association of Suicidology, New Orleans.

McIntosh, J. L. (1992). Methods of suicide. In R. W. Maris, A. L. Berman, J. T. Maltsberger, & R. I. Yufit (Eds.), *Assessment and prediction of suicide* (pp. 381–397). New York: Guilford.

McIntosh, J. L., & Hubbard, R. W. (1988). Indirect self-destructive behavior among the elderly: a review with case examples. *Journal of Gerontological Social Work, 13,* 37–48.

McIntosh, J. L., Hubbard, R. W., & Santos, J. F. (1980). Suicide among nonwhite elderly: 1960–1977. Presented at the meeting of the Gerontological Society of America, San Diego.

McIntosh, J. L., Hubbard, R. W., & Santos, J. F. (1981). Suicide among the elderly: a review of issues with case studies. *Journal of Gerontological Social Work, 4,* 63–74.

McIntosh, J. L., & Santos, J. F. (1981). Suicide among minority elderly: a preliminary investigation. *Suicide and Life-Threatening Behavior, 11*, 151–166.

McIntosh, J. L., & Santos, J. F. (1985–86). Methods of suicide by age: sex and race differences among the young and old. *International Journal of Aging and Human Development, 22*, 123–139.

McIntosh, J. L., & Santos, J. F. (1990). Similarities and differences in elderly and youth suicide. Presented at the meeting of the American Association of Suicidology, New Orleans.

Meerloo, J. A. M. (1968). Hidden suicide. In H. L. P. Resnik (Ed.), *Suicidal behaviors: diagnosis and management* (pp. 82–89). Boston: Little, Brown.

Menninger, K. (1938). *Man against himself.* New York: Harcourt, Brace & World.

Miller, M. (1979). *Suicide after sixty.* New York: Springer Publishing Co.

Moody, H. R. (1984). Can suicide on grounds of old age be ethically justified? In M. Tallmer, E. R. Pritchard, A. H. Kutscher, R. DeBellis, M. S. Hale, & I. K. Goldberg (Eds.), *The life-threatened elderly* (pp. 64–92). New York: Columbia University Press.

Morgan, A. C. (1989). Special issues of assessment and treatment of suicide risk in the elderly. In D. Jacobs & H. N. Brown (Eds.), *Suicide: understanding and responding: Harvard Medical School Perspectives* (pp. 239–255). Madison, CT: International Universities Press.

National Center for Health Statistics. (1989). Advance report of final mortality statistics, 1987. *NCHS Monthly Vital Statistics Report, 38*(5, suppl).

Nelson, F. L., & Farberow, N. L. (1977). Indirect suicide in the elderly, chronically ill patient. In K. Achte & J. Lonnqvist (Eds.), *Suicide research: proceedings of the seminars of suicide research by Yrjo Jahnsson Foundation 1974–1977* (pp. 125–139). Helsinki: Psychiatria Fennica (Supplementum 1976).

Nelson, F. L., & Farberow, N. L. (1980). Indirect self-destructive behavior in the elderly nursing home patient. *Journal of Gerontology, 35*, 949–957.

Nelson, F. L., & Farberow, N. L. (1982). The development of an indirect self-destructive behaviour scale for use with chronically ill medical patients. *International Journal of Social Psychiatry, 28*, 5–14.

Osgood, N. J. (1985). *Suicide in the elderly: a practitioner's guide to diagnosis and mental health intervention.* Rockville, MD: Aspen.

Osgood, N J., Brant, B. A., & Lipman, A. (1991). *Suicide among the elderly in long-term care facilities.* Westport, CT: Greenwood Press.

Osgood, N. J., & McIntosh, J. L. (1986). *Suicide and the elderly: an annotated bibliography and review.* Westport, CT: Greenwood Press.

Patterson, R. D., Abrahams, R., & Baker, F. (1974, November). Preventing self-destructive behavior. *Geriatrics, 29*, 115–118, 121.

Pollinger-Hass, A., & Hendin, H. (1983). Suicide among older people: projections for the future. *Suicide and Life-Threatening Behavior, 13*, 147–154.

Portwood, D. (1978). *Common sense suicide: the final right.* New York: Dodd, Mead.

Prado, C. G. (1990). *The last choice: preemptive suicide in advanced age.* Westport, CT: Greenwood Press.

Rico-Velasco, J., & Mynko, L. (1973). Suicide and marital status: a changing relationship? *Journal of Marriage and the Family, 35*, 239–244.

Sainsbury, P. (1962). Suicide in the middle and later years. In H. T. Blumenthal (Ed.), *Aging around the world: medical and clinical aspects of aging* (pp. 97–105). New York: Columbia University Press.

Sainsbury, P. (1963). Social and epidemiological aspects of suicide with special reference to the aged. In R. H. Williams, C. Tibbitts, & W. Donahue (Eds.), *Processes of aging: social and psychological perspectives* (Vol. 2; pp. 153–175). New York: Atherton Press.

Santos, J. F., & VandenBos, G. R. (Eds.) (1982). *Psychology and the older adult: challenges for training in the 1980s.* Washington, DC: American Psychological Association.

Shneidman, E. S. (1965, May). Preventing suicide. *Americal Journal of Nursing, 65*, 111–116.

Shneidman, E. S. (1969). Prologue: fifty-eight years. In E. S. Shneidman, (Ed.), *On the nature of suicide* (pp. 1–30). San Francisco: Jossey-Bass.

Shneidman, E. S. (1973). *Deaths of man*. New York: Quadrangle.

Shneidman, E. S. (1985). *Definition of suicide*. New York: John Wiley & Sons.

Shulman, K. (1978). Suicide and parasuicide in old age: a review. *Age and Ageing, 7*, 201–209.

Stack, S. (1980). The effects of marital dissolution on suicide. *Journal of Marriage and the Family, 42*, 83–91.

Stack, S. (1981). Divorce and suicide: a time series analysis, 1933–1970. *Journal of Family Issues, 2*, 77–90.

Stack, S. (1982). Suicide: a decade review of the sociological literature. *Deviant Behavior, 4*, 41–66.

Stenback, A. (1980). Depression and suicidal behavior in old age. In J. E. Birren & R. B. Sloane (Eds.), *Handbook of mental health and aging* (pp. 616–652). Englewood Cliffs, NJ: Prentice-Hall.

Stroebe, M. S., & Stroebe, W. (1983). Who suffers more? Sex differences in health risks of the widowed. *Psychological Bulletin, 93*, 279–301.

Trout, D. L. (1980). The role of social isolation in suicide. *Suicide and Life-Threatening Behavior, 10*, 10–23.

Trovato, F. (1986). The relationship between marital dissolution and suicide: the Canadian case. *Journal of Marriage and the Family, 48*, 341–348.

United States Bureau of the Census. (1988). Estimates of the population of the United States, by age, sex, and race: 1980 to 1987. *Current Population Reports*, Series P-25, No. 1022.

Veevers, J. E. (1973). Parenthood and suicide: an examination of a neglected variable. *Social Science and Medicine, 7*, 135–144.

Wasserman, I. M. (1984). A longitudinal analysis of the linkage between suicide, unemployment and marital dissolution. *Journal of Marriage and the Family, 46*, 853–859.

Weissman, M. M. (1974). The epidemiology of suicide attempts, 1960 to 1971. *Archives of General Psychiatry, 30*, 737–746.

Wenz, F. V. (1977). Marital status, anomie, and forms of social isolation: a case of high suicide rate among the widowed in an urban sub-area. *Diseases of the Nervous System, 38*, 891–895.

Wenz, F. V. (1981). Family size, depression and parent suicide risk. In J. P. Soubrier & J. Vedrinne (Eds.), *Depression and suicide: medical, psychological, and socio-cultural aspects* (pp. 310–315). Elmsford, NY: Pergamon Press.

Wexler, L., Weissman, M. M., & Kasl, S. V. (1978). Suicide attempts 1970–1975: updating a United States study and comparisons with international trends. *British Journal of Psychiatry, 132*, 180–185.

Wolff, K. (1970). Observations on depression and suicide in the geriatric patient. In K. Wolff (Ed.), *Patterns of self-destruction: depression and suicide* (pp. 33–42). Springfield, IL: Charles C Thomas.

PART III
Assessment

7

Psychological Testing for Potentially Suicidal Individuals

JAMES R. EYMAN AND SUSANNE KOHN EYMAN

A survey by the Risk Assessment Committee of the American Association of Suicidology revealed that most mental health practitioners evaluate both immediate risk and long-term vulnerability to suicide through interview techniques (Jobes, Eyman, & Yufit, 1990). Although clinicians found psychological testing more helpful than suicide scales, they used both instruments less frequently than interviews. This finding was true even among psychologists.

Many of the survey respondents were not familiar with the suicide scales currently available. Others were concerned about the psychometric properties of the scales, particularly the predictive capacity and likelihood of obtaining results that were misleading. Nevertheless, an additional concern was the rate of "false negative" results obtained by many suicide scales. Most respondents said that they would use suicide scales as part of an evaluation for suicide risk if the scales were more reliable and valid. Some suicide scales can be used effectively as screening instruments or as supplements to an interview and to psychological testing.

It is surprising that psychologists do not find psychological testing to be generally helpful for the assessment of suicide risk, particularly as the question of suicide potential is a common referral issue. Some psychologists are not trained to use testing to reveal suicidal potential. In-depth testing also requires a substantially greater investment of time and resources than is required by the typical interview.

Interview techniques are probably the most effective way to elicit information about current, acute suicide risk. Psychological tests do not provide some of the necessary information to make this assessment, such as presence and intensity of suicidal ideation, thoroughness of a plan, and access to a method. It is also difficult to predict behavior from psychological test data because any behavior is multiply determined. However, our clinical experience indicates that the use of a psychological test battery can supply information that is highly relevant to understanding an individual's vulnerability to engage in suicidal behavior.

We review briefly and evaluate the psychometric properties and usefulness of the best and most commonly used suicide scales. The literature related to suicide assessment with the Minnesota Multiphasic Personality Inventory (MMPI), the Rorschach, and the Thematic Apperception Test (TAT) is also briefly reviewed. For a more in-depth review of suicide scales and psychological tests, readers should refer to Eyman, Mikawa, and Eyman (1990), and Eyman and Eyman (1990, 1991). A discussion of the psychological characteristics that might contribute to suicide vulnerability emphasizes how psychological testing can elicit this information. Finally, a case example illustrates our approach to using psychological testing to assess suicide risk.

Suicide Scales as Screening Instruments

In a variety of settings, it is important to identify individuals who might be at risk for suicidal behavior. For nonclinical populations, such as those found in an academic environment, it would be both impractical and unnecessary to either interview or conduct psychological testing of everyone to assess their suicide potential. There are far too few suicidal individuals to justify such a massive effort. However, self- or group-administered suicide scales can be used economically to identify persons at risk. Those who are assessed as vulnerable can then be referred for further evaluation.

In clinical settings, suicide screening instruments can also help sensitize clinicians to patients who might be suicidal. A more in-depth evaluation of risk can then be initiated and a treatment approach planned. It is particularly important to use screening instruments with those patients whose demographic and diagnostic characteristics place them in a high-risk group, such as substance abusers or depressed persons.

No suicide scale is an ideal screening instrument. One must use a scale that has the best normative data for the population in question and that is oriented toward the particular information most needed (e.g., degree of hopelessness, severity of suicidal ideation, and suicide intent). Of the instruments available, there are five scales that are the most psychometrically sound and useful: (1) Hopelessness Scale; (2) Suicide Intent Scale; (3) Scale for Suicide Ideation; (4) Suicidal Ideation Questionnaire; and (5) Suicide Probability Scale.

Hopelessness Scale

The Hopelessness Scale is a 20-item true-or-false questionnaire used to measure pessimism about the future (Beck, Weissman, Lester, & Trexler, 1974b). A score of 1 is given to each item that the respondent marks in the direction of pessimism. The total score is the number of pessimistic responses endorsed.

The Hopelessness Scale is a psychometrically sound instrument with good reliability and validity. The internal consistency of the scale was found to be .93 (Beck et al., 1974b). Several studies have found that hopelessness, as measured by this scale, is a significantly better indicator of suicidal risk that depression, as measured by various depression scales (Beck, Steer, Kovacs, & Garrison,

1985; Drake, Gates, & Cotton, 1986; Dyer & Kreitman, 1984; Fawcett et al., 1987; Kovacs, Beck, & Weissman, 1975; Wetzel, Margulies, Davis, & Karam, 1980). The Suicide Hopelessness Scale was developed by using a principal component factor analysis with varimax rotation that obtained three factors: (1) "feelings about the future"; (2) "loss of motivation"; and (3) "future expectations." The first factor, feelings about the future, accounted for 41% of the variance. Items were associated with such affects as hope, enthusiasm, happiness, and faith. Factor two, loss of motivation, accounted for 6% of the variance and loaded on items that concerned giving up. Three percent of the variance was found in the third factor, future expectations, which included items regarding anticipation of what life will be like.

The Hopelessness Scale is a consistent measure of pessimism about the future that has been shown to be an excellent indicator of potential suicidal behavior among adults. It is recommended for use to both screen for adults at potential suicide risk and as part of a more comprehensive clinical battery. If an individual scores high on the Hopelessness Scale, the clinician can use that information to discuss with the individual the cause of the feeling and how the feeling might relate to suicidality. Research on this scale has focused primarily on an adult population, however, so the scale should not be used with children and adolescents. In addition, one study found that the Hopelessness Scale did not discriminate between minority female adolescent suicide attempters and a matched group of nonsuicidal psychiatric patients (Rotheram-Borus & Trautman, 1988). Thus the Hopelessness Scale might not be an indicator of suicidality among adolescent minority females.

Suicide Intent Scale

The Suicide Intent Scale (Beck, Schuyler, & Herman, 1974a), is a semistructured interview administered by a trained clinician to assess suicidal intent from information about the intensity of the attempter's wish to die at the time of the attempt. The scale contains 15 items rated in intensity from 0 to 2 that are divided into two sections. The first section, objective circumstances related to the suicide attempt, describes the individual's behavior and events surrounding the attempt. The second section records the individual's thoughts and feelings at the time of the attempt. Each item is rated on a 3-point scale of severity (0–2), and a total score is the sum of the item scores for questions 1 through 15. The Suicide Intent Scale has been found to have an interitem reliability ranging from .91 (Beck, Morris, & Beck, 1974) to .95 (Beck et al., 1974a), an odd–even internal consistency coefficient of .82 (Beck et al., 1974a), a split-half reliability of .80 (Beck, Kovacs, & Weissman, 1979), and an interrater reliability of .95 (Minkoff, Bergman, Beck, & Beck, 1973). This scale has been correlated with hopelessness and depression (Wetzel et al., 1980). Other studies have found some relation between age and suicidal intent score (Dyer & Kreitman, 1984; Lester, Beck, & Trexler, 1975; Pierce, 1977) or no relation at all (Beck et al.,

1979; Goldney, 1979; Minkoff et al., 1973; Silver, Bohnert, Beck, & Marcus, 1971).

The Suicidal Intent Scale appears to be a psychometrically sound and useful instrument that can help clinicians evaluate the seriousness of a past suicide attempt and to predict the potential lethality of any future attempt. With this scale, assessment of intent is derived from the information gathered about previous suicide attempts, so the scale can be used only with those people who already have made a suicide attempt. There have been no studies on the use of the Suicidal Intent Scale with adolescents.

Scale for Suicide Ideation

To quantify the intensity of current conscious suicidal thoughts and plans, Beck and colleagues (1979) developed a Scale for Suicide Ideation. This 19-item scale is rated by a trained interviewer on the basis of a semistructured interview. Three alternative statements are scored from 0–2, and the total score is the sum of the scores for each item. The Scale for Suicide Ideation covers five areas: (1) attitude toward living and dying; (2) suicide ideation or wish; (3) nature of contemplated attempt; (4) actualization of contemplated attempt; and (5) background factors. Internal consistency was found to be .89 and interrelater reliability .83. This scale discriminates among groups differing in degree of suicidal ideation.

Miller, Norman, Bishop, and Dow (1986) modified the Scale for Suicide Ideation so it could be administered by paraprofessionals. Their modified scale contains 25 items and has acceptable reliability and validity. Another version of the scale is a self-report instrument that can be administered by computer (Beck, Steer, & Ranieri, 1988). Both forms have adequate reliability and validity.

The scale for Suicide Ideation is a psychometrically sound instrument that correlates with clinicians' independent ratings of suicidal ideation and discriminates between suicide attempters and nonattempters. This scale is recommended for use only with adults because there has been little research on its use with an adolescent population.

Suicidal Ideation Questionnaire

To assess an adolescent's severity and recent frequency of suicidal ideation, Reynolds (1988) developed the Suicidal Ideation Questionnaire (SIQ). Forms for high school and junior or middle high school students were designed primarily as screening instruments to identify adolescents at risk of suicidal behavior. The high school version includes 30 items and the junior high school version 15 items.

The items consist of statements pertaining to thoughts of death, suicide, and self-injury. The student is asked to respond to each question according to how frequently a particular statement was "on my mind" during the past month, using a 7-point scale ranging from "almost every day" to "I never had this thought." Interpretation is based on a total score for severity of suicidal ideation that is a sum of the item scores, critical items related to specific thoughts and plans, and individual patterns. Reynolds suggested evaluating for suicidal risk adolescents who score above a cutoff score or those who endorse three critical items on the high school version and two critical items on the junior high school version.

The normative data for the SIQ was developed by studying 2180 adolescents in grades seven through twelve. Internal consistency was high throughout the grade levels, and the item/total score correlations were generally high. The test–retest correlation was .72. The SIQ also correlated well with common measures of depression. Spirito (1987) found a significantly higher mean score (69.6) among a group of adolescent suicide attempters than in a normative population (17.8). Although approximately two-thirds of these attempters were above the cutoff score, the false-negative rate of approximately 33% is somewhat high.

The carefully constructed SIQ is one of the best suicide screening instruments for an adolescent population, but it cannot be recommended for clinical use by itself (Lewinsohn, Garrison, Langhinrichsen, & Marsteller, 1989). Further information is needed about the scale's ability to distinguish adolescents at serious clinical risk (i.e., those who have made a serious suicide attempt or completed suicide) from those who make mild suicide attempts or do not engage in suicide behavior. In addition, research is needed about discriminate validity, including the ability of the scale to distinguish suicidal ideation from other psychological variables such as hopelessness. The instrument was developed with a sample of junior and senior high school students, so the scores from this general population are most likely inappropriate for adolescent psychiatric inpatients.

Suicide Probability Scale

To assess suicide risk in adolescents and adults, Cull and Gill (1986) designed the Suicide Probability Scale. This 36-item questionnaire asks respondents to rate the frequency of occurrence for each item of a 4-point Likert Scale. However, instructions are vague about whether the respondent is to base the rating on current or past experiences.

Hand-tabulated responses provide a total weighted score, a normalized total T-score, and a suicide probability score, which is the statistical likelihood that an individual might belong to a population of lethal suicide attempters. Four subscale scores are obtained: hopelessness, suicide ideation, negative self-evaluation, and hostility. The subscales, which are all highly correlated with each other and with the total score, were developed by factor-analyzing the responses of the original sample of 1158 subjects. Golding (1985) indicated that a factor analysis showed the scale items to be scattered among various factors and highly correlated so the subscales are not statistically sound or independent; hence they should be used with caution.

Interpreting the scale requires the clinician to assign a "presumptive risk" category by assessing how "at risk" the individual is in the type of setting where he or she is being seen. Thus in order to interpret an instrument whose goal is the assessment of suicide potential, the clinician is faced with the unusual task of determining how "at risk" the patient is. Although the internal consistency and split-half reliability on this scale was good, the scale misclassified 71% of the individuals in the low-risk presumptive group. Differential cutting scores resulted in a false-negative rate of 2% in the high-risk group and 17% in the intermediate presumptive risk group. Because the validity studies for adults were conducted for the same population on which the scale was developed, these results are suspect. In addition, Cull and Gill asserted that the scale can be used for individuals 14 years of age and over, but no validation studies exist for this group; hence the scale cannot be recommended for use with adolescents.

Psychological Testing

Psychological testing can yield a rich portrait of an individual's psychological life. The testing process allows the psychologist to tease out and understand the personality characteristics, ego functioning, and dynamic underpinnings that might fuel a person's suicidal urges. Given such an understanding, the clinician can then develop an appreciation for the circumstances under which persons might feel most hopeless about leading a reasonably rewarding life, causing them to consider suicide as a possible solution.

To develop such a robust picture of a person requires a full psychological test battery. No single test can supply such complete information because each instrument is oriented primarily toward eliciting various aspects of psychological functioning. Experienced clinicians are not limited to quantitative analysis; they also make use of content, the patient's approach to each task, and the relationship that develops with the patient. Using a battery of tests allows the clinical psychologist to generate hypotheses from one test that can be confirmed or disconfirmed by another.

Previous attempts to establish the usefulness of psychological testing to assess suicidality had a different perspective than the one we suggest. Numerous efforts have been made to find one Rorschach sign, or a constellation of Rorschach variables, and MMPI configurations or single scales that would indicate if a person is suicidal. Smith (1985) has identified several problems with such an approach. First, like most behavior, suicide is multiply determined by complex personal motivations and environmental circumstances. This intricate constellation of variables cannot possibly be captured for everyone by one indicator. The "sign approach" rests on the faulty assumption that all suicidal behavior has the same dynamic underpinnings. Second, the sign approach is based on the notion that suicidal urges are expressed the same psychologically by each person. For example, Blatt and Ritzler (1974) believed that most individuals who are seriously suicidal give transparent responses on the Rorschach. Third, Smith (1985) pointed out that sign approaches are oriented toward either–or thinking; that is, the person is or is not suicidal. Most researchers investigating the usefulness of psychological testing to evaluate suicide potential have not gone beyond this simplistic way of thinking. In reality, nearly everyone has had suicidal thoughts of varying intensity and could become actively suicidal under certain stressful circumstances.

In an extensive review of the use of psychological testing to predict suicide potential, Eyman and Eyman (1991) concluded that despite considerable research effort no MMPI item, scale, or profile configuration has been found to differentiate consistently between suicidal and nonsuicidal individuals. Although no research has been reported that evaluates the usefulness of the MMPI-2 for assessing suicide potential, it is doubtful that the restandardized version provides more valid indicators. The MMPI-2 has a new 37-item content scale to measure depression. Although the manual states that elevation on this scale is characteristic of individuals with possible suicidal thoughts or wishes to die, there is no empirical evidence of the validity of this assumption. The Koss-Butcher Critical Items Scale, which has not been found useful for assessing suicide potential, has been retained in the MMPI-2. All but five of the questions dealing with depressed and suicidal ideation are found in the Depression Content Scale. An experimental version of the MMPI-2, form-TX, for use with adolescents is currently being evaluated (Williams, 1990).

The few studies using only the TAT to assess suicidality have all failed to differentiate suicidal groups from nonsuicidal groups. More research has been attempted using the Rorschach, but the results have been disappointing. Research efforts have focused on investigating responses to the "vaginal" area of Card VII (Sapolsky, 1963), transparency responses (Blatt & Ritzler, 1974), and color and color-shading responses (Appelbaum & Colson, 1968; Appelbaum & Holzman, 1962; Colson & Hurwitz, 1973; Hertz, 1948). Although there has been some support for the transparency response and the color-shading response, consistent empirical support has not been obtained. Clinicians might be alerted by the presence of either transparency responses or color-shading responses to other evidence of potential suicidality. However, the absence of one of these signs should not be interpreted as evidence that the person is not suicidal.

From a configurational approach, the presence of a certain number of signs from a constellation of variables is necessary to determine if an individual is suicidal. Exner (1978, 1986; Exner & Weiner, 1982; Exner & Wylie, 1977), Hertz (1948, 1949), and Martin (1951, 1960) have all developed Rorschach configurational approaches. Although the variables of Hertz and Martin have not been consistently supported, they seem promising. Exner and Weiner (1982) reported that the Children's Suicide Constellation was not able to discriminate suicidal from nonsuicidal children and should not be used clinically. Similarly, we found in an earlier study (Eyman & Eyman, 1987) that the Adult Suicide Constellation was ineffective in identifying seriously suicidal individuals.

Relevant Factors Assessed Through Psychological Testing

To elicit the full spectrum of psychological functioning, we use a battery that includes the WAIS-R, Rorschach, and TAT (Rapaport, Gill, & Schafer, 1978). By looking at the test data in conjunction with information obtained from evaluating the examiner–patient relationship, we seek to understand the patient's conscious and unconscious experience of self in the world, his or her capacity to form relationships and the quality of those relationships, and his or her ability to experience and understand feelings and to delay acting on them when they are inappropriate. We also evaluate how the patient resolves intrapsychic conflict or keeps it from awareness, as well as his or her intellectual functioning, ability to maintain an accurate perception of reality, problem-solving skills, capacity to reflect on internal experiences and behavior, and psychological resources that influence engagement in treatment.

In addition to these variables, a number of characteristics related to suicidal behavior (Smith & Eyman, 1988) can be assessed by psychological testing (Eyman & Eyman, 1991; Richman & Eyman, 1990; Smith, 1985). According to this view, those individuals who cling to unrealistic, unobtainable, and conflicted ideas about themselves are the most vulnerable to engaging in seriously suicidal behavior. Similarly, they have unrealistic and conflicting expectations of how others will nurture them and help maintain their self-esteem. When these expectations are not fulfilled, a great deal of anger is generated, which makes the person uncomfortable. Life situations occasionally require us to readjust our self-perceptions and our expectations of others. Suicidally vulnerable people are less flexible and thus are less able to modify their views of self and others, and so are more subject to severe disappointment and feelings of hopelessness. They may begin to think seriously about death as a way to end their painful turmoil.

Research on this model has delineated four variables that can help identify seriously suicidal individuals: (1) conflicting high expectations of self; (2) conflictual dependent yearnings; (3) defensive stance toward aggression; and (4) anxious but serious attitude about death (Eyman & Eyman, 1991, Richman & Eyman, 1990; Smith, 1985; Smith & Eyman, 1988). Although this information might be apparent anywhere in the test battery, each psychological test can elicit certain information relevant to the patient's suicide potential. Each test has specific areas that merit particular attention.

WAIS-R. In addition to IQ scores, the WAIS-R can elicit a great deal of information. It allows the psychologist to attend to the clarity of the patient's thinking and reasoning, which is evident in articulation and in how each task is approached. In addition, a patient can give bizarre responses that indicate a lack of reality attunement. The more disorganized the patient, the poorer the judgment and the greater the risk for acting on suicidal urges. Impulsivity can also be assessed; a patient may be too hasty in approaching a task or in responding (e.g., "yell fire" to the movie question on the comprehension subtest). The patient's skills at problem-solving can be assessed particularly well on the comprehension, picture arrangement, and block design subtest. The more accurate, flexible, and extensive the coping repertoire, the better is the patient's ability to negotiate life's difficulties without resorting to suicidal behavior.

Clinicians can also obtain important information by paying careful attention to how patients react to inevitable failures on WAIS-R items. Of particular relevance is whether the patient can realistically appraise his or her own performance. The unrealistically high expectations of many suicidal patients may cause them to feel that they are not doing well enough, no matter what the reality. The more harsh and unaccepting an individual is of personal limitations, the poorer the self-esteem and the more risk for suicidal behavior.

The WAIS-R is also a good place to look for indications of prominent or disruptive affective experience. For example, depression and the associated paucity of mental energy can be inferred from a lack of elaboration on verbal items and by motoric slowness on performance tasks. Anxiety might be apparent in lapses of concentration, particularly on the arithmetic and digit span subtests. Clinicians should evaluate content for themes of anger, depression, guilt, dependency, and overtly suicidal thoughts and should attend to whether concerns about these issues interfere with the patient's capacity to correctly answer questions.

RORSCHACH. The best instrument for evaluating unconscious material and ego functioning is the Rorschach. It is of value in this area because the patient has limited knowledge about what responses convey to the examiner and because it is the least structured test in the battery. Thus it can elicit information about aspects of the personality that are farthest from the person's awareness. In addition, it can be used to assess a variety of ego functions relevant to suicide potential. The patient's reality testing becomes apparent through scoring how well the responses adhere to the reality of the blots. In general, the poorer the form level, the greater the suicide risk. Also, peculiar elaborations of responses, such as confabulations, contaminations, and bizarre logic (Rapaport et al., 1978), can indicate disorganization and possible thought disorder. Spoiled responses (an otherwise good response that is spoiled by a significant intrusion of affect-laden concern), such as "a butterfly whose wings are falling off because of the holes" on Card I, indicate both a crisis state in which remnants of good ego functioning are seen alongside deterioration, as well as the possibility of an internal experience in which good things are fragile and easily overwhelmed. Such responses may indicate that the person feels overwhelmed and out of control without the necessary coping resources.

Another aspect of ego functioning that can be assessed through the Rorschach is the discomfort patients feel about their emotional life. Prominent affective experiences and how the patient responds to them can be inferred by content analysis as well as by scoring the responses to colored areas. The more evidence there is of a defensive effort to minimize emerging affect, particularly anger and aggression, the greater is the suicide risk. Suicidally vulnerable individuals who function in the mid-borderline to neurotic level of ego organization may minimize the few aggressive responses they give; and color responses tend to be form-dominated, which indicates rigid over-control of anger. In contrast, suicidal individuals who function at the psychotic level give graphically aggressive responses and demonstrate a lack of control over aggressive feelings. Nevertheless, these individuals may be aghast at their responses and often seem concerned about what they mean. Regardless of the level of ego organization, any suicidal individual may produce content with depressive themes that are also evident on achromatic scoring. Of particular concern are those responses that convey a sense of deterioration and death.

Evaluating object relations is also an important aspect of suicide assessment. Rorschach content analysis can yield information about how patients experience themselves, as well as their hopes and ideals. Unrealistically high self-expectations in Rorschach percepts might suggest an exaggerated sense of self-importance that the person strives to maintain. Rorschach responses that portray this experience include percepts of a person in an idealized or perfect role, such as that of a king, angel, or priest; grand objects or famous people, such as a gold chalice or someone like Napoleon; and prized or expensive objects, such as diamonds on a crown. The capacity for relatedness is evident in the patient's ability to create human movement responses. The quality of the movement responses should be assessed. Are human percepts viewed as benign and positive, or as threatening or destructive interactions? Finally, the examiner should assess any prominent areas of conflict. These "core conflicts" often surface as repetitive themes on the Rorschach or as responses that elicit disorganization. Conflict areas can give clues about the circumstances under which individuals become suicidal.

THEMATIC APPERCEPTION TEST. The TAT is particularly useful for fleshing out interpersonal issues, resultant affects, and coping approaches that are more covert in other tests. Unrealistic and counterdependent relationship paradigms and the sense of disappointment and anger that are often characteristic of suicidal people can unfold in TAT stories. Such stories might revolve around a failure to obtain gratification from others, forcing the hero to be self-sufficient; or the hero might feel a lack of needed resources because others are unavailable. Stories that reflect a general disillusionment about relationships or those in which characters are always left alone and lonely can reflect the suicidal individual's sense of isolation and lack of hope that relationships will be gratifying. Unrealistically high self-expectations can be seen in stories in which the hero accomplishes more than what most people can, for example, becoming the world's greatest violinist, magically healing those who are sick, or being admired and adored by everyone. Unrealistically high self-expectations are also evident in stories with a hero who is never satisfied. Sometimes a fairytale, happy-ever-after ending is tacked onto a story, with no indication that the situation has been worked through and resolved. Suicidal patients may also tell stories in which the hero attempts to cope by withdrawing socially or by negating the help of others.

Because many TAT cards elicit themes of death, this test provides valuable information about the patient's attitude and feelings about death. When presented with the TAT cards, suicidal individuals often exhibit heightened anxiety and a sober regard for death. They may avoid telling stories about suicide or death, and when they do tell them it is without embellishment. The stories typically reveal anxiety because the patient tells them haltingly or often reworks them, as by abruptly changing the plot.

The TAT is useful for clarifying the circumstances under which a patient is most likely to feel overwhelmed with despair and then turn to suicide as an option. Situations that might elicit suicidal behavior include repetitive themes, content areas around which the patient disorganizes or refuses to develop or continue a story, or themes for which the patient ends the story with death or suicide.

EXAMINER-PATIENT RELATIONSHIP. Many psychologists believe that testing should be conducted in a sterile, impersonal manner, so as not to interfere with the patient's responses. Our approach differs in that we believe that the mere presence of the examiner makes testing an interpersonal process. Thus the relationship between the patient and the psychologist can be an important source of diagnostic information.

Some suicidal patients may have difficulty engaging in a collaborative relationship with the psychologist. Trying to hide their distress and turmoil, they are only superficially compliant with the task at hand. Other suicidal patients see the examiner as unhelpful, perhaps even sadistically intrusive, or as someone who can nurture the patient and magically make everything better.

Case of Mrs. P.

Mrs. P. is a middle-aged woman with a history of depression and suicidal ideation that began early in life. She had seen several psychotherapists, had been hospitalized many times, and had been treated with various types of psychotropic medication as well as electroconvulsive therapy (ECT). The youngest of four children, she had grown up in a home where there was "no love between my parents." Her mother was frequently depressed and bedridden. Even so, Mrs. P. had felt that, of all her siblings, she was the most special to her mother. Her father was cold and unavailable to the children and verbally abusive to his wife.

The patient's first significant depressive episode began at around age 18 when she was attempting to separate from home. This brief episode resolved itself after the patient met her future husband, who was "so kind and would do anything for me." A few years into their marriage, Mrs. P. again became depressed and sought psychotherapy, which she eventually terminated. On becoming more depressed, she entered a psychiatric hospital. After discharge, she entered outpatient psychotherapy but became sexually involved with her psychotherapist who allowed her to "do things I couldn't have otherwise done, like be a mother." She eventually terminated, then several years later this former psychotherapist died and Mrs. P. again became depressed and suicidal.

When Mrs. P.'s adolescent daughter moved away from home and married, the patient's depression worsened. She entered psychotherapy but again became sexually involved with her psychotherapist. The patient gradually became convinced that if she could have an orgasm with this psychotherapist it would permanently cure her depression. However, this psychotherapist also died— just before the session when the patient had thought, in retrospect, that she would have the "curing" orgasm. She then went on to several psychotherapists, asking each to have intercourse with her; each time her request was denied Mrs. P. became more depressed, helpless, and suicidal. She again entered a psychiatric hospital and this time underwent psychological testing.

OBJECT RELATIONS. Mrs. P. appeared conflicted about relationships. On one hand, the test data suggested that she harbored a wish to be cared for, guided, and nurtured without having to reach out overtly to others. The story she told for TAT Card 13MF ("looks like she's lying there waiting to be loved") poignantly expressed the passive quality of her interpersonal yearnings. Her Rorschach response on Card II ("baby elephants with their trunks together") illustrated her wish for a merger with another. Her wishes to be nurtured and "fed" were further apparent in Rorschach percepts of "chicken drumstick" (Card V) and "cotton candy" (Card IX). Mrs. P.'s dependency was also evident during the testing process because she repeatedly questioned the examiner about what she should do and if her answers were correct.

Despite these intense wishes to be cared for, there were no indications from the test that Mrs. P. had a capacity to engage others in meaningful relationships. There were no human movement responses on the Rorschach (no M's); and on Cards III and VII, where humans are typically seen, she saw a "monster with arms upraised, probably trying to attack something" and "a caterpillar and a snail." Her TAT stories were marked by an absence of any positive human interaction. Both her Rorschach and her TAT responses suggested that she experienced relationships as either disappointing or fearful.

ABSENCE OF LIGHT. Mrs. P.'s depressive experience of blackness and internal emptiness was particularly evident in her story about TAT Card 14. In that story, all the "light" (goodness and resources) is outside herself, and she can only gaze at it longingly: "It's so black on the inside. I can't visualize it ever being so black inside . . . maybe that's why he's looking outside, so he can look into the light." It is noteworthy that this patient disavowed her own depression and feelings of badness because she did not want others to know about her bleak, empty internal experience.

Accompanying Mrs. P.'s depression was a self-image bereft of internal resources. Her answer to the "forest" question on the comprehension subtest of the WAIS-R illustrated that she could effectively cope with situations, in that she gave an appropriate explanation about how to get out of the forest, but it also showed that she was most comfortable seeing herself as ineffectual. She seemed to resent being expected to muster her strength and rise to the challenge: "This is the daytime, and you would probably never get out of it if it was nighttime. If it was afternoon you could get out by knowing the sun was in the west and then it might take a long time depending upon the size of the forest. I wouldn't want to be in that situation." Mrs. P.'s unwillingness to use whatever ability she did possess was also evident in her guessing at the more difficult arithmetic problems and in her presenting herself as overwhelmed and inadequate even though she possessed the computational skills to correctly solve the problems.

Although Mrs. P. viewed herself as incapable, she, at times, tried to appear competent. She was, however, very critical of her abilities and intolerant of failure. To the vocabulary word "ominous," she replied, "It makes me angry. I should know it." Similarly, to "audacious," she stated, "I should know it. I've heard it." With the information question about *Hamlet,* she devalued the question and tried to rationalize her failure, "I never liked literature that well. I can't answer it."

REALITY TESTING. The test data revealed this patient's level of ego organization to be in the borderline range. She gave some responses that did not adhere to the reality of the plot and received several F− scores. Her F+% (Rapaport et al., 1978), the percentage of all form-dominated responses with good form level, was a low 44%. When emotionally aroused, she exhibited even greater impairment in her reality testing, as shown by her lower F+'% of 37%. She gave several Rorschach indicators of thought disorder: six peculiar verbalizations, one perseveration, and one autistic logic.

DISCOMFORT WITH ANGER. Although the test data showed Mrs. P. to be angry, bitter, and vengeful, she was uncomfortable with these emotions, trying to deny their existence and relevance to her. A story she told about a picture by Klee, used to supplement the TAT, illustrates this process.

This guy looks like he's angry, like he wants revenge on the other one. He probably felt that the other one did something to make him feel so angry. And the outcome, I don't know. If I were, well, I'd never do it like that, revengeful. But if I weren't, I don't know.

Similarly, Mrs. P.'s one aggressive response on the Rorschach (of a monster ready to attack) was her last response to a card, and she followed it with a disclaimer indicating she had difficulty making anything out of the next card. Although the reason for her long latency on this next card was not clear, her delayed response might have been needed so she could regroup and repress the angry emotions aroused by the preceding card.

ATTITUDE ABOUT DEATH. The patient gave only one response about death, which was in her TAT story to card 3BM.

Is that supposed to be a gun? It looks like it is—or won't you tell me? I'm supposed to figure that out for myself. . . . It looks like a gun to me. This person came here because they are really sad and down. . . . . I don't know what to do, whether he should take his life or live. . . . Probably will decide it is worth living. I'm supposed to say what he is feeling?. . . Devastated, crushed, cut to pieces.

Mrs. P. was uncomfortable with this story, especially the thought of suicide. She conveyed her internal concern by having to struggle to tell the tale, as exemplified by long pauses and indecision about the gun and suicide. She also tacked on a positive outcome, saying the man would decide to live but giving no convincing evidence of why this decision would be made. It was as if the story felt too real to her, too close to home; and she therefore needed to replace these upsetting thoughts and feelings with a happy ending.

SUMMARY OF TEST FINDINGS. The psychologist who tested Mrs. P. considered her to be vulnerable to making a serious suicide attempt. She showed many characteristics often found among seriously suicidal individuals. The discomfort she exhibited while telling the TAT story about suicide reflected the serious yet ambivalent attitude about death found among many suicidal individuals. She was uncomfortable about expressing her feelings, particularly anger, and did not want either herself or others to be aware of her emotional experiences. Her unwillingness to be emotionally expressive impeded her ability to resolve conflict and therefore distanced her even further from other people. Mrs. P.'s view of herself as without goodness or resources left her feeling incapable, vulnerable, and unable to achieve anything that would please herself or others. Although dependent, she felt conflicted about her yearnings for closer relationships and would not engage in meaningful interactions, which limited her opportunity for feeling cared for by others.

Even though Mrs. P. experienced increasing pessimism that relationships could be satisfying, she harbored the hope that someone would fill her with "light" and take away her "blackness." This hope, albeit unrealistic, was life-sustaining. She would then be most vulnerable to suicidal behavior when anything occurred to shatter that dream. At the same time, she would be left feeling totally alone, with no hope that someone else could make her life fulfilling.

OUTCOME. This patient entered the hospital with the expectation that it was not only the best place for her treatment but that it represented her last chance to relieve the depression. Throughout her hospitalization, Mrs. P. maintained the delusional idea that she could be cured if a male psychotherapist would help her to have an orgasm. The patient became increasingly despondent, hopeless, and helpless. She complied only superficially with treatment, finding it difficult to discuss with staff members her mood swings and her depressive, suicidal despair. After several months in treatment, Mrs. P. began to verbalize her belief that she could not stay in the hospital without showing some progress.

At her request, the treatment team discussed alternatives, including a nursing home or a state hospital. A few days later Mrs. P. killed herself.

Guidelines for Practice: Chapter Summary

1. Psychological testing is an important tool for obtaining information crucial to understanding suicidal individuals.
2. Suicide screening scales can help identify those potentially at risk among a large group of people.
3. Once a person has been identified as possibly at risk, a thorough battery of psychological tests can answer questions about the patient's personality makeup, characteristics that contribute to suicidality, and the circumstances that might lead to suicidal behavior. Important tests include the following:
   a. Wais-R
   b. Rorschach
   c. TAT

REFERENCES

Appelbaum, S. A., & Colson, D. B. (1968). A reexamination of the color-shading Rorschach test response and suicide attempts. *Journal of Projective Techniques and Personality Assessment, 32,* 160–164.

Appelbaum, S. A., & Holzman, P. S. (1962). The color-shading response and suicide. *Journal of Projective Techniques, 26,* 155–161.

Beck, A. T., Kovacs, M., & Weissman, A. (1979). Assessment of suicidal intention: the Scale for suicide ideation. *Journal of Consulting and Clinical Psychology, 47,* 343–352.

Beck, A. T., Schuyler, D., & Herman, I. (1974a). Development of suicidal intent scales. In A. T. Beck, H. L. P. Resnik, & D. J. Lettieri (Eds.), *The prediction of suicide* (pp. 45–56). Bowie, MD: Charles Press.

Beck, A. T., Steer, R. A., Kovacs, M., & Garrison, B. (1985). Hopelessness and eventual suicide: a 10-year prospective study of patients hospitalized with suicidal ideation. *American Journal of Psychiatry, 142,* 559–563.

Beck, A. T., Steer, R. A., & Ranieri, W. F. (1988). Scale for Suicide Ideation: psychometric properties of a self-report version. *Journal of Clinical Psychology, 44,* 499–505.

Beck, A. T., Weissman, A., Lester, D., & Trexler, L. (1974b). The measurement of pessimism: the Hopelessness Scale. *Journal of Consulting and Clinical Psychology, 42,* 861–865.

Beck, R. W., Morris, J. B., & Beck, A. T. (1974). Cross-validation of the Suicidal Intent Scale. *Psychological Reports, 34,* 445–446.

Blatt, S. J., & Ritzler, D. A. (1974) Suicide and the representation of transparency and cross-sections on the Rorschach. *Journal of Consulting and Clinical Psychology, 42,* 280–287.

Colson, D. B., & Hurwitz, B. A. (1973). A new experimental approach to the relationship between color-shading and suicide attempts. *Journal of Personality Assessment, 37,* 237–241.

Cull, J. G., & Gill, W. S. (1986). *Suicide Probability Scale (SPS): Manual.* Los Angeles: Western Psychological Services.

Drake, R. E., Gates, C., & Cotton, P. G. (1986). Suicide among schizophrenics: a comparison of attempters and completed suicides. *British Journal of Psychiatry, 149,* 784–787.

Dyer, J. A. T., & Kreitman, N. (1984). Hopelessness, depression and suicidal intent in parasuicide. *British Journal of Psychiatry, 144*, 127–133.

Exner, J. E. (1978). *The Rorschach: a comprehensive system* (Vol. 2). New York: John Wiley & Sons.

Exner, J. E. (1986). Structural data IV—special indices. *The Rorschach: a comprehensive system. Vol. 1. Basic foundation* (2nd Ed.); pp. 411–428. New York: John Wiley & Sons.

Exner, J. E. & Weiner, I. (1982). *The Rorschach: a comprehensive system* (Vol. 3). New York: John Wiley & Sons.

Exner, J. E., & Wylie, J. (1977). Some Rorschach data concerning suicide. *Journal of Personality Assessment, 41*, 339–348.

Eyman, J. R., & Eyman, S. K. (1990): Suicide risk and assessment instruments. In P. Cimbolic & D. Jobes (Eds.), *Youth suicide: issues, assessment and intervention* (pp. 9–32). Springfield, IL: Charles C Thomas.

Eyman, J. R., & Eyman, S. K. (1991) Personality assessment in suicide prediction. *Suicide and Life Threatening Behavior, 21*, 37–55.

Eyman, J. R., Mikawa, J., & Eyman, S. K. (1990). The problem of adolescent suicide: issues and assessment. In P. McReynolds, J. Rosen, & G. Chelune (Eds.), *Advances in psychology assessment* (Vol. 7, pp. 165–202). New York: Plenum Press.

Eyman, S., & Eyman, J. R. (1987). An investigation of Exner's suicide constellation. Presented at the meeting of the American Psychological Association, New York.

Fawcett, J., Scheftner, W., Clark, D., Hedeker, D., Gibbons, R., & Coryell, W. (1987): Clinical predictors of suicide and patients with major affective disorders: a controlled prospective study. *American Journal of Psychiatry, 144*, 35–40.

Golding, S. L. (1985). Suicide Probability Scale. In J. W. Mitchell (Eds.), *Ninth mental measurements yearbook* (pp. 1500–1501). Lincoln: University of Nebraska Press.

Goldney, R. D. (1979): Assessment of suicidal intent by a visual analog scale. *Australian and New Zealand Journal of Psychiatry, 13*, 153–155.

Hertz, M. R. (1948). Suicidal configurations in Rorschach records. *Rorschach Research Exchange, 12*, 1–56.

Hertz, M. R. (1949). Further study of "suicidal" configurations in Rorschach records. *Rorschach Research Exchange, 13*, 44–73.

Jobes, D., Eyman, J. R., & Yufit, R. (1990). Suicide risk assessment survey. Presented at the meeting of the American Association of Suicidology, New Orleans.

Kovacs, M., Beck, A. T., & Weissman, A. (1975). Hopelessness: an indicator of suicidal risk. *Suicide, 5*(2), 98–103.

Lester, D., Beck, A. T., & Trexler, L. (1975). Extrapolation from attempted suicide to completed suicides. *Journal of Abnormal Psychology, 84*, 563–566.

Lewinsohn, P., Garrison, C., Langhinrichsen, J., & Marstellar, F. (1989). The assessment of suicidal behavior in adolescents: a review of scales of suitable for epidemiologic and clinical research. Unpublished manuscript, Oregon Research Institute.

Martin, H. A. (1951). A Rorschach study of suicide. Doctoral dissertation, University of Kentucky.

Martin, H. A. (1960). A Rorschach study of suicide. *Dissertation Abstracts, 20*, 37–38.

Miller, I. W., Norman, W. H., Bishop, S. B., & Dow, M. G. (1986). The modified Scale for Suicidal Ideation: reliability and validity. *Journal of Consulting and Clinical Psychology, 54*, 724–725.

Minkoff, K., Bergman, E., Beck, A. T., & Beck, R. (1973). Hopelessness, depression, and attempted suicide. *American Journal of Psychiatry, 130*, 455–459.

Pierce, D. W. (1977). Suicidal intent in self-injury. *British Journal of Psychiatry, 130*, 377–385.

Rapaport, D., Gill, M., & Schafer, R. (1978). *Diagnostic psychological testing* (Rev. Ed.). New York: International Universities Press.

Reynolds, W. M. (1988). *Suicidal Ideation Questionnaire*. Odessa, FL: Psychological Assessment Resources.

Richman, J., & Eyman J. R. (1990). Psychotherapy of suicide: individual, group, and family approaches. In D. Lester (Ed.), *Current concepts of suicide* (pp. 139–158). Philadelphia: Charles Press.

Rotheram-Borus, M. J., & Trautman, P. D. (1988). Hopelessness, depression, and suicidal intent among adolescent suicide attempters. *Journal of the American Academy of Child Adolescent Psychiatry, 27*, 700–704.

Sapolsky, A. (1963). An indicator of suicidal ideation on the Rorschach test. *Journal of Projective Techniques and Personality Assessment, 27*, 332–335.

Silver, M. A., Bohnert, M., Beck, A. T., & Marcus, D. (1971). Relation of depression of attempted suicide and seriousness of intent. *Archives of General Psychiatry, 25*, 573–576.

Smith, K. (1985). Suicide assessment: an ego vulnerabilities approach. *Bulletin of the Menniger Clinic, 19*, 489–499.

Smith, K., & Eyman, J. R. (1988). Ego structure and object differentiation. In H. Lerner & P. Lerner (Eds.), *Primitive mental states and Rorschach* (pp. 175–202). Madison, CT: International Universities Press.

Spirito, A. (1987). Adolescent suicide attempters: SIQ scores. Unpublished data.

Wetzel, R. D., Margulies, T., Davis, R., & Karam, E. (1980). Hopelessness, depression, and suicide intent. *Journal of Clinical Psychiatry, 41*(5), 159–160.

Williams, C. (1990). Adolescents and the MMPI: present and future. *MMPI-2 News and Profiles, 1*, 5.

8

Structured Clinical Assessment of Suicide Risk in Emergency Room and Hospital Settings

ROBERT I. YUFIT AND BRUCE BONGAR

The emergency most frequently encountered by mental health professionals is suicide (Schein, 1976), with clinicians consistently ranking work with suicidal patients as the most stressful of all clinical endeavors (Deutsch, 1984). One of the most challenging assessment tasks faced by clinicians in emergency room and inpatient settings is the need to identify the latently suicidal person and then to accurately evaluate the suicide potential.

Although outpatient therapy is possible for persons at high suicidal risk, referral to an emergency, or facility hospitalization, is often more usual and prudent in such cases because the opportunities to control and anticipate suicide are greater in emergency room and inpatient settings (Yufit, 1988; Bongar, 1991). Motto (1979) noted that the first management decision when treating a suicidal patient is to determine the treatment setting, which includes consideration of characteristics of both the patient and therapist and a careful evaluation (including a clear definition of the risks and the rationale for the decisions one is making). The central reason for not utilizing an outpatient management approach is the clinician's judgment that "the patient is not likely to survive as an outpatient" (Motto, 1979, p. 3).

However, the clinician must never forget that each management decision is a result of both the characteristics of vulnerability and coping that are unique to the patient, his or her social support matrix, and the therapist's equally unique level of experience plus his or her capability and tolerance for stress and uncertainty (Yufit, 1992).

Thus one of the clinician's first tasks in determining appropriate patient populations for continued outpatient management is to distinguish between acute clinical states related to *DSM III-R* Axis I clinical syndromes and chronic suicidal behavior that is part of an Axis II personality disorder (Bongar, 1991; Goldsmith, Fyer, & Frances, 1990; Simon, 1988). One authority, Simon (1987, 1988), pointed out that the clinician who treats a suicidal outpatient and who makes a gross error in deciding not to seek commitment for the patient meeting these legal criteria may be held liable (Simon, 1988). In addition, Simon (1987, 1988) noted a number of considerations when evaluating a patient in an emergency situation and evaluating the need for hospitalization:

"(1) Acutely suicidal patients often suffer from *DSM-III-R* Axis I disorders, typically major affective or schizophrenic disorders that require immediate hospitalization and inpatient treatment. The suicidal risk may recede with the remission in the acute or recurrent episode of the illness. Clinicians must act promptly and affirmatively to hospitalize, further evaluate, and to monitor such patients.

(2) Those patients who are chronically suicidal can more typically be treated as outpatients and usually these patients meet the Axis II criteria for a personality disorder. However, inpatient care may be necessary when their suicidal impulses become acute and exacerbated due to a crisis [in] the patient's life, or if the patient also develops an Axis I clinical syndrome, typically an affective disorder. A patient's chronically suicidal behavior may be one way to deal with her/his inner sense of hopelessness or of rage, secondary to a sense of low self-esteem; or she/he uses the suicidal crisis as a means of escape from her/his life turmoil and attempts to regain control. When such crises occur, a real risk of self-harming behavior, if not self-destructive behavior, does exist and hospitalization may become necessary.

(3) If such chronically suicidal patients are not hospitalized, they should be continuously reevaluated for vulnerability to self-harm or self-destructiveness during each clinical contact. An important role for the outpatient clinician is to make sure that the patient sees her/him as readily available, the patient's medications must be closely monitored (both for compliance and for side effects), and the patient's interpersonal support systems must be mobilized to help [him or her] deal with the intense stresses of these chronic suicidal crises.

(4) A mental health professional who conducts treatment on an outpatient basis with a chronically suicidal patient must learn to be able to tolerate a certain level of chronic suicidal ideation in these patients in order to effectively continue psychotherapy. For clinicians who elect this management course, the well-documented risk/benefit analysis is sine qua non in the ongoing outpatient management plan."

Here it is vital to note Litman's cautionary dictum that most chronically suicidal patients have a history of contacts with mental health professionals; 20% of the suicides in this patient group were in treatment at the time of their death, and 50% were in treatment with psychiatrists (Litman, 1988). Litman stated that the actual ongoing treatment of these difficult patients calls for flexibility and the use of a variety of therapeutic modalities. One major complication to treatment is "therapist burnout." Other serious complications to providing optimal care are a sense of complacency on the part of the experienced psychotherapist and the intense anxiety experienced by the neophyte therapist who confronts the suicidal scenario for the first time.

In the emergent assessment of imminence and degree of lethal dangerousness, it also may be useful to conceptualize more generally the patient's "danger-to-self" thoughts, feelings, and behaviors along a thought–impulse–action continuum (e.g., attention seeking to self-harming to actual self-destruction behaviors). Shneidman (1989) found that a communication of intent is present in 80% of completed suicides. Although placing this communication of intent by patients at the 50 to 70% level, Fawcett (1988) discovered that high-risk suicidal patients tend to communicate their intent to their significant other only, whereas those in the moderate to mild risk group more frequently threaten suicide to doctors, other family members, and so forth. In addition, and except when contraindicated by toxic interpersonal matrices, the clinicians in the emergency room or the inpatient intake clinicians must avail themselves of all available sources of information through comprehensive information gathering and collaboration with the family and significant others, in addition to comprehensive direct assessment of the patient. Such an extended network of assessment information is a vital element in any successful assessment, management, or treatment plan. (This point is discussed further in Chapter 16.)

Initial Intake Evaluation Goals

A review of the case law and data on malpractice in psychology and psychiatry suggests that an acceptable standard of care requires an initial and periodic evaluation of suicide potential for *all* patients seen in clinical practice (Bongar, 1991). Furthermore, prudent practice activities dictate that reasonable clinical care requires that a patient who is "either suspected or confirmed to be suicidal" must be the subject of certain affirmative precautions (Simon, 1988, p. 85). The mental health professional who fails either to reasonably assess a patient's suicidality or to implement an appropriate management plan based on the detection of elevated risk is likely to be exposed to liability if the patient is harmed by a suicide attempt (Bongar, 1991; Simon, 1987, 1988). Simon (1988, p. 85) also noted that "the law tends to assume that suicide is preventable, in most circumstances, if it is foreseeable."

Litman (1982) found in a survey of suicides in the Los Angeles area that 1% of patients being treated in general medical-surgical or psychiatric hospitals committed suicide during their hospital treatment (approximately one-third of these suicides resulted in lawsuits against the hospitals). Psychiatric units were the targets of about half of these suits. Furthermore, suicidal phenomena are ubiquitous in the inpatient settings, and dangerousness to self is the most common reason for admission to a psychiatric unit (Friedman, 1989). In an important review on the management of suicide in inpatient settings, Friedman found that the frequency of completed suicide in patients with a history of psychiatric hospitalization is many times that of patients in the general population.

In addition to the assessment and hospitalization of patients who present as at high risk of suicide during the initial emergency room evaluation, once the patient is admitted to an inpatient facility, the clinician and staff must assess suicide risk when deciding issues of specific treatments, privileges, discharge, and so forth. Often the inpatient staff must also deal (sometimes continuously) with the presence of suicidal gestures, thoughts, and impulses within the context of the therapeutic milieu (Friedman, 1989; Peterson & Bongar, 1990). (A complete review of hospital assessment of the suicidal patient is beyond the scope of this chapter. The reader is directed to the extended discussion and review of this topic by Friedman [1989], as well as to the concept of the Suicide Assessment Team [Yufit, 1988]).

In 1957 Litman stated that in cases where the clinician conducting the assessment for risk determines that an outpatient is a high suicide risk—but that he or she can function as an outpatient—it is the clinician's responsibility to ensure that the risk is communicated to all concerned parties (i.e., the family and significant others). The assessing clinician must dispassionately provide the family and significant others with detailed, clear information as to the risks and benefits of both inpatient and outpatient treatment and obtain their informed consent (Sadoff, 1990). At this juncture in the assessment, the clinician may wish to consider routinely involving a senior colleague for a "second opinion" on this particular decision.

Bongar (1991), after reviewing the clinical and legal literature on the psychotherapeutic care of high-risk outpatients, found that there was a consensus (with one exception) that the outpatient environment exposes the patient to much greater danger because the patient is not under 24-hour monitored care. Bongar also pointed out that courts have typically seen a psychiatric decision to hospitalize as the more usual and customary one for a patient who is at high and imminent risk. Hospitals are usually deemed the environment where maximum protection can best be provided.

Klerman (1986) has cautioned mental health professionals that they should carefully examine their own policies and procedures for hospitalizing patients to ensure that their decisions are risk-benefit decisions that are focused on the optimal care and safety of the patient and not merely defensive reactions based on fear of litigation. (An automatic hospitalization of a patient at any level of suicide risk can be a significant iatrogenic effect of treatment and can significantly impair the outpatient therapy.) Klerman further warned that, because of the threat of a lawsuit, the clinician may become so worried about the question of hospitalization that important details in therapy are missed. In addition, the inability of the mental health professional to deal effectively with the pressure and strain of the situation may influence the course of the treatment—and not necessarily in ways that are beneficial to the patient.

Empirical Evidence on Risk Assessment Practices

The Risk Assessment Committee of the American Association of Suicidology (AAS) reported the findings of their survey on suicide risk assessment procedures among a random sample of practitioners in psychology, psychiatry, and social work (including members and nonmembers of the AAS). Although the data from this study should be interpreted with caution because of the moderate sample size ($n = 414$) and relatively low response rate (38%), the findings do provide at least an initial glimpse into the usual assessment practices of practitioners (Jobes, Eyman, & Yufit, 1990). Of particular interest for the purposes of the present chapter is the amount of assessment activity carried out by the respondents, many of whom appeared to be psychologists: They reported evaluating, on average, 1.9 suicidal adolescents and 2.8 suicidal adults per month.

More specifically, Jobes et al. found that

the more commonly known suicide assessment instruments [appear] to be used infrequently and most of the traditional instruments are rated as having a limited usefulness. For adolescents, the most frequently utilized instruments were the Beck Hopelessness Scale (BHS) (28%) and the Beck Suicide Intent Scale (SIS) (23%). For adults the most frequently utilized scales were the BHS (34%) and SIS (28%). The latter two scales were rated as "somewhat useful" for both adults and adolescents. While traditional psychological tests were found to be used more frequently, these instruments were not rated as very useful. The most commonly used psychological tests were the MMPI (47%), Beck Depression Inventory (46%), the Rorschach (44%), and the Thematic Apperception Technique (42%). Of these assessment tools, only the BDI received a rating of "somewhat useful."

In addition, with a limited number of exceptions, there were few major differences in the approaches used to assess acute versus chronic risk, and a limited difference in the assessment practices of clinicians who worked with adults compared to those clinicians who assessed adolescents. Instead, most of these mental health professionals seemed to rely primarily on some form of a clinical interview to assess suicide (specifically, on certain valued questions and observations).

While noting the typical practitioner's preference (and primary reliance) for interview and observational data, we attempt, in the remainder of this chapter, to present a structured approach to the assessment of risk in both emergency room and inpatient settings. More specifically, we discuss the use of a specialty assessment sequence, along with suicide scales and risk estimators. (For specific psychological, epidemiological, biological, cognitive, psychodynamic, and pharmacological findings on risk factors see Chapters, 1, 3, 9, 10, and 11.) Although the following section details specific structured assessment approaches, it is critical to state that this discussion must always be buttressed and supported by the formulation of clinical judgment via important clinical observations and key elements that are elicited in the clinical interview as integral components of a model of comprehensive clinical assessment and risk management.

Suicide Assessment Team and Battery

Yufit (1988) has stated that suicidal behavior is best assessed through the use of a Suicide Assessment Team (SAT). Such a team might be composed of a multidisciplinary staff of psychologists, social workers, nurses, and psychology graduate students, with specialty training in the use of a focused screening interview format and other assessment techniques for the identification and determination of suicide potential. The SAT assessment would involve three levels: the focused clinical interview (level I); this interview plus specialized rating scales (level II); and a more extended psychological assessment (level

III), including the above interviews and ratings as well as special psychological assessment techniques, termed the Suicide Assessment Battery (SAB).

The SAB would be used to make a more extended evaluation of suicide potential (level III) beyond specific interview ratings (level I) and the scores on the specialized suicide rating scales (level II). Before proceeding to level III, the clinician already would have seen the patient in a structured interview (level I) and utilized rating scales such as the Beck Hopelessness Scale or Suicide Intent Scale, as well as the Suicide Screening Checklist (Yufit, 1989), in order to screen the patient for suicide potential. As Yufit (1988, p. 26) has stated, level III techniques "would most likely be used with inpatients who make suicidal threats or attempts, or where suicide can be inferred" and is helpful for making a more extended evaluation of suicide potential.

A useful technique in the SAB is the Time Questionnaire (TQ). This instrument is a semiprojective personality assessment technique that has been found to correlate with suicide potential (Yufit & Benzies, 1973). The TQ has been given to "over 1500 persons, including clinical and non-clinical samples, as well as matched sample populations; it has consistently differentiated high lethal suicidal persons from lower levels of suicide lethality and nonsuicidal persons as well as a variety of psychiatric diagnostic groups, on the basis of uniquely different time perspectives . . . the TQ is a key technique in the SAB" (Yufit, 1988, p. 27). Yufit (unpublished data) has now incorporated the Coping Abilities Questionnaire, a 15-item instrument that measures the range of coping ability (from excellent to minimal). (For the complete Coping Abilities Questionnaire see Appendix I).

In addition to the use of the TQ and the Coping Abilities Questionnaire, Yufit has found that other possible key elements in an SAB might incorporate the following instruments.

(1) Suicide Screening Checklist (SSC), a 60-item checklist that provides a supplement to clinical judgment
(2) California Risk Estimator for Suicide, an empirically derived instrument using significant items from a large-scale prospective study of a clinic instrument used to assess suicide (Motto, Heilbron, & Juster, 1985)
(3) Experiential Inventory, in which the patient is asked to list 10 most important life experiences (past, present, future) (modified from Cottle, 1976)
(4) Sentence Completion, a 32-item form with sentence stems especially selected to elicit affect related to morbid thoughts, self-destruction, hope, trust, the future, etc.
(5) Draw-A-Person in the Rain, a variation of the "Draw-a-Person" projective technique in which rain is an ambiguous environmental stimulus. . . . "Scoring is subjective in nature, but the work of Machover and other exponents of DAP can be utilized." (Yufit, 1988, p. 30)
(6) The use of an autobiography, in which the patient is asked to write his or her life history in any way that they wish
(7) Rorschach Inkblot technique, primarily used for associational content
(8) Thematic Apperception Technique, analysis of story themes related to isolation, hopelessness, mistrust, morbid content, and future orientation
(9) Erikson Questionnaire, a multiple choice instrument giving scores on Erik Erikson's

developmental model of eight stages (especially those stages that are related to intimacy-isolation, trust-mistrust, autonomy-shame and doubt)

(10) The Phillipson Object Relations Technique (this is a variant of the TAT and usually gives more in-depth and elaborate data than the more structured TAT)

(11) Humor Test, a 104-item, objectively scored questionnaire which gives polar opposite scores on 13 factor-analytically-determined scales

(12) Q-sort set, a 22-item set representative by "descriptive items relating to variables often associated with suicidal behavior" which are keyed to Henry Murray's "need/press" framework (Yufit, 1988, p. 28). Two sorts are made, a sort for the actual self, and an additional sort for the "ideal" self (these are then intercorrelated) and repeated.

The list of the above 12 assessment techniques shows how a clinician might use a broad array of available and useful measures, from which an SAB can be derived: Usually the SAB includes five or six of the above measures (e.g., one possible SAB battery might include the Suicide Screening Checklist, Beck Hopelessness Scale, Coping Questionnaire, Time Questionnaire, and Q-sort).

Yufit (1988) also noted that the use of the above SAB, which includes a number of scales still in experimental form, "may be questioned, but they are considered very useful to trying to fill the lacunae and tap the nuances in the complex task of identification and assessment of suicide potential" (p. 32). Yufit concluded that

even in a psychiatric hospital setting, where psychiatric sophistication may be considered "deep" (i.e., in a more environmentally saturated context), there is a need for more *comprehensive* evaluation procedures of the complex behavior of self-harming and self-destructive behaviors. At this stage of development, these techniques are not necessarily conclusive, nor are they often objective, but they very often do serve as important *guidelines* to assist in the identification and the assessment of the components of suicide potential. They should supplement clinical judgment, not substitute for it. [p. 33]

In an effort to formalize and distill the elements of risk detection (as well as to try to enhance the accuracy of diagnosing suicide risk), Yufit (1989) has begun development and field testing of a new, more integrated instrument cited previously, the Suicide Screening Checklist (SSC). The SSC can be used by a clinician or interviewer to assess an individual for the purposes of iden-tifying suicidal potential. Here suicidal potential (or suicide risk) refers to the likelihood that such a person will engage in behavior that directly or indirectly leads to his or her self-destruction. The SSC is a screening tool constructed from empirical data and uses known and presumed correlates of suicidal be-havior. Like many suicide scales and risk estimators, it is designed to comple-ment and improve "the validity of clinical judgment" (Yufit, 1989, p. 139). Thus the use of specific instruments such as the SSC (as well as batteries such as the SAB) allow clinicians to supplement their own clinical intuition with a systematized approach to collecting assessment information. (The SSC is con-tained in Appendix II.)

Yufit further suggested that an accurate (and widely accepted) model for the assessment of elevated risk in the suicidal patient may well require a future research endeavor that would entail a large-scale collaborative multicenter study that was designed from the start to evaluate *all* of the existing assessment procedures for both efficacy and significance. The assessment practices questionnaire data (Jobes et al., 1990), though only the first step in understanding clinical practices with regard to suicide, is of great help in seeing how practicing clinicians use interview, observational and rating scale material.

Bongar (1991) has constructed a comprehensive and integrated decision checklist that has drawn on and recognized the important contributions of Beck and his colleagues' empirical model (e.g., hopelessness, helpless, negative cognitions), Fawcett and colleagues' empirical model (e.g., chronic versus acute, the social matrix, communication of intent), Hirschfeld and Davidson's epidemiological model, Simon's checklist model, Shneidman's psychological model, and Yufit's SAT/SAB/SSC protocols.

This clinical and legal formulation of a standard of care for the assessment of elevated risk (Bongar, 1991) follows the decision-checklist tradition and presents a large number of specific steps and detection-decision points.* However, we suggest a cautionary note: Such approaches often offer a possible set of known risk factors and are often not meant to be definitive or exhaustive. Mental health professionals can almost certainly tailor the assessment of elevated risk to what Motto (1989) has shown to be the uniqueness of every decision on suicide probability.

Guidelines for Practice: Chapter Summary

Yufit (1988), based on many years of conducting clinical assessments of suicidal patients, noted that the assessment of suicide potential is often accomplished through the use of the initial focused interview (based on a referral from the emergency room, inpatient service, or outpatient clinician) and the experienced psychotherapist's clinical judgment. The level I (focused interview) would explore and rate the

1. Patient's conscious intent (and ambivalence) of ending his or her own life
2. Risk of rescue or possible interruption during suicide attempt
3. Degree of planning
4. Behavior level (e.g., threats, ideation, gestures, overt attempt)
5. Lethality of attempt made
6. Extent of physical injury and/or toxicity
7. Precipitant factors
8. Intensity of current life stress and pain
9. History of previous attempts, gestures, threats, and ideation
10. Degree of depression

---

*For a complete presentation of this comprehensive checklist approach the reader is directed to the material on detection of elevated risk in Bongar (1991, Ch. 3).

11. Intense observation of the patient
    a. Ability to relate to the examiner during the interview
    b. Overt dress and grooming
    c. Posture
    d. Degree of agitation
    e. Ability to discuss the problem
12. Changes in the patient's behavior during the initial contact
13. Overall psychological status which should focus on balance between coping and vulnerability (Yufit, 1992)

When there is more time available than in an emergency situation, it behooves clinicians to use data from the specific risk factors for suicide, knowledge of the general formulation of clinical judgment, and their own clinical experience and training, combined with common sense, to estimate the risk; the clinician may also wish to consider Yufit's suggestion to continue the evaluation via a level II/III (SAT/SAB) specialty referral. (The appendices to this chapter contain selected examples of screening tools that might prove useful for conducting a level II/III evaluation.)

If the patient has exhibited any previous suicidal behavior, Motto (1989) noted that such behaviors can demonstrate a breach of resistance to pain, as a suicide attempt must be considered an indication of increased vulnerability. Moreover, Clark, Gibbons, Fawcett, and Scheftner (1989, p. 42) suggested that when considering patients with moderate to severe affective disorders the clinician "should not interpret the absence of any recent suicide attempts to mean that the patient is at relatively low risk for attempting suicide in the future. . . . Suicide attempts made many years ago may have equal value to recent attempts when estimating an individual's predisposition to non-lethal attempts in the future. In addition, Shneidman (1987, 1989) stressed evaluation of perturbation, lethality, and environmental "press" as critical factors in the estimation of risk—all of which should be carefully considered during the assessment of risk in inpatient and outpatient settings.

Finally, Pokorny (1983) noted that the identification and care of the suicidal patient in clinical practice is made up of a *sequence of small decisions*, a point we wish to underscore. Murphy (1988, p. 53) expanded on this dictum by noting that

the first decision might be based on some alerting sign or clinical configuration, and the decision would be to investigate further. After further investigation, one might stop, if no additional alerting or confirming indicator were found. Or one might decide to explore the situation even further; perhaps even to hospitalize, for example. "In each case, the decision is not what to do for all time, but rather what to do next, for the near future (Pokorny, 1983). . . . There is continuing opportunity for feedback, and thus for modification of risk assessment and intervention.

REFERENCES

Bongar, B. (1991). *The suicidal patient: clinical and legal standards of care.* Washington, DC: American Psychological Association.

Clark, D. C., Gibbons, R. D., Fawcett, J., & Scheftner, W. A. (1989). What is the mechanism by which suicide attempts predispose to later suicide attempts? A mathematical model. *Journal of Abnormal Psychology, 98*(1), 42–49.

Cottle, T. J. (1976). *Perceiving time.* New York: John Wiley & Sons.

Deutsch, C. J. (1984). Self-report sources of stress among psychotherapists. *Professional Psychology: Research and Practice, 15,* 833–845.

Fawcett, J. (1988). Interventions against suicide. In D. G. Jacobs and J. Fawcett (chairmen), Suicide and the psychiatrist: clinical challenges. Presented at a symposium sponsored by the Suicide Education Institute of Boston in collaboration with The Center of Suicide Research and Prevention at the American Psychiatric Association Annual Meeting, Montreal.

Freidman, R. S. (1989). Hospital treatment of the suicidal patient. In D. G. Jacobs & H. N. Brown (Eds.), *Suicide: understanding and responding: Harvard Medical School perspectives on suicide* (pp. 379–402.). Madison, CT: International Universities Press.

Goldsmith, S. J., Fyer, M., & Frances, A. (1990). Personality and suicide. In S. J. Blumenthal & D. J. Kupfer (Eds.), *Suicide over the life cycle: risk factors, assessment, and treatment of suicidal patients* (pp. 155–176). Washington, DC: American Psychiatric Press.

Jobes, D. A., Eyman, J. R., & Yufit, R. I. (1990). Suicide risk assessment survey. Presented at the annual meeting of the American Association of Suicidology, New Orleans.

Klerman, G. L. (Ed.) (1986). *Suicide and depression among adolescents and young adults.* Washington, DC: American Psychiatric Press.

Litman, R. E. (1957). Some aspects to the treatment of the potentially suicidal patient. In E. S. Shneidman & N. L. Farberow (Eds.), *Clues to suicide* (pp. 11–118). New York: McGraw-Hill.

Litman, R. E. (1982) Hospital suicides: lawsuits and standards. *Suicide and Life-Threatening Behavior, 12*(4), 212–220.

Litman, R. E. (1988). Treating high-risk chronically suicidal patients. In D. G. Jacobs and J. Fawcett (chairmen), Suicide and the psychiatrist: clinical challenges. Presented at a symposium sponsored by the Suicide Education Institute of Boston in collaboration with The Center of Suicide Research and Prevention at the American Psychiatric Association Annual Meeting, Montreal.

Maris, R., Berman, A., Maltsberger, J. T., & Yufit, R. I. (1992). *Assessment and prediction of suicide.* New York: Guilford Press.

Motto, J. A. (1979). Guidelines for the management of the suicidal patient. *Weekly Psychiatry Update Series Lesson 20, 3,* 3–7 (available from Biomedia, Princeton, NJ).

Motto, J. A. (1989). Problems in suicide risk assessment. In D. G. Jacobs & H. N. Brown (Eds)., *Suicide: understanding and responding: Harvard Medical School perspectives on suicide* (pp. 129–142). Madison, CT: International Universities Press.

Motto, J. A., Heilbron, D. C., & Juster, R. P. (1985). Development of a clinical instrument to estimate suicide risk. *American Journal of Psychiatry, 142,* 680–686.

Murphy, G. E. (1988). The prediction of suicide. In S. Lesse (Ed.), *What we know about suicide behavior and how to treat it* (pp. 47–58). Northvale, NJ: Jason Aronson.

Peterson, L. G., & Bongar, B. (1990). Training physicians in the clinical evaluation of the suicidal patient. In M. Hale (Ed.), *Teaching methods in consultation-liaison psychiatry* (pp. 89–108). Basel: Karger.

Pokorny, A. D. (1983). Prediction of suicide in psychiatric patients. *Archives of General Psychiatry, 40,* 249–257.

Sadoff, R. L. (1990). Argument for the plaintiff—expert opinion: death in hindsight. In R. I. Simon (Ed.), *Review of clinical psychiatry and the law* (pp. 331–335). Washington, DC: American Psychiatric Association.

Schein, H. M. (1976). Obstacles in the education of psychiatric residents. *Omega, 7,* 75–82.

Shneidman, E. S. (1987). A psychological approach to suicide. In G. R. VandenBos & B. K. Bryant (Eds.), *Cataclysms, crises, and catastrophes: psychology in action* (pp. 147–183). Washington, DC: American Psychological Association.

Shneidman, E. S. (1989). Overview: a multidimensional approach to suicide. In D. G. Jacobs & H. N. Brown (Eds.), *Suicide: understanding and responding: Harvard Medical School perspectives on suicide* (pp. 1–30). Madison, CT: International Universities Press.

Simon, R. I. (1987). *Clinical psychiatry and the law*. Washington, DC: American Psychiatric Press.

Simon, R. I. (1988). *Concise guide to clinical psychiatry and the law*. Washington, DC: American Psychiatric Press.

Yufit, R. I. (1988). Manual of procedures—assessing suicide potential: suicide assessment team. Unpublished manual (available from R. I. Yufit, PhD, Department of Psychiatry and Behavioral Sciences, Division of Clinical Psychology, Northwestern University Medical School, Chicago, IL.)

Yufit, R. I. (1989). Developing a suicide screening instrument for adolescents and young adults. In Alcohol, Drug Abuse, and Mental Health Administration, *Report of the Secretary's Task Force on Youth Suicide* (Vol. 4; pp. 4–129 to 4–144). DHSS Publ. No. (ADM) 89–1624. Washington, DC: U.S. Government Printing Office.

Yufit, R. I., & Benzies, B. (1973). Assessing suicide potential by time perspective. *Suicide and Life-Threatening Behavior, 3*(4), 270–282.

Coping Abilities Questionnaire

Patient Name: _____

Date: _____

Rater: _____

	Weight	Yes	No	Uncertain
1. Flexible approach in dealing with adversity (failure, loss, rejection).	×5	___	___	___
2. Firm belief in oneself; feels capable	×2	___	___	___
3. Sense of trust in self	×3	___	___	___
4. Sense of trust in others	×3	___	___	___
5. Deals with change adequately	×2	___	___	___
6. Well-developed sense of humor (enjoys fun)	×3	___	___	___
7. Has continuity of the self, i.e., has an identity	×3	___	___	___
8. Maintains adequate perspective in stress situations; not easily overwhelmed	×3	___	___	___
9. Values life in a meaningful manner	×3	___	___	___
10. Has a developed future time perspective	×5	___	___	___
11. Available external support system		___	___	___
12. Has current close relationships	×4	___	___	___
13. Has control over energy level		___	___	___
14. Enjoys work; achievement oriented; has generativity	×2	___	___	___
15. Can give, be loving	×2	___	___	___
Sum	___	___	___	___

TENTATIVE SCORING KEY

Circle one:
31–40	Excellent coping abilities
24–30	Good coping abilities
15–23	Fair coping abilities
Below 14	Minimal coping abilities

Level of Confidence: High Low

Reasons for low rating:

_____

_____

_____

APPENDIX II

Suicide Screening Checklist

Patient Name: _____

Date: _____

Rater: _____

Directions: Score each item on basis of interview responses or chart information. Verify doubtful information with family members when possible. If no parentheses after item, score +1 for each "yes" or use listed weighted score in parentheses. "No" or "uncertain" scores = 0. Try to minimize "uncertain" scores. Sum all scores and categorize as indicated.

SUICIDE HISTORY (maximum section score = 18)

	Yes	No	Uncertain
1. Prior attempt	___	___	___
2. Two or more prior attempts in past year (highly lethal* = ×2)	___	___	___
3. Prior suicide threats or ideation	___	___	___
4. Suicidal attempts in family (×2)	___	___	___
5. Completed attempts in family (×3)	___	___	___
6. Current suicidal preoccupation, threats, attempt (×2); detailed, highly lethal plan (×2); access to weapon, medication in home (×4); all three "yes" = 8	___	___	___
7. Ongoing preoccupation with death	___	___	___

PSYCHIATRIC HISTORY (maximum score = 13)

	Yes	No	Uncertain
8. Drug, alcohol abuse: past (×2); present (×3)	___	___	___
9. Dx of mental disorder (×2); Dx of schizophrenia or manic-depressive (×3)	___	___	___
10. Poor impulse control (current = ×3)	___	___	___
11. Explosive rage episodes (circle: chronic, single recent, single past)	___	___	___
12. Accident-proneness (frequency, examples)	___	___	___
13. Panic attacks: (recurrent = ×2)	___	___	___

SCHOOL (when relevant) (maximum score = 8) or   JOB (maximum score = 8)

14. Grade failure	14. Demotion ($\times 2$)
15. Rejection, poor social relations	15. Rejection, poor social relations
16. Probation or school dropout ($\times 2$)	16. Fired ($\times 2$)
17. Disciplinary crisis ($\times 2$)	17. Laid off
18. Anticipation of severe punishment	18. Disciplinary crisis
19. Unwanted change of schools	19. Unwanted change

FAMILY (maximum score = 26)

20. Recent major negative change, e.g., loss (death, divorce, serious health problem); (irreversible loss = $\times 3$) (divorce = $\times 3$); both "yes" = 6
21. Lack of emotional support, estranged
22. Loss of employment (parent or self)
23. Major depression in parent, sibling ($\times 2$)
24. Alcoholism other drug use in family member ($\times 2$)
25. Psychiatric illness in family member ($\times 2$)
25a. If $23 + 24 + 25 = 6$, add 6 more
26. History of pysical or sexual abuse (both = $\times 2$)

SOCIETAL (maximum score = 3)

27. "Contagion" suicide episode
28. Economic down-shift in community
29. Loss of major support system (family, job career)

PERSONALITY AND BEHAVIOR; COGNITIVE STYLE (maximum score = 56)

30. Anger, rage (held in = $\times 4$)
31. Depression (intensely depressed = $\times 2$; agitated depressed = $\times 4$)
32. Hopelessness ($\times 2$)
32a. If $30 + 31 + 32 = 10$, add 10 more
33. Mistrust (paranoid = $\times 2$)
34. Disgust or despair (both = 2)
35. Withdrawn, isolated (loneliness = $\times 2$)
36. Low "future time" perspective ($\times 3$)
37. High "past" orientation ($\times 3$)

(continued)

	Yes	No	Uncertain
37a. If 36 + 37 = 6, add 6 more	___	___	___
38. Rigidity, perfectionism ($\times 4$)	___	___	___
39. Lack of belonging ($\times 3$)	___	___	___
40. Indifference, lack of motivation (boredom = $\times 2$)	___	___	___
41. Worthlessness, no one cares	___	___	___
42. Shame or guilt (both = $\times 2$)	___	___	___
43. Helplessness	___	___	___
44. Inability to have fun ($\times 2$)	___	___	___
45. Extreme mood or energy fluctuation (both = $\times 2$)	___	___	___
46. Giving away valuables	___	___	___

PHYSICAL (maximum score = 12)

	Yes	No	Uncertain
47. Male ($\times 2$); Caucasian ($\times 2$); both "yes" = 4	___	___	___
48. Delayed puberty	___	___	___
49. Recent injury leading to deformity, impairment (if permanent = $\times 2$)	___	___	___
50. Loss of appetite	___	___	___
51. Marked weight loss (>10 lb) ($\times 2$)	___	___	___
52. Sleep disturbance (onset, middle, early awake); hypersomnia	___	___	___
53. Ongoing physical pain	___	___	___

INTERVIEWER BEHAVIOR (maximum score = 16)

	Yes	No	Uncertain
54. Noncommunicative, encapsulated	___	___	___
55. Negative reaction of patient to interviewer ($\times 2$)	___	___	___
56. Negative reaction of interviewer to patient	___	___	___
57. Increasing "distance" during interview ($\times 3$)	___	___	___
58. Increasing hostility, noncooperation ($\times 3$)	___	___	___
59. Highly self-critical, self-pitying ($\times 2$)	___	___	___
60. Discusses death, suicide ($\times 4$)	___	___	___

Sum of pages 1 + 2 + 3

SUICIDE POTENTIAL RISK
GUIDELINES

Current Intention

Attention _____
Escape pain _____
Punish: self/others _____
Self-harm _____
To die _____

Confidence level: High Low
Reasons for low confidence rating

Tentative Scoring
Very high risk = 110–150
High risk = 70–109
Moderate risk = 40–69
Low risk = 39 and below
Immediate risk _____
Long-term risk _____

*Highly lethal defined as low risk for rescue, low chance of reversibility, e.g., comatose, serious medical injury.
© R.I. Yufit, 1988.

159

9

Psychodynamic Approaches to the Assessment and Management of Suicide

CHRISTOPHER G. LOVETT AND JOHN T. MALTSBERGER

No one kills himself who did not want to kill another, or at least wish death to another.

Wilhelm Stekel, 1910

Only he who is wished dead by someone else kills himself.

Paul Federn, 1929

Suicide brings no solution, only cessation of struggle.

D. W. Winnicott, 1971

In one of the first psychoanalytic works to explore the problem of suicide, Karl Menniger (1938) divided the motive for suicide into the wish to kill, the wish to be killed, and the wish to die. Such suicidal wishes can be conscious, pre-conscious, or unconscious. Through the use of certain defensive operations the potentially self-destructive person may succeed in hiding from himself and others the extent of his hopelessness and sense of inner danger. Each year many such individuals die by suicide, often only days after discharge from an inpatient psychiatric unit. Despite their assurances to the hospital staff, the resolution of their most noxious symptoms, and in many cases their own conscious sense of safety from deadly impulses, these patients fall victim to the destructive forces within them.

The importance of understanding the underlying motivations of suicidal acts is often overlooked in these cases, as is the dynamic relation between such impulses and other elements of the unconscious intrapsychic processes that promote or inhibit their expression through self-destructive action. As Rangell (1988, p. 32) pointed out, when examining the potential for suicide it is important to bear in mind that "the moment of decision always rests on deeper states."

A task confronts every clinician engaged in the estimation of suicide risk. The task requires integration of information derived from a psychiatric history and mental status examination with an estimation of the danger posed by inferrable unconscious factors. To complete this task and arrive at an accurate appraisal of suicide potential, we employ a method of psychodynamic for-

160

mulation (Maltsberger, 1986). It emphasizes understanding the specific emotional stresses in the patient's life that can promote suicide and the resources, both within the personality and in the external world, that would work to counteract such destructive forces.

Confronted by the task of determining how much at risk for suicide any single patient may be, clinicians must organize their observations, their previous knowledge of the patient, and their emotional responses to the patient's predicament. This step is necessary to intervene effectively, to ensure the patient's safety, and to further plan the patient's treatment. Burdened by the pressures inherent in making a life-and-death decision, the clinician is often tempted to seize some single element of the patient's presentation, and to rely on this one aspect as the sole reliable ("bottom line") indicator of the patient's immediate suicidal potential.

Often one relies on the patient's diagnosis and conclusions drawn from the mental status examination to settle questions of suicide risk. Because suicide and affective illness are frequently associated with one another, the process of clinical decision-making comes to hinge on an estimation of how depressed the patient seems to be. Depressive affect, however, like anxiety, plays some part in all psychic conflict (Brenner, 1982). Depression may be prominent, mild, or even absent in a patient's overt symptomatology despite its playing a major role in the underlying conflict. Similarly, the urge toward suicide is often independent of the patient's observable mental status, and it can remain intense despite the clearing of other symptoms of mental disturbance, including depression.

Like the affects of anxiety or depression, the vulnerability to suicide is found at all levels of descriptive psychopathology. Though many patients who commit suicide are seriously depressed, many others are not. Some are driven to attempt suicide because they cannot bear depressive affect of even mild severity. An experience of shame or humiliation arising from a narcissistic injury prompts a sudden act of self-destruction in others (Chasseguet-Smirgel, 1975/1985; Rangell, 1988).

The discovery of a depressive syndrome, a history of suicide attempts, or even a history of depression and suicide in the patient's parents or other relatives are valuable data. Clearly this information should be sought and taken into account as part of any thorough assessment. No descriptive or historical information, however, can answer the question confronting the clinician at the moment of clinical decision: At how much risk for suicide is this patient at this moment, or over the next 24 hours, or over the coming weekend?

For just such reasons some clinicians blind themselves, eschew diagnostic or historical information, and make decisions about suicide risk by intuition. Sometimes intuitive decision making is called "empathic judgment": "I just sense that this patient would not kill himself," such people say. We acknowledge that some intuitions may, in fact, be correct. One may rightly believe that the patient with whom he sits will never kill himself even if his emotional pain were to swell to agony. When not balanced by a systematically formulated clinical assessment, however, such intuitions must leave us dangerously unsure. Otherwise we cannot tell whether they reflect what the patient actually feels, what we wish he would feel, or merely what we, because of some unconscious resistance on our part, can allow ourselves to imagine or accept that he feels.

We agree with those who emphasize that empathic understanding of patients is essential when approaching the assessment and treatment of suicidal states. Empathy is the basis of all clinical sensitivity. Our need to understand how patients feel, or what their symptoms mean to them symbolically, and to make some live, emotional contact with their innermost wishes and fears is not limited to work with suicidal patients. A therapist's empathic immersion into the patient's inner life plays a vital therapeutic role. It may prove to be life-saving as well in the specific case of the suicidal patient. Empathic understanding conveys to the patient that another person can understand and accept his hopelessness or self-hatred (Ornstein & Kay, 1983). Well-informed empathy—that takes into account such psychodynamic factors as the importance of unconscious fantasy in prompting suicidal acts—may also provide the groundwork for a thorough evaluation of those aspects of the patient's inner life that may make the difference between life and death.

Psychodynamic Formulation of Suicide Risk

A psychodynamic formulation of suicide risk provides the clinician with a more thorough, systematic method for assessing suicide danger than those methods based principally on a mental status examination or clinical intuition (Maltsberger, 1986). It is grounded in the belief that the urge to suicide is primarily a psychodynamic matter, largely independent of the patient's observable mental state. The formal elements of mental illness work only to intensify the suicidal impulse, alter the shape of its expression, or compensate for its presence. A suicidal state arises when internal or external resources vital for a sense of self-worth and inner safety have been lost and the patient cannot replace them (Buie & Maltsberger, 1989).

By integrating the presenting clinical material from the patient's history, present illness, and mental status examination with various dynamic considerations (e.g., the stability of the patient's identity or his or her proneness to malignant regressions), the psychodynamic formulation illuminates the different vulnerabilities and strengths of the patient who threatens suicide. It provides a means of assessing and integrating the influences, both internal and external, that drive patients toward self-destruction or protect them from it.

By asserting that suicide is a psychodynamic process, we emphasize that, like any symptom, it emerges as a compromise formation out of unconscious conflicts between unacceptable wishes and the defenses against them. The patients' conscious intentions are often a misleading reflection of their underlying motivations, including the unconscious wish to live or to die.

In addition, the patients' self-reports seldom provide an accurate picture of their vulnerability to dangerous regression in which self-destruction emerges as the only conflict resolution the patient can consider. We agree with Rice (1985) that suicide represents a regressive transformation into action of an unconscious fantasy. Exploring the nature of such fantasies, the underlying conflicts that give rise to them, and the integrity of certain crucial ego functions that protect against such regressions constitute important aspects of the psychodynamic assessment of suicide risk.

In brief, there are five components in the psychodynamic case formulation for patients when there is a question of suicide potential.

1. *Assessment of the patient's past responses to emotional stresses, particularly losses.* The latter may be the losses inherent in the normal developmental process, such as the movement during adolescence away from the parents of childhood, or they may be more unexpected, traumatic losses, such as the death of a loved one. We examine the patients' capacity to tolerate experiences of separation, grief, disappointment, disillusionment, or failure and their capacity to engage in mourning. We pay close attention to the characteristic effects of such experiences on the patients and to their usual reactions, preferred defenses, and vulnerability to regression.

2. *Assessment of the patient's vulnerability to three varieties of affect closely associated with morbid self-destructiveness.* Aloneness, intense self-hatred or contempt, and intense, murderous rage each may crescendo into intolerable psychic pain. Self-hate and murderous rage are better known and more commonly associated with suicide than aloneness, about which we offer further comment. It was described by Adler and Buie (1979) as a subjective state of empty, cold isolation experienced as hopeless, beyond the relief of love or comfort from within or without, and accompanied by a panicked feeling of terror, or horror. It is close in kind to annihilation anxiety, and patients so afflicted feel they are dying.

This second aspect of the psychodynamic formulation represents a special case of a more general inquiry into the patient's capacity for tolerating painful affect. It takes into account any tendency to experience certain events or situations in ways that evoke painful states of mind. For instance, the tendency to feel aloneness, self-hatred, or murderous rage often derives from a sensitivity to narcissistic injury. We look to see affective crises in the context of events wherein they arise. When patients might be prone to such experiences, as well as the factors that enhance their capacity to tolerate them without turning to self-destruction, provide essential information.

3. *Assessment of the sustaining aspect of the patient's object relations.* We try to determine the patient's sustaining resources, both internal, in the form of "good" internalized objects or soothing introjects, and external, in the form of what Kohut (1971) referred to as "self-objects." Self-objects, or exterior sustaining resources, may enable patients to quiet incipient panic and to ward off self-hate, murderous rage, or aloneness if their internal resources do not suffice. In other words, an assessment is needed of the origin and nature of patients' narcissistic supplies and the degree to which the regulation of their narcissistic equilibrium depends on internal or external sources of affirmation and admiration.

The psychodynamic formulation estimates patients' autonomy versus their dependence on self-objects for survival. In addition, because it is the loss of such sustaining resources that triggers a patient's particular dynamic motivations for suicide, the essential elements of any treatment plan require identification of lost vital resources, their restoration or replacement through the provision of encouragement and support, and mourning (when possible) the losses that are irretrievable.

4. *Assessment of the emergence and emotional importance of death fantasies.* Unconscious fantasies play an enormously important role in the precipitation and shaping of suicidal acts. Once grasped by the clinician, they provide a crucial window into the underlying motives and the dynamic structure of the person's experience of longing for self-destruction.

The regressed suicidal patient can transform death into a powerful, feared, longed-for, or even erotized object with whom the patient is engaged in an intense relationship, sometimes to the exclusion of those in the external world. In these instances, death can be unconsciously personified by the suicidally preoccupied patient, who seeks the "Grim Reaper." This fantasy may also assume the form of a conscious, delusional belief held by severely ill patients. Death, or in some cases "fate," may serve as a "screen" object for an ambivalently loved person who is felt to be lost or unavailable (Asch, 1980), for a figure associated with some early trauma in the patient's life, or for the split-off destructive aspects of the self.

Some patients believe that Death stalks them as a felt presence, similar to the hooded figure of Ingmar Bergman's film *The Seventh Seal*. Patients who say the dark Angel of Death awaits them provides another example of such personifications. Other patients believe that to die is to travel somewhere. Others look to rejoin some loved person who has died and now seems to them to offer the love, safety, or the redemption they desperately seek.

5. *Assessment of the patient's ego capacities in areas that may contribute to a vulnerability to suicide.* We assess patients' impulsiveness, their capacity for reality testing, the stability of their identity, and their characteristic mechanisms of defense and functional effectiveness.

Assessment of the patient's defensive organization requires attention to the degree to which self-destructive impulses are dissociated from the core self, which we all experience as the dynamic center of initiative and experience. Passive suicidal longings, the wish that one might die, unaccompanied by consciously, active impulses to hurt oneself are usually less indicative of immediate risk than a distinct urge to injure or kill oneself. Passive wishes, however, sometimes augur impulsive, unexpected, self-destructive acts. Their presence must be regarded seriously and explored thoroughly with the patient.

The therapist further needs to know how realistic patients can be about the worth and availability of the sustaining resources that remain accessible to them. This area requires us to estimate the patient's capacity for reality testing, under normal circumstances as well as during times of stress and regressed functioning. When death fantasies command a delusional power over the patient's world (death, for example, can be experienced as the all-forgiving mother who pleads with her child to die and join her) the risk for suicide is great.

No single element of the psychodynamic formulation stands alone. A regressive loss of reality testing, for instance, does not by itself indicate great risk for suicide; it may be offset by the experienced, felt presence of sustaining resources that ensure the patient's continued attachment to life. Alternatively, the presence of intense, murderous rage or self-hatred does not lead invariably to suicide if, for example, the afflicted patient understands such feelings as a habitual but temporary reaction to the frustration of self-object needs. Such patients are able to maintain some perspective, engage in reality testing, and have faith in their capacity to reestablish connections once the emotional storm has passed.

Psychoanalytic Perspectives on Suicide

Due in large part to the influence of Freud's (1917/1957) formulations on melancholia, there has been a trend in psychoanalytic thinking to link suicide exclusively with the dynamics of depression (Meissner, 1977). The classical psychoanalytic model of suicide emphasizes the internalization of a lost, ambivalently loved object, resulting in hostile attacks on the self in the form of self-reproaches, self-devaluation, and self-destructive acts. This concentration on the dynamics of hostile, self-punitive impulses of "depressive suicide" has obscured the passive, libidinal, or narcissistic aims expressed in some suicidal acts (Asch, 1980, 1985; Rice, 1985). Freud's (1919/1955, 1920/1957) other writings that dealt with longings for death or impulses toward self-destruction nevertheless suggested just such a focus on longings to reestablish primitive narcissistic states or to reunite with a lost object through masochistic surrender.

For suicidal patients the psychodynamic formulation is focused on the patient's capacity (or incapacity) to attain and sustain stable states of inner security and safety (Buie, unpublished). The feeling of safety, normally a background to everyday experience, is essential to the experience of aliveness (Sandler, 1987). Any individual requires some minimum amount of safety feeling in order to tolerate staying alive. Initially derived from states of bodily well-being, the feeling of safety later becomes localized in the self and colors the self-experience through feelings of self-esteem and the sense of the self as intact and cohesive. Experiences of anxiety or self-disorganization lower the level of safety feeling. In extreme cases, where self-disintegration is threatened, the patient experiences the terror of annihilation anxiety, a condition of extreme self-danger.

The need to restore the sense of integration and stability that accompanies the feeling of safety takes precedence over the fulfillment of wishes usually regarded as regulated by the pleasure principle. Freud (1920/1957) wrote that an activity that leads to pleasure, whether libidinal or aggressive in quality, may be sacrificed or defensively distorted in order to prevent the shattering, dis-

organizing anxieties experienced when the feeling of safety is severely threatened. Flarsheim (1975) described the importance of such feelings in the treatment of severely self-destructive patients:

When a patient suffers from severe anxiety about survival, everything in his life is
subservient to the need for reassurance. Even an overt suicidal act can have the meaning
of taking destructiveness into one's own hands rather than feeling vulnerable to uncon-
trolled destructiveness. Under such circumstances there is no room for seeking pleasure
for its own sake. Pleasure and pain, as all other experiences, are pressed into the service
of reassurance against overwhelming anxiety. [p. 163]

Although it seems paradoxical, some patients believe they can achieve safety
through dying. Their suicidal wishes may arise from the loss of certain crucial
sustaining resources in life and at the same time express the belief that it is
only through death that the loss can be restored. It may be evident in fantasied
self-object relationships with idealized images of death. Such treasured fantasy
relationships, conscious or unconscious, seem to offer the love and affirmation
no person in the real world ever has. Death in such cases can promise to restore
permanently a lost holding, soothing resource. The patient anticipates that in
death the longed-for reunion or merger with a loving, powerful object will
reestablish self-worth and abolish all psychic pain.

During early development, experiences of loving handling and optimal responsiveness by the mother and other caregivers assist in the development of
a "basic sense of entitlement" (Bacal & Newman, 1990). It represents a healthy,
confident expectation of responsiveness to one's needs, including a sense of
comfortable expectation that self-object functions will be available from others
when needed. An essential aspect of a basic sense of healthy entitlement is that
one is entitled to survive, which is taken for granted and unquestioned in any
conscious way.

For suicide-vulnerable patients, however, the fundamental sense that one
has "the right to a life" (Modell, 1984) is often not well established. The narcissistic pathology of many suicide-prone individuals drives them to desperate,
ruthless behavior because their basic right to survive seems threatened (Adler
& Buie, 1979). Such patients often function according to an unconscious, or
even conscious, belief that they must have their wishes met because their frustration provokes unbearable states of fragmentation or aloneness that feel like
impending death.

Klein (1935) viewed suicide as an expression of the "death instinct" turned
against the self. Freud's speculative notion of a death instinct (1920/1957) grew
out of his interest in narcissism and narcissistic libido. Freud's conceptualization
of the death instinct includes a wish to remain in or return to a state of continuing primary narcissism. It constitutes an aim of achieving a primitive state
of existence prior to self-awareness and the frustrations of separateness. Freud
defined it as the aim of achieving inactivity, passivity, or sleep—in other words,
the peaceful oblivion of an eternal, undifferentiated homeostasis. From the
point of view of the death instinct, the aim of all life is psychic death and a

return to an inanimate state. This tendency is opposed by all of the individual's strivings for autonomy, individuation, and connection with others.

Within the context of the death instinct, however, the principal role of the individual's various ego functions is to ensure that the organism follows its own path through life to death. Explicitly referred to by Freud (1910/1957) at various times as the "self-preservative instincts," these psychic functions are responsible for protecting the organism from any traumatic disruption of its unfolding course of development. It is in this sense that Freud's notion of the death instinct overlaps with wishes for self-preservation and feelings of safety. Many of the ideas contained within Freud's speculations about the death instinct, such as his description of narcissistic regressions and wishes to preserve the inherited potential of the organism, were expanded in later psychoanalytic writings on narcissism and the early development of the self.

Kohut (1971), for instance, described a nuclear self, the central narcissistic sector of the personality, that includes its basic program of action. According to Kohut, it is the earliest cohesive, core configuration of the self. The content and form of the nuclear self are determined by the combined forces of the individuals' specific biological inheritance and the selective self-object responses of the parents. The nuclear program of action, the creative-productive potential of the self, is an intrinsic design for development and action that is realized over the course of the life-span if not disrupted by profound self-object failure or some other traumatic influence.

Similarly, Winnicott (1965, 1971) introduced the term "true self" to indicate an inherited potential that found its expression in spontaneous action. He emphasized that a "good enough" early environment allows the maturational processes the individual inherits to be set in motion and to continue in an unbroken line of living growth. In contrast, the infant who is forced to react to "impingements" by the environment is disturbed out of a state of "being." When reacting, an infant is not "being." The spontaneous processes of the self are interrupted, and a deep sense of futility and hopelessness develops. It is accompanied by a temporary loss of identity, a loss of the feeling of aliveness, and a loss of the sense that one possesses a personal life that is worth living.

From Winnicott's point of view, trauma involves disruption and fragmentation of the personal continuity of experience. As a result, primitive defenses arise to defend against a dreaded repetition of the early experience of disintegration, loss of self, and hopelessness. It may entail reliance on compliance and the establishment of a "false self," wherein all that is original and creative within the person is hidden.

Certain unmet needs and wishes may also persist as essential parts of Kohut's nuclear self. These needs and wishes generate powerful, unconscious fantasies that serve as guiding forces in the person's life. These underlying, defended-against wishes are not limited to sexual and aggressive ones but include longings for psychic stability, safety, and narcissistic repair (Shane & Shane, 1990).

The accumulating evidence for the importance of early, preoedipal stages of development and new understanding of the pathological effects of early interferences in the development of the self have contributed to a broadening of the psychoanalytic understanding of suicide. Self-punishment for unconscious hatred is no longer viewed as the sole unconscious dynamic in the psychogenesis of suicide. It is now acknowledged that suicidal acts may be motivated by a need to restore a feeling of safety and of self-cohesion through death, to rescue the person's last remaining traces of self-esteem by merging with a powerful, loved object in unconscious fantasy, to confirm a sense of identity, or to destroy a false self in the hope of a rebirth of a more authentic self (Kay, 1989; Litman, 1989; Meissner, 1977). Many suicides are not prompted by melancholic guilt and the dynamics that accompany it. Many suicides derive from unconscious complexes giving rise not to guilt but to feelings of unbearable emptiness, inner deadness, and intense shame (Kohut, 1972).

The clinician must establish genuine empathic contact with the subjective experience of the suicidal patient. In doing so one must bear in mind that for the suicidal patient death may not be the worst imaginable fate (Rangell, 1988). Instead, suicide may be sought after as the fulfillment of some libidinal or defensive aim associated with the idea of death (Fenichel, 1945). The concept of death may unconsciously symbolize separation, castration, object loss, or loss of love; and thus it is readily associated with unconscious conflict (Brenner, 1982). Suicide is a symptomatic act and contains within its structure both manifest and latent content (Rice, 1985). A conscious wish for death is part of the symptom's manifest content, but underneath the wish to die there may reside wishes to undo separation anxiety, restore past losses, or repair narcissistic injuries.

The importance of unconscious fantasies in the psychodynamic understanding of suicide is based on their role as organizers of psychic reality. Suicidal persons aim to resolve their intrapsychic conflicts (Asch, 1980). Fantasies provide us a useful means to understand the nature and composition of the unconscious conflicts that lead to symptomatic acts such as suicide attempts (Abend, 1990; Rice, 1985). The total destruction of the self is probably never the unconscious aim of a suicidal act (Asch, 1985). Usually one can discern multiple aims, such as the wish to destroy the hated internalized object or an unacceptable part of the self, the wish to fuse with the good, loving internal objects, or the wish to disintegrate or disappear in order to be reborn (Maltsberger & Buie, 1980). In addition, many suicide fantasies have a dyadic quality in the organization of unconscious object relationships (Asch, 1980). The fantasy is often sadomasochistic, and the patient may play complementary roles, both passive and active, victim and executioner, and thus enact all three of the wishes outlined by Menninger (1938): to kill, to be murdered, and to die by dissolving into a blissful sleep.

Sustaining Resources and Survival

One of the central components of the psychodynamic formulation merits further comment: the patient's relations with both external sustaining self-objects and internalized objects. Separation, individuation, and a reliable sense of autonomy become possible only when emotional resources for maintaining narcissistic equilibrium develop internally. The capacity to sustain a sense of self-worth, self-cohesion, and self-esteem depend on the development of these resources. The capacity to sustain narcissistic injury, perhaps a disappointment or rejection, without "going to pieces" depends on their development. The sustaining "self-maintenance functions" (Buie, unpublished) of the personality are based on internal "good" objects, or soothing introjects, which in unconscious fantasy are thought to provide these functions. The development of such capacities is contingent on the adequate phase-appropriate provision of self-maintenance functions during childhood by parental figures. They ultimately derive from images of the loving parents, and in this sense good parents are built into the inner world from the earliest stages of development onward.

"Good" internal objects develop as the growing child experiences consistent caring, soothing, and holding at the hands of those who love him. Within the inner world they stand in contrast to the "bad" objects, composed of images of frustrating, hated, and hateful early objects. When these self-maintenance functions are properly provided by others over the course of development, the individual grows in his capacity to provide them to himself through processes of internalization.

A large proportion of persons vulnerable to suicide have not developed an inner world colored primarily by "good" objects; and in keeping with the relative failure of development in this area, they possess too few inner resources to sustain an internal sense of coherence, well-being, and goodness. Lacking such reliable "good" introjects, in order to hold together they must perforce turn to external resources. Reliance on exterior sustaining resources is their only way to achieve any sense of comfort or to maintain self-cohesion and self-esteem. Without external resources such patients are prone to experience feelings of aloneness, self-hatred, and narcissistic rage.

So long as external sustaining resources are consistently and dependably available to the suicide-vulnerable individual, intolerable psychic pain is avoided and suicide is not a threat. The loss of a stabilizing external resource, however, may well precipitate a flood of deadly affect and plunge the patient into a suicidal regression. At such moments such patients' inner world and scanty remaining "good" objects threaten to disintegrate or are unconsciously felt to be destroyed. They believe that their internal "bad" objects have overpowered their inner world, creating a barren, cruel inner landscape that Freud (1923/1961) described as a "pure culture of the death instinct" (p. 53). Any lingering sense of goodness, self-worth, or value washes away, and suicide seems the only escape from the sense that the self is dying or is already dead.

The patient's history usually indicates the sort of external sustaining resource he or she requires to maintain emotional integrity. There are three classes of external sustaining resource: other people, work, and special, intensely narcissistically invested aspects of the self.

Suicide-vulnerable persons may depend on others to support their self-esteem, to protect them from the incessant internal pressures of a cruelly punishing superego, or to soothe them when the threat of abandonment begins to escalate. Other patients may depend on a highly valued career to refuel their narcissism. Yet others rely on some aspect of the self—physical prowess or intellectual capacities—prized as vital for equilibrium and well-being. The psychodynamic formulation is crucial for planning treatment of the suicidal patient, because it shows what resources for self-soothing and self-worth must be restored to regain narcissistic equilibrium.

Case Example

The dynamic interplay between external and internal sources of narcissistic sustenance can be illustrated by the case of Julie, a 17-year-old girl who came for treatment because of feelings of depression and apathy.

Julie was a popular, gifted student who wrote inspired, though emotionally turbulent, poetry. She complained that she had never enjoyed herself and had never felt real in any way. Although her outward adaptation was excellent, she reported that over the previous several months she had been retreating more and more frequently to an elaborate, competing internal world she referred to as "The Cyclinder." This region was populated by only four inhabitants: (1) a sadistic, brutal figure referred to as "The Violator"; (2) a timid, abused woman, "The Abdicator," who sulked about not being known by anyone and who knew no other way of being except to submit to The Violator as part of an endless, masochistic trial; (3) "The Eye," a creature chained to the wall of the cave in which the three of them lived, and who was forced to watch, helplessly, the cruelty inflicted by The Violator upon the woman, whom the patient readily acknowledged was herself; (4) "The Comforter," a benign but slow-moving sloth who lived in a different cave on the other side of the cylinder.

Julie imagined The Cylinder to be a hollow tube with no bottom or top, stretching on forever, providing no exit, no release—it was an internal chamber of horrors. External reality was classified as "The Blue World"; it was dealt with by an empty, shallow part of herself referred to as "Bomb," The Abdicator's double. When Julie first came to treatment The Comforter had no means of getting across the cylinder to the woman. In other words, the patient felt she literally had no access to her "good" internal objects. Instead, she was left at the mercy of the sadistic elements of her inner world; and, in telling of this dilemma, Julie provided a graphic representation of her self-hatred and feelings of helplessness.

Julie was suicidal at the time she began her treatment. Her involvement in the world of The Cylinder was beyond that of a sustaining fantasy. It reflected the depths of her narcissistic regression to a point where internal objects had become more important to her than people in the real world. She described fantasies of killing herself by carbon monoxide poisoning in the garage when her parents were scheduled to meet with her therapist. Because Julie saw her treatment as a source of hope, however, and had begun to recognize that her suicidal wishes stemmed from angry despair over a recent disappointment with a boyfriend, the therapist decided to work with the patient intensively as an outpatient.

During a very early session the therapist's concerns about Julie's safety were allayed somewhat when she related two developments within The Cylinder. She first reported that The Comforter had somehow found a vine that stretched across the cylinder. Now he could reach the woman's cave and help her. She did not know, she said, whether this vine had not been there before or The Comforter simply had not noticed it. In addition, she reported that the therapist had moved into The Cylinder and now had his own cave and his own name, appropriately, "The One Who Waits."

This title bestowed on the therapist within the patient's inner world possessed great significance, because despite the desperation felt by the patient the therapist had not taken any precipitous action. The experience of not being pressured or controlled from sources located either outside or inside was completely novel to Julie and contributed to her hope that her true self, her own intrinsic sense of aliveness, could be allowed to emerge spontaneously and be nurtured.

This example highlights the interdependence of internal sources of hope, nurturance, and comfort, and the availability of external resources. At times such exterior sustaining resources must replace what the patient cannot provide for himself or herself. They serve to rectify a deficit, to offset what in a suicidal patient can be a life-threatening lack. In other cases, such self-object functions as might be provided by a therapist are required only to refuel the patient's own "good" internal objects and self-regulatory capacities. In the case of Julie, intensive treatment, lasting only a few months, resulted in dismantling of The Cylinder as part of a spontaneous development on her part. She no longer required its presence as a place of refuge from the deprivations she experienced in her experiences with others. Instead, she began to feel more alive and real in her actual life outside.

In Julie's clinical assessment the nascent presence of hopelessness and the indications of a belief, however frustrated at the time, in the potential usefulness of sustaining resources played a major part in the therapist's decision to treat the patient intensively as an outpatient. Also important, however, were the indications that the patient did not have an idealized or inviting fantasy of death but, instead, saw herself as fighting to be truly alive. She had demonstrated a good capacity to appraise the therapy and the therapist realistically as potential but not magical sources of, as she said it, "patient support."

We believe this brief case summary highlights the importance of understanding unconscious conflicts and the internalized object relationships through which they are subjectively experienced. Julie's inner life was dominated by sado-masochistic relationships, unconscious beating fantasies, and a sense that her everyday interactions and involvement with the world were false and unreal. For treatment, it was first important to provide support of the patient's fragmented, devalued self through an attitude of empathic holding. It allowed her to utilize the sustaining functions of self-soothing, self-worth, and self-cohesion offered by the therapist that she could not provide to herself. Only when this holding function was solidly in place, as signaled by the development of an idealizing transference and a sense that her internal "good" objects had been revived and had again become accessible to her, was a more exploratory form of treatment possible.

Guidelines for Practice: Chapter Summary

The psychodynamic formulation of suicide risk represents a systematic approach to the evaluation of those aspects of a patient's personality and current life situation that may precipitate or prevent self-destructive acts. It is carried out by appraising several aspects of patients' history and current functioning: their usual responses to loss or narcissistic injury; their vulnerability to painful affect states associated with suicide; the quality of their object relations, specifically respecting the nature and availability of external and internal resources necessary to the maintenance of feelings of self-worth and self-cohesion; the presence and relative influence within their psychic economy of death fantasies; and the intactness of those ego capacities that insulate the personality from regressive losses of control over self-destructive impulses.

Among the dangers encountered in the assessment and treatment of a potentially suicidal patient is the clinician's countertransference hatred (Maltsberger & Buie, 1980) or feelings of helplessness (Adler, 1972). At times these feelings promote the therapist's emotional withdrawal. Such withdrawal is usually unconsciously, or even consciously, experienced by patients as a rejecting reinforcement of their own self-destructive wishes (Flarsheim, 1975). In addition, therapists' desperate resort to the use of ultimata, their demands that the patient "contract" for safety, or their failure to take a suicide "gesture" seriously may be experienced by the patient as empathic failures. The patients' sense of despair then deepens and their urges toward self-destruction intensify. Despite the often obvious sadomasochistic and provocative attitudes of some suicidal patients, it is often useful to view such phenomena as also reflecting the patient's wish to survive and, ultimately, to thrive.

There is an intrinsic connection between the patients' needs for external sustaining resources and their sensitivity to narcissistic injury within the treatment setting. The psychodynamic formulation is central not only to the assessment of suicide risk but also to the process of planning for its ongoing treatment. Determining what sustaining resources must be restored and provided during treatment also demands that the clinician maintain an alertness to indications by the patients that their sense of self-worth or self-cohesion has been disrupted by "failure" on the part of the therapist to provide a vital self-object function. Calling attention to apparent shifts in the patient's attitudes within a session, even a first evaluation session, and noting with the patient any changes in feelings of worthlessness and self-hatred, murderous rage, and, in particular, aloneness allows an invaluable window into the person's vulnerabilities that give rise to suicidal states.

The assessment and treatment of suicidal patients requires more of the emotional strength of the clinician than does the treatment of most, if not all, other categories of patients. The experience of failure and the frustration of the wish to heal, especially of someone so obviously in need as the self-destructive patient, is one many find foreign to their experience and to their *amour propre*. Kohut (1972) related suicide to the need for an unconditionally responsive self-object, yet all of us bring certain conditions to the clinical encounter with patients through the vehicle of our own personalities.

The therapeutic aim must not be to avoid any empathic disruption with suicidal patients. Their narcissistic vulnerabilities and brittle self-organizations ensure that such disjunctions are a part of any therapeutic endeavor. As with any psychodynamic treatment approach, however, the clinician's intent must be to try to understand the patient's subjective experience of such failures, both conscious and unconscious. It is through the repeated working through of empathic failure that those vulnerable to suicide are able to acquire through internalization the functions necessary to sustain their sense of safety, restore their sense of self-worth and self-cohesion in the face of the loss of sustaining resources, and revive their feelings of aliveness when formerly death would have seemed an inevitable choice.

REFERENCES

Abend, S. M. (1990). Unconscious fantasies, structural theory, and compromise formation. *Journal of the American Psychoanalytic Association, 38*, 61–73.

Abram, H. S., Moore, G. I., & Westervelt, F. B. (1971). Suicidal behavior in chronic dialysis patients. *American Journal of Psychiatry, 127*, 1199.

Adler, G. (1972). Helplessness in the helpers. *British Journal of Medical Psychology, 45*, 315–326.

Adler, G., & Buie, D. H. (1979). Aloneness and borderline psychopathology: the possible relevance of child development issues. *International Journal of Psycho-Analysis, 60*, 83–96.

Asch, S. S. (1980). Suicide and the hidden executioner. *International Review of Psycho-Analysis, 7*, 51–60.

Asch, S. S. (1985). Suicide—fantasies of internal object relations. *International Journal of Psychoanalytic Psychotherapy, 11*, 269–274.

Bacal, H. A., & Newman, K. M. (1990). *Theories of object relations: bridges to self psychology*. New York: Columbia University Press.

Barraclough, B., Bunch, J., Nelson, B., & Sainsbury, P. (1974). A hundred cases of suicide: clinical aspects. *British Journal of Psychiatry, 125*, 355–373.

Berman, A. (1986). A critical look at our adolescence: notes on turning 18 (and 75). *Suicide and Life-Threatening Behavior, 16*(1), 1–12.

Blumenthal (1990). An overview and synopsis of risk factors, assessment, and treatment of suicidal patients over the life cycle. In S. J. Blumenthal & D. J. Kupfer (Eds.), *Suicide over the life cycle: risk factors, assessment, and treatment of suicidal patients* (pp. 685–734). Washington, DC: American Psychiatric Press.

Bongar, B., & Harmatz, M. (1989). Graduate training in clinical psychology and the study of suicide. *Professional Psychology: Research and Practice, 20*(4), 209–213.

Bongar, B., Peterson, L. G., Harris, E. A., & Aissis, J. (1989). Clinical and legal considerations in the management of suicidal patients: an integrative overview. *Journal of Integrative and Eclectic Psychotherapy, 8*(1), 53–67.

Brenner, C. (1982). *The mind in conflict.* New York: International Universities Press.

Brown, H. N. (1989). Patient suicide and therapists in training. In D. G. Jacobs & H. N. Brown (Eds.), *Suicide: understanding and responding: Harvard Medical School perspectives on suicide* (pp. 415–434). Madison, CT: International Universities Press.

Buie, D. H. (unpublished). The psychotherapy of suicidal patients.

Buie, D. H., & Maltsberger, J. T. (1989). The psychological vulnerability to suicide. In D. Jacobs & H. N. Brown (Eds.), *Suicide: understanding and responding: Harvard Medical School perspectives on suicide* (pp. 59–71). Madison, CT: International Universities Press.

Chasseguet-Smirgel, J. (1975/1985). *The ego ideal: a psychoanalytic essay on the malady of the ideal* (P. Barrows, trans.). New York: Norton.

Chemtob, C. M., Hamada, R. S., Bauer, G. B., Kinney, B., & Torigoe, R. Y. (1988). Patient suicide: frequency and impact on psychiatrists. *American Journal of Psychiatry, 145,* 224–228.

Chemtob, C. M., Hamada, R. S., Bauer, G. B., Torigoe, R. Y., & Kinney, B. (1988). Patient suicide: frequency and impact on psychologists. *Professional Psychology: Research and Practice, 19*(4), 421–425.

Deutsch, C. J. (1984). Self-report sources of stress among psychotherapists. *Professional Psychology: Research and Practice, 15,* 833–845.

Dorpat, T. L., & Ripley, H. S. (1960). A study of suicide in the Seattle area. *Comprehensive Psychiatry, 1,* 349–359.

Dorpat, T. L., Anderson, W. F., & Ripley, H. S. (1968). The relationship of physical illness to suicide. In H. L. P. Resnik (Ed.), *Suicidal behaviors* (pp. 209–219). Boston: Little, Brown.

Fenichel, O. (1945). *The psychoanalytic theory of neurosis.* New York: Norton.

Flarsheim, A. (1975). The therapist's collusion with the patient's wish for suicide. In P. L. Giovacchini (Ed.), *Tactics and techniques in psychoanalytic therapy, Vol. 2. Countertransference* (pp. 155–195). New York: Jason Aronson.

Freud, S. (1910/1957). The psycho-analytic view of psychogenic disturbance of vision. In J. Strachey (Ed. & Trans.), *The standard edition of the complete psychological works of Sigmund Freud* (Vol. 11; pp. 209–218). London: Hogarth Press.

Freud, S. (1917/1957). Mourning and melancholia. In J. Strachey (Ed. & Trans.), *The standard edition of the complete psychological works of Sigmund Freud* (Vol. 14; pp. 243–258). London: Hogarth Press.

Freud, S. (1919/1955). A child is being beaten. In J. Strachey (Ed. & Trans.), *The standard edition of the complete psychological works of Sigmund Freud* (Vol. 17; pp. 179–204). London: Hogarth Press.

Freud, S. (1920/1957). Beyond the pleasure principle. In J. Strachey (Ed. & Trans.), *The standard edition of the complete psychological works of Sigmund Freud* (Vol. 18; pp. 7–64). London: Hogarth Press.

Freud, S. (1923/1961). The ego and the id. In J. Strachey (Ed. & Trans.), *The standard edition of the complete psychological works of Sigmund Freud* (Vol. 19; pp. 12–59). London: Hogarth Press.

Henn, R. F. (1978). Patient suicide as part of psychiatric residency. *American Journal of Psychiatry, 135,* 745–746.

Hirschfeld, R., & Davidson, L. (1988). Risk factors for suicide. In A. J. Frances & R. E. Hales (Eds.), *American Psychiatric Press Review of Psychiatry* (Vol. 7; pp. 307–333). Washington, DC: American Psychiatric Press.

Kapantais, G., & Powell-Griner, E. P. (1989). *Characteristics of persons dying from AIDS: preliminary data from the 1986 National Mortality Followback Survey.* Advance data report no. 173. Washington, DC: National Center for Health Statistics.

Kay, J. (1989). Self-psychological perspectives on suicide. In A. Goldberg (Ed.), *Dimensions of self experience: progress in self psychology,* (Vol. 5; pp. 169–186). Hillsdale, NJ: Analytic Press.

Kleespies, P. M., Smith, M. R., & Becker, B. R. (1990). Psychology interns as patient suicide survivors: incidence, impact, and recovery. *Professional Psychology: Research and Practice, 21*(4), 257–263.

Klein, M. (1935). A contribution to the psychogenesis of manic-depressive states. *International Journal of Psychoanalysis, 16,* 145–174.

Kohut, H. (1971). *The analysis of the self.* New York: International Universities Press.

Kohut, H. (1972). Thoughts on narcissism and narcissistic rage. *Psychoanalytic Study of the Child, 27,* 360–400.

Litman, R. (1989). Suicides: What do they have in mind? In D. Jacobs & H. Brown (Eds.), *Suicide: understanding and responding.* Madison, CT: International Universities Press.

Maltsberger, J. T. (1986). *Suicide risk: the formulation of clinical judgment.* New York: New York University Press.

Maltsberger, J. T., & Buie, D. H. (1980). The devices of suicide: revenge, riddance, and rebirth. *International Review of Psycho-Analysis, 7,* 61–72.

Maris, R. W., Dorpat, T. L., Hathorne, B. C., Heilig, S. M., Powell, W. J., Stone, H., & Ward, H. P. (1973). Education and training in suicidology for the seventies. In H. L. P. Resnick & B. C. Hathorne (Eds.), *Suicide prevention in the seventies* (pp. 23–44). Washington, DC: U.S. Government Printing Office.

Meissner, W. W. (1977). Psychoanalytic notes on suicide. *International Journal of Psychoanalytic Psychotherapy, 6,* 415–447.

Menninger, K. A. (1938). *Man against himself.* New York: Harcourt, Brace & World.

Modell, A. H. (1984). *Psychoanalysis in a new context.* New York: International Universities Press.

Murphy, G. E. (1986). The physician's role in suicide prevention. In A. Roy (Ed.), *Suicide* (pp. 171–179). Baltimore: Williams & Wilkins.

Ornstein, P. H., & Kay, J. (1983). Ethical problems in the psychotherapy of the suicidal patient. *Psychiatric Annals, 13,* 322–340.

Peterson, L. G., & Bongar, B. (1989). The suicidal patient. In A. Lazare (Ed.), *Outpatient psychiatry: diagnosis and treatment* (2nd Ed.; pp. 569–584). Baltimore: Williams & Wilkins.

Peterson, L. G., & Bongar, B. (1990). Training physicians in the clinical evaluation of the suicidal patient. In M. Hale (Ed.), *Teaching methods in consultation-liaison psychiatry* (pp. 89–108). Basel: Karger.

Rangell, L. (1988). The decision to terminate one's life: psychoanalytic thoughts on suicide. *Suicide and Life-Threatening Behavior, 18,* 28–46.

Rice, E. (1985). The role of the oedipal fantasy in masturbatory and suicidal phenomena. *International Journal of Psychoanalytic Psychotherapy, 11,* 243–267.

Robins, E. (1986). Completed suicide. In A. Roy (Ed.), *Suicide* (pp. 123–133). Baltimore: Williams & Wilkins.

Sandler, J. (1987). *From safety to superego.* New York: Guilford Press.

Schein, H. M. (1976). Obstacles in the education of psychiatric residents. *Omega, 7,* 75–82.

Shane, M., & Shane, E. (1990). Unconscious fantasy: developmental and self-psychological considerations. *Journal of the American Psychoanalytic Association, 38,* 75–92.

Winnicott, D. W. (1965). *The maturational processes and the facilitating environment.* New York: International Universities Press.

Winnicott, D. W. (1971). *Playing and reality.* London: Tavistock Publications.

PART IV
Treatment

10

Guidelines for Handling the Suicidal Patient: A Cognitive Perspective

ROBERT J. BERCHICK AND FRED D. WRIGHT

This chapter describes in detail a model for assessing and treating suicidal patients in an outpatient setting. The model is derived from 20 years of work by Aaron T. Beck, M.D., and his colleagues at the Center for Cognitive Therapy, University of Pennsylvania. It integrates basic research, assessment techniques, and therapeutic strategies.

The Center for Cognitive Therapy has three basic functions: treatment, research, and training. Approximately 450 patients are evaluated each year, and roughly 200 are in active treatment weekly. These patients are heterogeneous in terms of diagnosis; however, depression, panic disorders, anxiety disorders, substance abuse, marital discord, and personality disorders are the most frequently represented diagnostic groups. As we have seen in other chapters of this book, diagnosis alone does not predict suicide; however, depression, schizophrenia, and alcohol abuse are seen as being highly represented diagnostic groups for suicide ideators, attempters, and completers (Beck, Steer, Kovacs, & Garrison, 1985; Fawcett et al., 1987; Pokorny, 1983). In a more recent study Weissman, Klerman, Markowitz, and Ouellette (1989) also reported panic disorder as a diagnostic group at increased risk for suicide. The Epidemiologic Catchment Area study found that subjects with panic disorder made more suicide attempts than subjects with other psychiatric disorders.

The Center treats a large number of patients in the suicide high-risk groups. This setting has provided an opportunity to conduct numerous research projects on suicide ideators, attempters, and completers. The following is an overview of these projects.

Research

Beck's research on suicide spans several decades. A discussion of all his research is beyond our scope and purpose here. We do review projects that serve as a framework for understanding a cognitive therapy perspective for working with suicidal outpatients.

The revised Beck Depression Inventory (BDI) consists of a 21-item instrument designed to assess the severity of depression in adolescents and adults (Beck, Rush, Shaw, & Emery, 1979b). Beck and Steer (1987, p. 1) noted that during the "last 26 years, the BDI has become one of the most widely accepted instruments in clinical psychology and psychiatry for assessing the intensity of depression in psychiatric patients, and for detecting possible depression in normal populations (Steer, Beck, & Garrison, 1986)."

A BDI total score can provide an estimate of the severity of the overall depression, and Beck and Steer (1987) stressed that it is also clinically important to attend to specific item content. Beck et al. (1985) have pointed out that the BDI's pessimism item (Item 2) was nearly as predictive of eventual suicide in 211 suicide ideators as the 20-item Hopelessness Scale (HS) (Beck, Weissman, Lester, & Trexler, 1974). The authors concluded that patients admitting to suicide ideation (Item 9) and hopelessness (Item 2) with ratings of 2 or 3 should be closely scrutinized for suicide potential.

It is also important to observe the overall pattern of the depression symptoms the patient is describing. The BDI reflects not only cognitive and affective symptoms but also somatic and "vegetative" symptoms. For example, some suicidal patients do not express suicidal ideation but have actually stopped eating and sleeping (Beck & Steer, 1987, pp. 7–8).

The Scale for Suicide Ideation (SSI) was developed by Beck, Kovacs, and Weissman (1979a) as a clinically administered scale to quantify and assess suicidal intention. The scale consists of 19 items: The first five tap into the individual's wishes and reasons for living and dying, as well as if the individual has any desire to make an active or passive suicide attempt. If persons indicate any desire to make an attempt on their life, the remaining 14 items are administered. Items 6 and 7 assess the duration and frequency of the suicide ideation, respectively. Item 8 elicits the individual's attitude toward the ideation, and Item 9 explores the individual's sense of control regarding acting on the ideation. Item 10 explores deterrents to an active attempt, and Item 11 investigates the reason or motive for the contemplated attempt. The next two items deal with specificity of plans for an attempt, as well as the availability or opportunity for the contemplated attempt. Item 14 explores the individual's sense of capability or courage to carry out the attempt. Item 15 assesses the individual's expectancy of carrying out the attempt, and Item 16 reviews if any preparation for the contemplated attempt has been made. Item 17 asks if a suicide note has been prepared, and the next item explores if any final acts in anticipation of death had been thought about or performed. The final item is a clinical judgment question concerning whether any deception or concealment of the contemplated suicide took place.

In terms of psychometric properties, Beck and colleagues (1979a) reported that the SSI has been shown to be internally consistent (alpha coefficient .89). Interrater reliability coefficient was .83 ($p < .0011$). Numerous estimates of concurrent validity, discriminative validity, and construct validity have also been favorable. Studies regarding the predictive validity of the SSI are ongoing.

In addition to the established potential of the SSI for research, it is helpful to the clinician for systematically gathering and quantifying information relevant to the individual in question's thoughts, plans, and wishes about suicide.

The Beck Hopelessness Scale (Beck et al., 1974) consists of 20 true-false questions assessing pessimism about the future. Beck et al. reported that the coefficient alpha values demonstrated for the Hopelessness Scale across diverse clinical and nonclinical populations are typically in the .80s, and its correlations with clinical ratings of hopelessness are in the .70s.

The Hopelessness Scale has been shown to be sensitive for predicting eventual suicide of both inpatients (Beck & Steer, 1989) and outpatients (Beck, Brown, Berchick, Stewart, & Steer, 1990). A group of 207 inpatients admitted because of suicidal ideation but with no recent attempts were entered into a long-term follow-up program. The minimum time in the follow-up program for inclusion in the preliminary report was 5 years. During that time 14 patients committed suicide. The Hopelessness Scale and Item 9 of the BDI predicted the eventual suicide. A score of more than 10 on the Hopelessness Scale correctly identified 91% of eventual suicides (Beck et al., 1989).

Special Procedures for Suicidal Patients

Because most patients at our Center for Cognitive Therapy carry a diagnosis of depression, hopelessness and suicidal ideation are frequently ongoing therapeutic issues. When suicidal ideation is a continuous (or intermittent), low- to moderate-level *purely ideational* symptom, it is dealt with as any other depressive symptom—primarily by the treating therapist (Beck et al., 1979b).

It is especially important that the therapist maintain close monitoring of *all* suicidal thoughts as well as the concomitant levels of hopelessness and depression. Regular administration of the first five items of the SSI (and the rest of the scale if any of these items are positive), the HS and the BDI at each session is strongly recommended, because these scales provide a reliable means of monitoring any changes in these variables over time.

Specific indicators of significant suicidal risk potential are HS scores over 10, and BDI Item 2 (pessimism) and Item 9 (suicide) scores of 2 or higher. If any of these scores are accompanied by SSI scores of 6 or higher, the option of hospitalization must be considered seriously.

Keep in mind, though, the major issue involved in the decision to have a patient hospitalized depends on whether the patient can cope throughout the crisis without being in a *protected* setting. Consideration regarding the advantages and disadvantages of hospitalization for *every* individual case must be made. Below are some highlighted criteria to be considered.

1. High suicide intent level or significant acute risk, typified by significant degrees of

   a. Specificity of method
   b. Planning of actual attempt

    c. Low impulse control

    d. Desire to make attempt.

    e. Absence of deterrents

    f. Absence of environmental support/pressure or antagonism from significant others

    g. New and pronounced life stressors

    h. Overt verbalizations of above

2. Sudden increase/recurrence of significant suicidal ideation

3. Evidence of recent attempts (or "gestures") or indications of imminence of an attempt from the patient's or others' reports

4. Patient with strong positive history of suicide attempts appears increasingly depressed (or elated) under unusual external stresses

5. In addition to obviously suicidal self-harm, one must also monitor anorexic self-harm as a potential criterion for hospitalization.

A model standard operating procedure such as the one used at the Center for Cognitive Therapy for any such suicidal *emergencies* is as follows.

1. The treating therapist should deal fully and directly with suicidal issues during the scheduled session, in an emergency session, or on the phone. A phone session should be followed by an emergency face-to-face session as soon as possible. The session should focus on elucidating the dimensions of the problem (i.e., administering the SSI, assessing risk factors) and treating the problem (dispelling the hopelessness, generating alternative solutions: use ingenuity to keep patients involved in treatment until the suicide crisis is resolved). The option of hospitalization should be discussed when appropriate.

2. *Immediate* consultation with the supervisor who is following the case. If the therapist's interventions have sufficiently alleviated the problem so the therapist is certain the acute risk is eliminated, the therapist then need only immediately inform the supervisor of the situation, discuss the case, and obtain the supervisor's approval.

3. If there is still any doubt about risk or if the therapist's own interventions have been unsuccessful, the therapist should immediately contact the Clinical Director or other covering senior staff member for emergency consultation with the patient. The Clinical Director can consult with the therapist and see the patient as is indicated by the situation.

To review, the chain of contact is the clinical director and then other covering senior staff member.

Treatment

All therapists should be familiar with *Cognitive Therapy of Depression* (Beck et al., 1979b), particularly Chapters 10 and 11.

1. All therapists should have available and be familiar with the list of common drugs and their lethal doses.

2. Therapists should also be aware of the various suicide crisis centers available.

In *Cognitive Therapy of Depression*, Beck and colleagues (1979b) devoted Chapters 10 and 11 to the suicidal patient. These chapters form the basis of the model used at the Center for Cognitive Therapy for therapeutic strategies involving the suicidal patient.

Of primary importance, as we have previously discussed, is assessment. The BDI should be administered routinely and checked by the clinician prior to the session. In particular, the two items shown to be sensitive to suicide must be carefully reviewed. If this instrument shows any suggestion of suicidal ideation or if other signs exist, the HS and SSI should be administered.

It is important for the therapist to attempt to elicit *collaboration* from the patient to work on the target *symptom* of suicide. If the patient is depressed, it is often helpful to explain that

if you look at the broad spectrum of emotional disorders which many individuals in our society experience, depression is one of the groups with the highest probability of recovery, especially when one considers spontaneous remission, psychotherapies such as cognitive therapy and interpersonal psychotherapy which have been shown to be efficacious, as well as the numerous medications that have been reported to be effective. Therefore, depression can be cured, as well as the sense of hopelessness behind your thoughts of suicide, by one or a combination of treatments.

This explanation sometimes helps to undermine their view of *suicide as a desirable alternative*.

A primary target in the treatment should be to help the patient see *more reasons for living than dying*. As the patient may be unable to "see" any reasons for living at the moment, the therapist should ask the patient to recall reasons for living prior to the depressive period and then have the patient explore whether these or similar reasons exist at the present time. If patients are hopeless, their basic *problem-solving skills* are often blocked; therefore the therapist must be active in generating the list. However, the therapist should use questioning and active listening to avoid the sense of trying to "talk the patient out of suicide."

The therapist should thoroughly explore the advantages for living as well as the disadvantages. It is often helpful to ask the patient to rate on a 0 to 10 scale C how certain or what is the probability that this advantage (or disadvantage) will come to fruition. Also, the patient should rate on a 0 to 10 scale I how important is this advantage (or disadvantage). The therapist should then question the patient or offer problem-solving strategies to see if it is possible for the patient to increase the qualitative ratings on the advantage side and decrease the values on the disadvantage side. The same technique, directed at the opposite objective, should be completed for the *advantages/disadvantages to dying*.

After the above strategies are employed, the therapist should then directly challenge the patient's sense of hopelessness. Too frequently therapists forget that few problems are insoluble and "buy into" the patient's sense of hopelessness. It is helpful to remind oneself, as well as the patient, that the sense of hopelessness may result from a distortion (e.g., *dichotomous thinking*) or from lack of effective problem-solving strategies. In the former case, the therapist should (1) ask for evidence; (2) look for alternative explanations; and (3) de-catastrophize—what is the worst that can happen/what is the best that can happen/what is most likely to happen. In the latter case, various *problem-solving* approaches should be considered.

During the period of high suicidal thought, family members should be consulted and activities for distraction assigned. Also, it is important for the therapist to see the patient often, as well as have daily check-in calls.

Guidelines for Practice: Chapter Summary

1. The BDI is administered at every therapy session.
2. Regardless of the total BDI score, Items 2 (hopelessness) and 9 (suicide ideation) are reviewed. If either item has a rating of 2 or 3, the SSI is administered.
3. The Hopelessness Scale is also administered at every therapy session. A score of 10 or above alerts the therapist to possible suicidal behavior.
4. Patients with suicidal ideation are monitored.
5. When assessing the possibility of suicidal behavior, the following high risk variables are taken into consideration.

   a. Does the patient have a method?
   b. Has the patient demonstrated low impulse control?
   c. Are there any deterrents?
   d. Does the patient have environmental support?
   e. Are there any new life stressors?
   f. Is there a sudden increase or recurrence of significant suicidal ideation?
   g. Is there any evidence of a recent attempt or "gesture," or are there indications of an imminent attempt?
   h. Does the patient have a history of previous attempts?

6. For an imminent suicidal emergency use the following guidelines.

   a. Focus on elucidating the dimensions of the problems through the use of the SSI and assessment of high-risk variables.
   b. Focus on the immediate problem (i.e., fired from job, breakup of a relationship) and attack the hopelessness by generating possible alternative solutions to the problem(s).
   c. Discuss hospitalization as an appropriate option.
   d. Discuss the situation with a colleague and supervisor if appropriate.

7. For long-term treatment of suicidal behavior, there are five areas of focused intervention.

a. If the patient is depressed, focus on symptom relief. Even minor relief can help to build hope in the patient. Standard cognitive therapy of depression (Beck et al., 1979b) is used.

b. Suicidal patients tend to have poor problem-solving ability. Problem-solving training is taught and practiced in an out of the therapy session.

c. Cognitive distortions tend to be more rigid among this group of patients. The therapist and patient collaborate to aggressively identify, test, and respond to these distortions. In addition, the therapist teaches the patient to label his or her habitual errors in thinking, such as arbitrary inference, selective abstraction, overgeneralization, personalization, and, most importantly, dichotomous thinking.

d. Hopelessness is attacked using a variety of methods. First, have the patient focus on alternative interpretations of his or her life situation. Second, look for current nonadaptive behavior and brainstorm other choices of behavior. Third, identify the cognitive distortion that leads to the current symptom of hopelessness.

e. Work at changing the patient's view of suicide as a desirable alternative. Focus on the advantages and disadvantages of suicide and point out the advantages and disadvantages of other solutions to their problems.

REFERENCES

Beck, A. T., Brown, G., Berchick, R. J., Stewart, B. L., & Steer, R. A. (1990). Relationship between hopelessness and ultimate suicide: a replication with psychiatric outpatients. *American Journal of Psychiatry, 147*, 190–195.

Beck, A. T., Kovacs, M., & Weissman, A. (1979a). Assessment of suicidal intention: the scale for suicidal ideation. *Journal of Consulting and Clinical Psychology, 47*, 343–352.

Beck, A. T., Rush, A. J., Shaw, B. F., & Emery, G. (1979b). *Cognitive therapy of depression.* New York: Guilford.

Beck, A. T., & Steer, R. A. (1987). *Manual for the revised Beck Depression Inventory.* San Antonio, TX: Psychological Corporation.

Beck, A. T., & Steer, R. A. (1989). Clinical predictors of eventual suicide: a 5- to 10-year prospective study of suicide attempters. *Journal of Affective Disorders, 17*, 203–209.

Beck, A. T., Steer, R. A., Kovacs, M., & Garrison, B. (1985). Hopelessness and eventual suicide: a ten-year prospective study of patients hospitalized with suicidal ideation. *American Journal of Psychiatry, 142*, 559–563.

Beck, A. T., Weissman, A., Lester, D., & Trexler, L. (1974). The measurement of pessimism: the hopelessness scale. *Journal of Consulting and Clinical Psychology, 42*, 861–865.

Fawcett, J., Scheftner, W., Clark, D., Hedeken, D., Gibbons, R., & Coryell, W. (1987). Clinical predictors of suicide in patients with major affective disorders: a controlled prospective study. *American Journal of Psychiatry, 144*, 35–40.

Pokorny, A. D. (1983). Prediction of suicide in psychiatric patients. *Archives of General Psychiatry, 40*, 249–257.

Steer, R. A., Beck, A. T., & Garrison, B. (1986). Applications of the Beck Depression Inventory.

In N. Sartorious & T. Ban (Eds.), *Assessment of depression* (pp. 123–142). New York: Springer-Verlag.

Weissman, M. M., Klerman, G. L., Markowitz, J. S., & Ouellette, R. (1989). Suicidal ideation and suicide attempts in panic disorder and attacks. *New England Journal of Medicine, 321,* 1209–1214.

11

Psychopharmacotherapy of Suicidal Ideation and Behavior

ANDREW EDMUND SLABY AND LARRY E. DUMONT

Suicide and other self-destructive behaviors require a differential diagnosis that calls for diagnosis-specific treatment in much the same way fever, seizures headache, depression, and anxiety suggest a differential diagnosis. It is true that in most instances individuals who attempt suicide do not want to die. They usually simply want to end their pain. They feel hopeless with no perceived means of ameliorating the torment except through self-inflicted death. It is hopelessness more than depression that kills. We all feel depressed or anxious at times. When we believe there is no way of mollifying the intensity or duration of the anxiety or depression, however, we experience hopelessness. Hopelessness in a person predisposed to suicide due to a number of causative and contributory factors leads to death. An abbreviated differential diagnosis of self-destructive behavior is portrayed in Table 11.1. Table 11.2 lists a number of contributing factors (Slaby, Lieb, & Tancredi, 1985).

Contributing factors interact with causative factors to enhance the risk of suicide. This situation may be mediated by biological mechanisms involving, in some instances, aberrations in catecholamine and indoleamine metabolism for which there are psychopharmacologically specific interventions. For instance, learning disorders and homosexual orientation are associated with increased risk of suicide. In the former case, early recognition of a learning problem leads to remedial measures to ensure that the full potential of someone so afflicted (and often with superior intelligence) can be achieved. Comparably, a person who feels alone and rejected by family and members of his or her community for sexually feeling more responsive to same-sex partners feels less alone and disconnected if those of us who believe we all have a part of our personality that is homoerotic speak out and provide support. Lack of recognition of a learning disorder can, however, lead to a sense of hopelessness such that it will never be possible to achieve scholastically despite all efforts. So, too, lack of support or overt rejection of a person who is predominantly homosexual leads to a sense that the world will never be safe for him or her, and that it will be impossible to live and love openly, as can those who are predominantly heterosexual. If a person has the biological predisposition to impulsivity in either instance, a fatal suicide attempt may occur.

187

Table 11.1. Differential Diagnosis of Self-Destructive Behavior

Adjustment disorders
Anxiety diorders
Bipolar disorder
Brief reactive psychosis
Delusional (paranoid) disorder
Depressive disorder
Impulse control disorder
Organic mental disorders associated with physical disorders or conditions
Personality disorders
Posttraumatic stress disorder
Psychoactive substance-induced organic mental disorders
Schizoaffective disorder
Schizophrenia

Source: The Handbook of Psychiatric Emergencies, 3rd ed., by A. E. Slaby, J. Lieb, and L. R. Tancredi, 1985, New York: Elsevier. By permission.

The often discussed association between suicide and affective illness and creativity (Coryell et al., 1989; Goodwin & Jamison, 1990) may relate to the fact that in the case of creativity and in the case of suicide, impulsivity leads to rapid decisions. In one case it may mean to go with an idea that is different—thinking of a usual thing in an unusual way, which is sometimes the essence of creative genius. In the other instance, an impulsive need to end the pain when depressed or anxious without hope leads to suicide.

Other contributing factors may operate alone or in concert with psycho-pharmacologically responsive causative variables. Alcoholics who attempt suicide have a greater incidence of lifelong psychiatric diagnoses of major depression, panic disorder, phobic disorder, and generalized anxiety disorder (Roy et al., 1990). In other cases alcohol or drugs may impair judgment, leading to self-destructive behavior without a major psychiatric disorder that requires

Table 11.2. Self-Destructive Behavior: Contributing Factors

Access to means of suicide	Learning disorders
Age	Marital status
Anomie	Occupation
Command hallucinations	Physical and psychological disabilities
Family history	Physical illness
Gender	Previous attempt
Generation in country	Religion
Homicide	Race
Homosexual orientation	Socioeconomic status
Hypochondriasis	Stressful life events
Insomnia	Suicide of others
Intelligence	Substance abuse
Lack of future plans	Unemployment
Lack of social support	Urban versus rural origin

Source: The Handbook of Psychiatric Emergencies., 3rd ed., by A. E. Slaby, J. Lieb, and L. R. Tancredi, 1985, New York: Elsevier. By permission.

treatment. Unemployed women (Hawton, Fagg, & Simkin, 1988), Vietnam veterans (Pollock et al., 1990), and adolescents who have been abused as children (Deykin, Alpert, & McNamara, 1985) are groups with both increased suicide risk and risk of substance abuse. In some instances posttraumatic stress disorder, affective illness, or an anxiety disorder is present, and the individual is self-medicating a psychiatric disorder. In others, substance abuse alone may be impairing judgment, leading to self-destructive and other destructive behavior. When guns are present (Rich et al., 1990) or high places are available, the acting out of the impulse to die is accelerated owing to the availability of the means to do so. It is well known that homocide and other assaultive behavior is correlated with self-destructive behavior (Alessi et al., 1984; Griffith & Bell, 1989; Kerkhof & Bernasco, 1990). This relation suggests that there may be a subpopulation of suicidal patients in whom a medication-responsive impulsivity leads to creativity in some but to homocide and suicide in others. The challenge to the biopsychiatric psychiatrist is how to save those who are creative, so they may enjoy the self-esteem of the actualization of their creativity without suffering the pain of an affective illness or anxiety disorder or in those so predisposed to death by suicide or homicide.

The lifetime likelihood of death by suicide is estimated to be approximately 1%. In actuality, the figure would probably be considerably greater if single car "accidents," "accidental" overdoses (particularly in the elderly), and mishaps such as falls from tall buildings, pedestrian train and car fatalities, and drownings not recorded as suicide were included (Slaby & Tancredi, 1991).

Affective illness (major depression, bipolar illness, and schizoaffective disorder) is the most common diagnosis among suicide completers, accounting for 60 to 70% (Bulik et al., 1990) of deaths. If other illnesses with a strong depressive component that can be managed with antidepressants are included, such as depressions associated with dysthymic disorder, cyclothymic disorder, narcissistic personality, and borderline personality disorder, the number may be as great as 80%. Many of those affectively ill are dually diagnosed. They self-medicate the mania or depression with drugs or alcohol, thereby decreasing judgment, increasing hopelessness, decreasing self-esteem, and enhancing the risk of impulsive self-inflicted death, particularly if the means to do so are available.

Risk of suicide for individuals with affective illness is estimated to be 30 times that of those not suffering from the disorder (Bulik et al., 1990). Approximately 15% of individuals with major depression take their own life (Bulik et al., 1990). Ten percent of schizophrenics are said to die by suicide during the first 10 years of the illness (Cohen, Test, & Brown, 1990) or, put another way, suicide of schizophrenics occurs at a rate 20 times that of the normal population (Breier & Astrachan, 1984). Lifetime prevalence of suicide of those with schizophrenia is 15% (Cohen et al., 1990). The incidence of suicide is so increasing among young schizophrenics (perhaps owing to lack of compliance with medication due to more abbreviated hospitalization with less indoctrination about the need for psychopharmacotherapy and to concurrent substance abuse) that it is now said to be the leading cause of death in this group. Symptoms of depression, especially hopelessness, are the most important variable in identifying those schizophrenics at most risk.

Panic disorder, which occurs in about 1.5% of the population at some time during their life, comparably carries a high risk of morbidity and mortality if not identified early and treated aggressively. Panic attacks themselves that do not comport to the diagnostic criteria for panic disorder are even two to three times more prevalent (Weissman et al., 1989). Of those with panic disorder approximately 20% make a suicide attempt, and of those with panic attacks about 12% (Reich, 1989; Weissman et al., 1989). Panic disorder, like affective illness, is associated with specific neuroendocrine changes and requires specific psychotropic intervention to reduce risk of suicide.

Even when contributing factors are examined for their role in increasing the risk of suicide, it is found that depression, anxiety, bizarre thinking, and hopelessness are variables discriminating those at greatest risk. Substance abuse may impair judgment and enhance impulsivity, but it also may be sought to self-medicate a panic disorder or affective illness, as individuals with these symptoms are the most suicidal. An untreated depression during adolescence can give genesis to depression as an enduring personality trait, creating a picture of double depression when major depression occurs at the time of a stressful life event later in life. The original depression, which may have gone unrecognized during adolescence and was therefore untreated, may have been a response to feeling hopeless about achieving better in school because of an unrecognized learning disorder or of making a meaningful love relationship with another human being because of adolescent (and adult) discrimination against gay youth. There probably also exists double anxiety when an untreated anxiety disorder during adolescence leads to an enduring part of the personality that, coupled with the onset of a panic disorder as an adult, leads to double anxiety.

Suicide attempters do not all die by suicide. Of those who attempt, only about 1% die by suicide during the 10 ensuing years (Reich, 1989). Of those who commit suicide, 35% have made a previous attempt (Reich, 1989). Attempts have psychological and physical cost. The parasuicidal individual frightens friends, lovers, and relatives, who then withdraw, creating more aloneness and hopelessness. Unsuccessful violent attempts can leave a person paraplegic, quadriplegic, or brain-damaged. The same factors—biochemical, psychological, and sociological—that are correlated with violent suicide are correlated with violent parasuicide (Slaby & Tancredi, 1991). Attempters who repeat, compared to those who do not, feel more depressed and hopeless about attempts to resolve their problems (Sakinofsky et al., 1990). Studies of adolescents who, like older people, have higher rates of suicides have indicated that some risk factors for attempts are the same as those for completed suicide: suicide attempts or completed suicides in the family, depression, and stressful life events (McHenry et al., 1982; Slaby & Tancredi, 1991; Tishler, McKenny, & Morgan, 1981).

Neurochemical Basis of Violent Behavior

Diagnosis-specific treatment of suicidal behavior has achieved its current status as a key factor in monitoring quality endurance of care provided to suicidal patients owing to (1) the evolving understanding that disturbances of mood, thought, and behavior have a differential diagnosis; and (2) the awareness of the neuroendocrine basis of major psychiatric illness. Today we know (Slaby & Tancredi, 1991) that anxiety, depression, thought disturbances, and suicidal behavior are sometimes due to organic mental disorders, anxiety disorders, affective disorders, schizophrenia, and other major psychiatric illnesses that dictate specific therapeutic interventions to reduce anxiety, depression, disturbed thinking, and violent behavior. The locus ceruleus in the brain has been demonstrated to be critical to understanding the etiology of anxiety and the mechanisms by which benzodiazepines such as alprazolam (Xanax) and lorazepam (Ativan) effect the reduction of anxiety. Electroshock, it appears, specifically alters the aberrations of indoleamine and catecholamine metabolism found with depression, restoring mood and thereby eliminating the single greatest risk factor for suicide: depression. Violent behavior with affective illness has been shown to relate specifically to changes in serotonin metabolism explored by Asberg (1986) and others. Data implicating the role of biological factors in suicide come from studies of attempters and completers (Asberg, 1986). The factors explored include measures of indoleamines and catecholamines and their metabolites in the brains of completers and in the cerebrospinal fluid of attempters, receptor binding, and other neuroendocrine measures (e.g., cortisol levels), and measure of the thyroid-releasing hormone (TRH).

Amines and Their Metabolites

Measures of monoamine transmitters in the brains of suicide victims have indicated a trend toward decreased amounts of serotonin (5-hydroxytryptamine) in the brains of completers (Asberg, 1980, 1986; Asberg, Bertilsson, & Martensson, 1984; Asberg, Nordstrom, & Traskman-Bendz, 1990; Cohen, Winchel, & Stanley, 1988; Shaw, Camps, & Eccleston, 1967; Stanley & Stanley, 1990; Traskman-Bendz, Asberg, Nordstrom, & Stanley, 1989; Traskman-Bendz, Asberg, & Schelling, 1990). There are, however, some conflicting reports. Autopsy studies indicate that serotonergic, but not dopaminergic or noradrenergic changes are associated with suicide (Traskman-Bendz et al., 1989). The areas of the brain affected include the raphe nuclei, hippocampus, frontal cortex, and hypothalamus (Asberg et al., 1987; Traskman-Bendz et al., 1990). Changes in receptor site activity provide further evidence for an anatomically selective change in adrenergic functioning in impulsively suicidal patients.

Low cerebrospinal fluid 5-hydroxyindoleacetic acid (CSF 5-HIAA), the principal metabolite of serotonin, is reported in those who attempt suicide regardless of diagnosis (Asberg, 1986; Asberg et al., 1990; Asberg, Traskman, Thoren, 1976b; Braunig, Rao, & Fimmers, 1989); DeLeo & Marazziti, 1988; Stanley &

Stanley, 1990). Violent attempts (e.g., jumping, neck slashing, hanging, shooting) are reported more frequently in those with lower CSF 5-HIAA levels who also have alcoholism, adjustment disorders, major depression, personality disorders (Brown et al., 1982), and schizophrenia (van Praag, 1983). The same low levels of 5-HIAA are reported in the CSF of violent criminal offenders (Linnoila et al., 1983). Bipolar disorder may represent an exception to this observation (Lidberg et al., 1985). In one study (Asberg, 1980) 40% of patients with a low level of CSF 5-HIAA attempted suicide in contrast to 15% of those with high 5-HIAA levels. This low level of 5-HIAA relates more specifically to suicide attempts *regardless of whether a patient reports depression* and more specifically to more *violent* attempts (Asberg et al., 1984; DeLeo & Marazitti, 1988; Traskman et al., 1981). Comparably lower blood serotonin levels have been associated with previous suicide attempts in studies of schizophrenics (Braunig et al., 1989).

Suicide attempters with a low concentration of CSF 5-HIAA have been reported to be more than ten times as likely to die by suicide than those who do not have the decrease (Asberg, 1986). In general, low CSF 5-HIAA and homovanillic acid (HVA), the breakdown product of the catecholamine dopamine, is associated with suicidal behavior (Asberg, 1986). Depressed patients with low HVA levels tend to exhibit psychomotor retardation in addition to increased suicide risk (Traskman et al., 1981; Traskman-Bendz et al., 1989; van Pragg & Korf, 1988). Patients with low 5-HIAA levels seem prone not only to more violent suicide attempts but also to attempts that are less premeditated (Asberg et al., 1976b) and more frequent (Asberg et al., 1976b; Cohen et al., 1988).

Receptor Binding

Postmortem studies indicate that suicide is associated with reduced activity of the serotonergic system, as indicated by decreased imipramine binding at presynaptic receptor sites in the frontal cortex (Asberg, 1986; Braunig et al., 1989; Stanley, Virgilio, & Gershon, 1982) and increased density of postsynaptic serotonin receptors (DeLeo & Marazitti, 1988; Mann et al., 1986; Stanley & Mann, 1983; Stanley & Stanley, 1990; Traskman-Bendz et al., 1990). The increase in postsynaptic B-adrenergic receptor binding represents a compensatory postsynaptic adaptive response to decreased presynaptic activation (Beigan & Israeli, 1988; Mann et al., 1986).

A study of serotonin platelet uptake (Modai et al., 1989) in adolescent patients who had either attempted suicide or had shown aggressive behavior prior to admission indicated that there were fewer binding sites for serotonin on platelets obtained from depressed patients. This finding suggests that serotonin platelet uptake may be a marker not related to merely symptoms of aggression, suicidality, or conduct disturbances (as may be true for CSF 5-HIAA) but, rather, to specific diagnostic disorders.

Other Neuroendocrine Measures

Concentrations of serum and of urinary 17-hydroxycorticosteroids have been found to be elevated in patients with depression, and there is an association between the inability to suppress cortisol secretion after dexamethasone administration and suicidal behavior. This finding is particularly pronounced in *suicidal* depressed patients (Asberg, 1980, 1986; Asberg et al., 1984, 1986a,b; Bunney & Fawcett, 1965; Cohen et al., 1988). Traskman et al. (1980) found elevated CSF cortisol in patients with endogenous depression but normal concentrations in suicide attempters; there was no correlation with CSF 5-HIAA. It is interesting to note that nonsuppression is also associated with increased state anxiety (Asberg et al., 1990). Serum cortisol and 5-HIAA changes appear as independent markers, enhancing their collective predictive value (Asberg et al., 1984).

Although the relation to self-destructive behavior is unclear, there is a blunted response of production of thyroid-stimulating hormone (TSH) to thyroid-releasing hormone (TRH) with antidepressant-responsive depressions (Asberg, 1986; Asberg et al., 1984).

Combining a number of biochemical measures, such as low CSF 5-HIAA, low HVA, and nonsuppression of the dexamethasone suppression test (DST), rather than any individual neuroendocrine measure maximizes the prediction of suicide (Traskman-Bendz et al., 1989). It appears that both low CSF 5-HIAA and activation of the hypothalamic-pituitary-adrenal axis are associated with increased suicide risk (Asberg et al., 1990).

In addition to increased suicidal behavior, there is a significant inverse correlation between CSF 5-HIAA and overt aggression (Asberg et al., 1990; Brown et al., 1982; Cohen et al., 1988; Lidberg, Asberg, & Sundquist-Stenmann, 1984; Lidberg et al., 1985; Linnoia et al., 1983; Traskman-Bendz et al., 1990) including murder (particularly impulsive homicide) and to increased anxiety (Asberg et al., 1984). The relation appears to relate to *overt* behaviors (e.g., hostile acts and suicide attempts) and not to self-reported affective states (e.g., thoughts of suicide or hostile feelings) (Asberg et al., 1984). The relation of low serotonin to both depression and expressed anger, social dominance, and fearlessness (Asberg, 1986) supports the literature from Freud to the present (Asberg et al., 1984) associating anger turned inward and outward to depression. The incidence of suicide among murderers in some European countries is higher than in any other risk group, exceeding in some instances 30% (Asberg et al., 1984).

It is noteworthy that administration of 5-hydroxytryptophan, a precursor of serotonin, has reversed aggressive behavior in animals induced by transection of the olfactory bulbs (DiChiara, Camba, & Spano, 1971).

In Asberg et al.'s studies (1984) of murderers, 5-HIAA levels that were comparable to those of suicide attempters were found in those who killed someone with whom they had an intense relationship, such as a paramour or spouse. It is this group who also is most likely to commit suicide after the homicide. Parents who attempt suicide after killing their own children (one of the most extreme violations of human bonding) have also been found to have lower CSF 5-HIAA levels (Lidberg et al., 1984).

In summary, impulsive violent behavior is more associated with disturbances of serotonin metabolism than are disorders of mood. The serotonergic system is involved with sexual drive, feeling, and fight and flight behavior (Linnoila et al., 1983; Traskman-Bendz et al., 1990), all of which obviously at times occur impulsively and sometimes are destructive to self. These changes are also reported with eating disorders, suicide, rape, assaults, and homicide (Cohen et al., 1988).

In addition to depression and aggression, it is known that schizophrenia, obsessive-compulsive disorder, and panic attacks are known to relate to serotonergic dysfunction (Van de Kar, 1990).

Although many suicide patients exhibit low CSF 5-HIAA, many healthy people who never will attempt suicide show a decrease (Asberg et al., 1990), indicating that in addition to this vulnerability other items such as situational factors (e.g., lack of social support, stress, and life events) and psychological variables (e.g., hopelessness, helplessness, and early loss of parents through death or divorce, relating a more stressful interpretation of later adult life events) play critical roles in making a person suicidal (Asberg et al., 1990). This fact indicates that the psychopharmacotherapy of impulsive suicidal behavior must be accompanied by psychotherapy and sociotherapy in order to maximize its ability to reduce self-destructive behavior.

Management

The first step in the psychopharmacotherapy of suicidal behavior is the identification of when a psychopharmacological intervention is necessary and will be efficacious. Obviously, if a person feels hopeless and in psychic pain because there is an unidentified learning disorder or a lack of social support because he or she is homosexual, a behavioral intervention is more appropriate if neurovegetative signs of a depression are not present. In the former case, it is identification of the nature of the learning disorder and provision of the appropriate treatment to allow compensation for the disorder. In the latter case, social support and appropriate mature homosexual role models are needed. It is possible in both of these instances, however, that one or two affective disorders exist individually or concurrently (so-called double depression) or an anxiety disorder or disease such as hypothyroidism or cardiac arrhythmia coexists with depression. In the last instance, a medical disorder may have been triggered by the stress of low self-esteem and rejection, stimulating a depressive or anxiety disorder. Depression and suicidal thoughts not treated appropriately with psychopharmacotherapy during adolescence can lead to the depression and suicidal thoughts becoming an enduring part of the adult personality.

When a stressful life event sets off a major depression with a suicide attempt later in life, one obtains a clinical history of a sustained depression since adolescent and an acute depression with the neurovegetative symptoms of sleep, appetite and sexual interest disturbance, and diurnal variation of mood (feeling worse in the morning). One drug may work for the acute depression, but psychotherapy with or without another antidepressant may be required for the dysthymia that has endured through the years since adolescence (Keller, 1990; Perez-Stable et al., 1990).

Only about one-third of affective illness that is antidepressant-responsive is treated despite the fact that it is the single greatest risk factor for suicide and the fact that those whom it affects are as a group found to be more creative and talented than those not so affected (Clayton, 1985; Slaby & Tancredi, 1991). The risk of suicide with those who are schizophrenic is about the same as major depression if untreated psychopharmacologically (viz. 15%) (Slaby & Tancredi, 1991), but the absolute number affected is much smaller. The risk of suicide attempts for those with panic disorder is about 20%. Panic disorder also requires diagnosis-specific treatment (Clayton, 1985) with anxiolytics or antidepressants.

The fact that depression and impulsive suicidal behavior are found to relate to specific defects in amine metabolism indicates the need for antidepressant medication that works to enhance available neurotransmitter amine compounds. As indicated earlier, suicide and outwardly aggressive acts appear to relate to deficiencies in serotonin production. This finding supports a more specific role in some cases for use of medication that increases availability of serotonin and that reduces impulsive episodic behavior: lithium carbonate, valproic acid, carbamazepine, β-blockers, and calcium channel blockers.

Benzodiazepines such as alprazolam (Xanax) and some antidepressants work specifically to counter panic disorder and other instances of episodic and chronic anxiety. Neuroleptics alone may serve to manage chronic schizophrenia, neuroleptics coupled with lithium or antidepressant medication for schizoaffective disorder, and neuroleptics with benzodiazepine therapy e.g., haloperidol (Haldol) or thiothixine (Navane) and lorazepam (Ativan) for acute psychotic episodes.

Fluoxetine is a selective serotonin uptake inhibitor that increases the amount of serotonin available at the neuronal synapse. It is equivalent or more effective than imipramine, the tricyclic antidepressant, in terms of its potency as an antidepressant (Asberg, 1986). The ability of most antidepressants to enhance serotonin function after prolonged treatment is believed to relate to a number of mechanisms depending on the nature of the antidepressant therapy (Blier, deMontigny, & Chaput, 1990; Price et al., 1990). Monoamine oxidase inhibitors (MAOIs) are believed to increase available releasable serotonin. Serotonin reuptake inhibitors and monoamine oxidase are thought to desensitize the neural firing-rate-inhibiting somatodendrite autoreceptors. Tricyclic (TCAs) and tetracyclic antidepressants and electroconvulsive therapy (ECT) are assumed to sensitize postsynaptic receptors and serotonin reuptake inhibitors and to desensitize serotonin-release-inhibiting terminal autoreceptors. Clinical findings further suggest (Price et al., 1990) that serotonin function is enhanced after long-term treatment with TCAs, MAOIs, and serotonin reuptake inhibitors. Trazadone (Asberg, 1986) has serotonin uptake blocking properties in addition to having other effects on serotonin neurons.

The fact that some antidepressants selectively block neuronal reuptake of serotonin is particularly appealing in the psychopharmacotherapy of depression associated with self-destructive behavior because drugs with fewer pharmacological effects are likely to cause fewer side effects and to be less toxic in case of overdose than more broad-spectrum antidepressants (Asberg et al., 1986a,b).

It appears at this time in the history of understanding suicidal behavior that serious questions could be raised regarding the quality of care provided a self-destructive patient if medical, surgical, and psychological causes of suicidality are not considered. Moreover, specific anxiety disorders, bipolar illness, major depression, schizophrenia, schizoaffective disorder, and other psychiatric, medical, and surgical disorders must be identified if diagnosis-specific interventions are not offered to the patient. It does not mean that a patient always accepts the treatment that will help, but it does offer a patient and their family the highest level of care medicine may offer.

Not all major depression is responsive to medication. It can in fact lead to a misimpression regarding psychopharmacotherapy. For instance, it was reported (Teicher, Glod, & Cole, 1990) that fluoxetine (Prozac) caused emergence of obsessional suicidal thoughts. Although it may be possible, there are a number of other explanations for the emergence of suicidal thoughts while on fluoxetine or any other antidepressant for that matter. Not every depressed patient referred for psychopharmacotherapy responds immediately or to the first drug used (or in some instances at all) any more than all patients who have hypertension referred to the most prominent internist has treatment-responsive high blood pressure. Some of the latter patients proceed to stroke and heart attacks, just as some moderately depressed patients do not respond to antidepressant therapy and become suicidal and in some cases commit suicide. Second, it has long been known that suicide risk may increase early in the course of treatment with antidepressants, as energy often returns more rapidly than the feelings of hopelessness disappear. This change creates a situation where a patient who may have suffered lethargy and lack of energy with depression and hopelessness regains energy before the depression lifts, providing the ability to act on a desire to end the pain through self-inflicted death. Finally, drugs such as fluoxetine (Prozac) and imipramine (Tofranil) may make a patient feel nervous and anxious, much as does a patient with an untreated anxiety disorder. This increased nervousness or anxiety as a side effect may enhance a feeling of hopelessness, as the condition may seem to be getting worse. This situation then may lead to a desire to commit suicide. Patients and significant others should be informed of potential side effects in order to enhance compliance with therapy and allow early intervention when a side effect requiring cessation of the drug arises.

Psychopharmacotherapy must be supplemented by a number of other therapeutic interventions in order to maximize the reduction of self-destructive behavior in those instances where psychotherapeutic intervention is indicated.

1. Psychotherapy directed at amelioration of the learned helplessness and low self-esteem that evolves with an untreated depression. Cognitive therapy is of particular help (Hollen, 1990). In instances where drug therapy is not indicated, cognitive therapy may be the principal modality of treatment. In other cases, it may be comparable in effectiveness to tricyclic antidepressant therapy (Hollon, 1990). In cases of a biologically based depression, it is an important conjunctive therapy in the acute state and is used to minimize the likelihood of exacerbation. Its success is predicated on the assumption that maladaptive information-processing and erroneous information play a causal role in the onset of some depressions, exacerbations of depression, and the persistence of learned helplessness after the neurovegetative signs of a depression remit.

2. Electroconvulsive therapy may be indicated for an acutely suicidal patient to prevent self-harm before psychopharmacotherapy can take effect and in instances where depression is resistant to treatment or a patient wishes rapid remission of depressive symptoms while awaiting therapeutic levels of antidepressant therapy. There are a number of myths regarding the use of ECT and its side effects. It is helpful to have a prospective patient talk to someone who has had course of ECT treatment to assuage fears regarding its use.

3. Because suicide ideation does not immediately respond to drugs or because an anxiety or depressive disorder may be treatment-resistant, it is necessary to provide appropriate safeguards to prevent a suicide attempt until self-destructive behavior is reduced. In its extreme form it includes 1:1 observation at arm's length. Other measures include removing from the patient anything that may be used to inflict harm and providing an environment as free as possible of access to means to commit suicide.

4. Psychopharmacologically responsive anxiety disorders, depression, and thought disorders may be self-medicated by a patient with alcohol and recreational drugs, impairing judgment, enhancing suicidal risk, and complicating the diagnostic presentation (so-called dually diagnosed patients). These patients require both specific treatment of the chemical dependence problem and awareness by clinicians not to use a psychotropic substance to which the patient may comparably become addicted if he or she is identified as having a chemically dependent personality. An alcoholic, for example, may substitute benzodiazepine for the alcohol used and become dependent—diazepam (Valium) dependency disorder. In some cases, not providing a person with appropriate benzodiazepine therapy if it is required may cause that individual, who does not have a genetic predisposition to alcoholism, to seek alcohol to reduce the pain of the anxiety in the same way they may attempt suicide to reduce the pain.

5. Obviously, truly predictable impulsive suicidal individuals require hospitalization (Roy, 1989); that is, if someone is profoundly depressed, feels hope-

less, and wants to kill himself or herself, the protection of hospitalization is indicated. There are, however, characterologically disordered patients with borderline or narcissistic personality disorders who are chronically depressed with no immediate plans to die and no predictable pattern of suicide. Hospitalization of some of these patients may reinforce manipulative suicidal threats rather than mollify them. Outpatient trials of antidepressants such as the MAOIs are sometimes effective against chronic dysthymia. Drugs such as lithium, valproic acid, diphenylhydantoin, clonazepam, β-blockers (e.g., propranolol), calcium channel blockers (e.g., verapamil), buspirone (Buspar), and carbamazepine (Tegretol), which are used for episodic outbursts, may be effective.

6. Given the fact that there is an excess of creative people who are suicidal and affectively ill, it is important to explain to them, their families, lovers, friends, and anyone else they deem critical to the success of their treatment that psychopharmacotherapy can prevent suicide of the creative person, and that there is no evidence of demonstrable attenuation of creativity with talented patients in treatment.

7. Supportive psychotherapy is critical early in the course of psychopharmacotherapy of the suicidal patient to enhance compliance with treatment and to counter the pain, loneliness, and hopelessness felt. External support in terms of family, lovers, and friends should be rallied to enhance the support.

8. Patients, family, lovers, and friends should be informed of side effects of therapy, including that at first the risk of suicide could increase and that in some instances mania may emerge as depression subsides. Specific effects, such as the possibility of addiction when benzodiazepines are used for anxiety disorders and the need for monitoring serum levels (as for lithium, valproate, and carbamazepine) should be reviewed together, emphasizing the need for monitoring potential side effects through physical examination (as with tardive dyskinesia) with patients who use neuroleptics and through laboratory studies of renal and thyroid functioning in patients taking lithium.

9. Patients who appear to be drug resistant may not be so but, rather, not have attained sufficient serum levels to have an impact. Serum levels provide evidence of noncompliance and indicate insufficient dosage for a specific patient. If serum levels are adequate, the response is enhanced in some cases with lithium augmentation, by the use of thyroxine, or by combining antidepressants or switching to another antidepressant that may have an impact through a different neurochemical mode of action. Some drugs, such as nortriptyline, have a specific "therapeutic window," a range above and below which there is decreased responsiveness.

Guidelines for Practice: Chapter Summary

Not all suicidal behavior is psychopharmacologically responsive. Some is due to deeply rooted psychological problems that may be more effectively treated by psychotherapy. At other times there are underlying medical or surgical disorders (e.g., antihypertensive medication side effects, adrenocortical dysfunction, tumors in the brain or elsewhere, AIDS) that can make a person feel self-destructive. These situations require specific medical or surgical intervention with or without pharmacotherapy or psychotherapy. Sometimes there exists social isolation because of extreme talent or lack of it, lack of sharing interests of a predominant subculture, gender identity issues, or other reasons that require sociotherapy to help increase a sense of belonging and to diminish anomie. In still other instances there is an existential aloneness, requiring help to define a *why* to live, so a patient may find a *how* to live.

Sometimes an illness exists that is making a person suicidal. Underlying neurotransmitter dysfunction may be manifested as affective illness, or an anxiety disorder, impulse disorder, or thought disorder. In these cases diagnosis-specific psychopharmacotherapy should be offered to diminish the manifestations of the disease and the impulse to die. Failure to offer this therapy, where indicated, and lack of effort to provide other requisite diagnosis-specific therapies falls short of providing the patient the highest quality of care that state-of-the-art medicine can provide.

REFERENCES

Alessi, N. E., McManus, M., Brickman, A., et al. (1984). Suicidal behavior among serious juvenile offenders. *American Journal of Psychiatry, 141,* 286–287.
Alvarez, D. (1973). *The savage god: a study of suicide.* New York: Bantam Books.
Asberg, M. (1980). Biochemical abnormalities in depressive illness. In G. Curzon (Ed.), *The biochemistry of psychiatric disturbance.* New York: John Wiley & Sons.
Asberg, M. (1986). Biochemical aspects of suicide. *Clinical Neuropharmacology, 9*(suppl. 4), 374–376.
Asberg, M., Bertilsson, L., & Martensson, B. (1984). CSF Monoamine: metabolites, depression, and suicide. In E. Usdin et al. (Eds.), *Frontiers in biochemical and pharmacological research in depression.* New York: Raven Press.
Asberg, M., Eriksson, B., Martensson, B., et al. (1986a). Therapeutic effects of serotonin uptake inhibitors in depression. *Journal of Clinical Psychiatry, 40*(4, suppl.), 23–35.
Asberg, M., Nordstrom, P., & Traskman-Bendz, L. (1986b). Biochemical factors in suicide. In A. Roy (Ed.), *Suicide.* Baltimore: Williams & Wilkins.
Asberg, M., Nordstrom, P., & Traskman-Bendz, L. (1990). Cerebrospinal fluid studies in suicide: an overview. *Annals of the New York Academy of Sciences, 487,* 243–255.
Asberg, M., Schalling, D., Traskman-Bendz, L., et al. (1987). Psychobiology of suicide, impulsivity, and related phenomena. In H. Y. Meltzer (ed.), *Psychopharmacology: a third generation of progress.* New York: Raven Press.
Asberg, M., Thorel, P., Traskman, L., et al. (1976). "Serotonergic depression" . . . a biochemical subgroup within the affective disorders. *Science, 191,* 478–480.
Asberg, M., Traskman, L., & Thoren, P. (1976b). 5-HIAA in the cerebrospinal fluid: a biochemical suicide predictor? *Archives General Psychiatry, 33,* 1193–1197.
Ballenger, J. R., Goodwin, F. K., Major, L. F., et al. (1979). Alcohol and serotonin metabolites in man. *Archives of General Psychiatry, 86,* 224–227.

Benfield, P., Hell, R. C., & Lewis, S. P. (1986). Fluoxetine: a review of its pharmacokinetic properties and therapeutic efficacy in depressive illness. *Drugs, 32*, 481–508.

Beskow, J. (1979). Suicide and mental disorder in Swedish men. *Acta Psychiatrica Scandinavica Supplementum, 277*, 5–138.

Biegan, A., & Israeli, M. (1988). Regionally selective increases in beta-adrenergic receptors density in the brains of suicide victims. *Brain Research, 442*, 199–203.

Blier, P., deMontigny, C., & Chaput, Y. (1990). A role for the serotonin system in the mechanism of action of antidepressant treatments: preclinical evidence. *Journal of Clinical Psychiatry, 51*(suppl.), 4–20.

Braunig, P., Rao, M. L., & Fimmers, R. Blood serotonin levels in suicidal schizophrenic patients. *Acta Psychiatrica Scandinavica, 79*, 86–189.

Breier, A., & Astrachan, B. M. (1984). Characterization of schizophrenic patients who commit suicide. *American Journal of Psychiatry, 141*, 206–209.

Brown, G. L., Ebert, M. H., Goyer, P. F., et al. (1982). Aggression, suicide and serotonin: relationship to CSF amine metabolites. *American Journal of Psychiatry, 139*, 741–746.

Buchsbaum, M. S., Coursey, P. A., & Murphy, D. L. (1976). The biochemical high risk paradigm: behavioral and genetic correlates of low platelet monoamine oxidase activity. *Science, 19*, 339–341.

Bulik, C. M., Carpenter, L. L., Kupfer, D. J., et al. (1990). Features associated with suicide attempts in recurrent major depression. *Journal of Affective Disorders, 18*, 29–37.

Bunney, W. E., Jr., & Fawcett, J. A. (1965). Possibility of a chemical test for suicidal potential. *Archives of General Psychiatry, 13*, 232–239.

Bunney, W. E., Jr., Fawcett, J. A., Davis, J. M., et al. (1969). Further evaluation of urinary 17-hydroxycorticosteroids in suicidal patients. *Archives of General Psychiatry, 21*, 138–150.

Cheetham, S. C., Crampton, C., Czudek, R., et al. (1989). Serotonin concentrations and turnover in brains of depressed suicides. *Brain Research, 502*, 332–340.

Clayton, P. J. (1983). A further look at secondary depression. In P. Clayton & J. Bennett (Eds.), *Treatment of depression: old controversies and new approaches*. New York: Raven Press.

Clayton, P. J. (1985). Suicide. *Psychiatric Clinics of North America, 8*, 203–214.

Clayton, P. J., Marten, S., Davis, M., et al. (1980). Mood disorder in women professionals. *Journal of Affective Disorders, 2*, 37–46.

Cohen, L. J., Test, M. A., & Brown, R. L. (1990). Suicide and schizophrenia: data from a prospective community treatment study. *American Journal of Psychiatry, 147*, 602–697.

Cohen, L. S., Winchel, R. M., & Stanley, M. (1988). Biochemical markers of suicide risk and adolescent suicide. *Clinical Neuropharmacology, 2*, 423–435.

Coryell, W., Endicott, J., Keller, M., et al. (1989). Bipolar affective disorder and high achievement: a familial association. *American Journal of Psychiatry, 146*, 983–988.

DeLeo, D., & Marazziti, D. (1988). Biological prediction of suicide: the role of serotonin. *Crisis, 9*, 1009–118.

Deykin, E. Y., Alpert, J. J., McNamara, J. J. (1985). A pilot study of the effect of exposure to child abuse or neglect on adolescent suicidal behavior. *American Journal of Psychiatry, 142*, 1299–1303.

Dichiara, G., Camba, R., & Spano, P. E. (1971). Evidence for inhibition by brain serotonin of mouse killing behavior in rats. *Nature, 233*, 272.

Goodwin, F. K., & Jamison, K. R. (1990). *Manic-depressive illness*. New York: Oxford University Press.

Gottfries, C. G., Oreland, L., & Wiberg, A. (1975). Lowered monoamine oxidase activity in brain from alcoholic suicides. *Journal of Neurochemistry, 25*, 667–673.

Gottfries, G. C., Von Knorrins, L., & Oreland, L. (1980). Platelet monoamine oxidase activity in mental disorders. 2. Affective psychoses and suicidal behavior. *Progress in Neuro-Psychopharmacology, 4*, 185–192.

Griffith, E. E. H., & Bell, C. C. (1989). Recent trends in suicide and homicide among blacks. *Journal of the American Medical Association, 262*, 2265–2269.

Guze, S. B., & Robins, E. (1970). Suicide and primary affective disorders. *British Journal of Psychiatry, 117*, 432–438.

Hagnell, O., & Rorsman, B. (1979). Suicide in the Lundby study: a comparative investigation of clinical aspects. *Neuropsychobiology, 5*, 61–73.

Hawton, K., Fagg, J., & Simkin, S. (1988). Female unemployment and attempted suicide. *British Journal of Psychiatry, 152*, 632–637.

Hollon, S. D. (1990). Cognitive therapy and pharmacotherapy for depression. *Psychiatric Annals, 20*, 249–258.

Keller, M. B. (1990). Depression: underrecognition and undertreatment by psychiatrists and other health care professionals. *Archives of Internal Medicine, 150*, 946–948.

Kerkhof, A., & Bernasco, W. (1990). Suicidal behavior in jails and prisons in The Netherlands: incidence, characteristics and prevention. *Suicide and Life-Threatening Behavior, 20*, 123–130.

Khuri, R., & Akiskal, H. S. (1983). Suicide prevention: the necessity of treating contributory psychiatric disorders. *Psychiatric Clinics of North America, 6*, 193–207.

Lidberg, L., Asberg, M., & Sundquist-Stenmann, J. (1984). 5-Hydroxyindoleacetic acid levels in attempted suicides who have killed their children. *The Lancet*, October 20, 928.

Lidberg, L., Tuck, J. R., Asberg, M., et al. (1985). Homicide, suicide and CSF 5HIAA. *Acta Psychiatrica Scandinavica, 71*, 230–236.

Linnoila, M., Virkkunnen, M., Scheimin, M., et al. (1983). Low cerebrospinal fluid 5-hydroxyindoleacetic acid concentration differentiates impulsive from non-impulsive violent behavior. *Life Sciences, 33*, 2609–2614.

Mann, J. J., & Stanley, M. (1984). Postmortem monoamine oxidase enzymes kinetics in the frontal cortex of suicide victims and controls. *Acta Psychiatrica Scandinavica, 69*, 135–139.

Mann, J. J., Stanley, M., McBride, P. A., et al. (1986). Increased serotonin 2 and beta 1 receptor binding in the frontal cortex of suicide victims. *Archives of General Psychiatry, 43*, 954–959.

Martin, R. L., Cloninger, C. R., Guze, S. B., et al. (1985). Mortality in a follow-up of 500 psychiatric outpatients. I. Total mortality. *Archives of General Psychiatry, 42*, 47–54.

Martin, R. L., Cloninger, C. R., Guze, S. B., et al. (1985). Mortality in a follow-up of 500 psychiatric outpatients. II. Cause specific mortality. *Archives of General Psychiatry, 42*, 58–66.

McKenny, P. C., Tishler, C. C., Kelly, C. (1982). Adolescent suicide: a comparison of attempters and nonattempters in an emergency room population. *Clinical Pediatrics, 21*, 266–270.

Modai, I., Apter, A., Meltzer, H., et al. (1989). Serotonin uptake by platelets of suicidal and aggressive psychiatric inpatients. *Neuropsychobiology, 21*, 9–13.

Morrison, J. R. (1982). Suicide in a psychiatric practice population. *Journal of Clinical Psychiatry, 43*, 348–350.

Paul, S. M. (1986). Serotonin reuptake in platelets and human brain: clinical implications. Presented at the WPP Regional Symposium, Copenhagen.

Perez-Stable, E. J., Miranda, J., Munozrf, A., et al. (1990). Depression in medical outpatients: underrecognition and misdiagnosis. *Archives of Internal Medicine, 150*, 1083–1088.

Plath, S. (1963). *The bell jar*. London: Faber & Faber.

Pollack, D. A., Rhodes, P., Boyle, C. A., et al. (1990). Estimating the number of suicides among Vietnam veterans. *American Journal of Psychiatry, 147*, 772–776.

Price, L. H., Charney, D. S., Delgado, P. L., et al. (1990) Clinical data on the role of serotonin in the mechanism(s) of action of antidepressant drugs. *Journal of Clinical Psychiatry*, *51*(4, suppl.), 44–50.

Rehavi, M., Paul, S. M., Skolnick, P., et al. (1980). Demonstration of high affinity bonding sites for ^{3}H-imipramine in human brain. *Life Sciences*, *26*, 2273–2279.

Reich, P. (1989). Panic attacks and the risk of suicide. *New England Journal of Medicine*, *321*, 1260–1261.

Rich, C. L., Young, J. G., Fowler, R. C., et al. (1990). Guns and suicide: possible effects of some specific legislation. *American Journal of Psychiatry*, *147*, 342–346.

Roy, A. (1989). Suicide. In H. Kaplan & B. Sadock (Eds.), *Comprehensive textbook of psychiatry* (Vol. 5, 5th Ed.). Baltimore: Williams & Wilkins.

Roy, A., Lamparski, D., DeJong, J., et al. (1990). Characteristics of alcoholics who attempt suicide. *American Journal of Psychiatry*, *147*, 761–763.

Sakinofesky, I., Roberts, R. S., Brown, Y., et al. (1990). Problem resolution and repetition of parasuicide: a prospective study. *British Journal of Psychiatry*, *156*, 395–399.

Shaw, D. M., Camps, F. E., & Eccleston, E. G. (1967). 5-Hydroxytryptamine in the hindbrain of depressive suicides. *British Journal of Psychiatry*, *113*, 1407–1411.

Slaby, A. E., Lieb, J., & Tancredi, L. R. (1985). *The handbook of psychiatric emergencies* (3rd Ed.). New York: Elsevier.

Slaby, A. E., & Tancredi, L. R. (1991). *The handbook of psychiatric emergencies* (4th Ed.). New York: Elsevier.

Stanley, M., & Stanley, B. (1990). Postmortem evidence for serotonin's role in suicide. *Journal of Clinical Psychiatry*, *51*(4, suppl.), 22–28.

Stanley, M., & Mann, J. J. (1983). Increased serotonin-2 binding sites in frontal cortex of suicide victims. *The Lancet*, *i*, 214–216.

Stanley, M., Virgilio, J., & Gershon, S. (1982). Tritiated imipramine binding sites are decreased in the frontal cortex of suicides. *Science*, *216*, 1337–1339.

Teicher, M. H., Glod, C., & Cole, J. O. (1990). Emergence of intense suicidal preoccupation during fluxetine treatment. *American Journal of Psychiatry*, *147*, 207–210.

Tishler, C. L., McKenny, P. C., & Morgan, K. C. (1981). Adolescent suicide attempts: some significant factors. *Suicide and Life-Threatening Behavior*, *11*, 86–92.

Traskman, L., Asberg, M., Berilsson, L., et al. (1981). Monoamine metabolites in CSF and suicidal behavior. *Archives of General Psychiatry*, *38*, 631–636.

Traskman, L., Tybring, B., Asberg, M., et al. (1980). Cortisol in the CSF of depressed and suicidal patients. *Archives of General Psychiatry*, *37*, 761–767.

Traskman-Bendz, L., Asberg, M., Nordstrom, P., & Stanley, M. (1989). Biochemical aspects of suicidal behavior. *Progress in Neuropsychopharmacology and Biological Psychiatry*, *13*, 335–344.

Traskman-Bendz, L., Asberg, M., & Schalling, D. (1990). Serotonergic function and suicidal behavior in personality disorders. *Annals of the New York Academy of Sciences*, *487*, 168–174.

Tsuang, M. T. (1978). Suicide in schizophrenics, manic depressives, and surgical controls. *Archives of General Psychiatry*, *35*, 153–155.

Tsuang, M. T., Dempsey, G. M., & Fleming, J. A. (1979). Can ECT prevent premature death and suicide in schizoaffective patients? *Journal of Affective Disorders*, *1*, 167–171.

Van de Kar, L. D. (1990). Neuroendocrine aspects of the serotonergic hypothesis of depression. *Neuroscience and Biobehavioral Reviews*, *13*, 237–246.

Valzell, L. (1981). *Psychobiology of aggression and violence*. New York: Raven Press.

Van Praag, H. M. (1983). CSF 5-HIAA and suicide in nondepressed schizophrenics. *Lancet*, *2*, 977.

Van Praag, H. M., & Korf, J. (1988). Endogenous depressions with and without disturbances in the 5-hydroxytryptamine metabolism: a biochemical disturbance? *Psychopharmacologia, 19*, 148–152.

Weissman, M. M. (1974). The epidemiology of suicide attempts, 1960–1971. *Archives of General Psychiatry, 30*, 737–746.

Weissman, M. M., Klerman, G. L., Markowitz, J. S., et al. (1989). Suicidal ideation and suicide attempts in panic disorder and attacks. *New England Journal of Medicine, 321*, 1209–1214.

12

Decision to Hospitalize and Alternatives to Hospitalization

BETSY S. COMSTOCK

The decision to hospitalize or not to hospitalize often emerges as the first order of business when individuals present with acute suicidal ideation, threats of suicide, or recent suicide attempt. Three problems are associated with the decision to hospitalize. First, there never are enough hospital beds to accommodate all of the suicidal individuals encountered in treatment settings. Second, mental health professionals repeatedly have been demonstrated not to be capable of predicting future events; thus the prediction of suicide is fraught with errors. Third, hospitalization may not offer the most efficacious treatment for the suicidal individual and indeed may impose barriers to effective treatment.

The decision for hospitalization in the event of suicide risk presents any psychotherapist with a conflict of values, where the benefits of hospitalization must be weighed against the disadvantages of hospitalization plus the benefits of ambulatory care. To this equation must be added the disadvantages of ambulatory care, although generally these factors are so obvious as to remain unstated.

This chapter reviews what is known about acute suicide risk states and the related functions of hospitals, consideration of case volume and the problems related to it, the methodology of risk assessment, recommendations for minimal indications for hospitalization, and a description of alternative treatments available when hospitalization is not used.

Acute Suicide Risk

Clinicians faced with individuals presenting with acute suicide risk need to understand the likelihood that the individual will commit suicide in the near future. Unfortunately, most research on this question is based on long-term follow-up studies rather than on examination of short-term suicide rates among those admitted versus those not admitted for hospital care. Obviously, it is difficult to construct randomized treatment protocols for individuals at risk for suicide. One exception has appeared in the literature, reported by Waterhouse and Platt (1990). In Great Britain there existed until 1984 a requirement that patients seen because of overdose or "parasuicide" be referred for general hospital admission and psychiatric evaluation. Because such admissions became the most common cause for acute admission for hospitalization for women and the second most common reason for men (Farmer, 1987) this policy was re-evaluated during an interval when house officers were allowed to make a distinction between patients requiring either medical hospital care or psychiatric referral and those considered of lesser risk and potentially dischargeable. A random assignment of brief hospitalization, generally less than 24 hours, or outpatient referral was established. The follow-up at 1 week and 16 weeks revealed no differences between the two groups of parasuicides with respect to a large number of measures of psychological functioning as well as of repeat overdoses.

Hawton (1987) provided a cogent summary of problems associated with the assessment of suicide risk. As was pointed out also by Pokorny (1983), suicide is a rare event. Even if clinicians could achieve 80% accuracy there would exist an unacceptably high false-positive rate in assignment of high risk. Further-more, virtually all risk assessment and outcome studies have dealt with groups of individuals studied on the basis of various risk factors and have not focused on the identification of outcome for discrete individuals, obviously a far more challenging task. The third problem of risk evaluation, already referred to, relates to the problem that individuals identified as being at risk inevitably have received some treatment, with the possible exception of the low-risk group identified above in the Waterhouse and Platt study. Data on nontreated pop-ulations simply do not exist. In general, risk factor studies make the assumption that factors being evaluated are stable over time, whereas in fact many of the important factors—for example, social isolation, substance abuse, or situational stress—clearly are unstable.

Within the United States the rationale for decisions to hospitalize seems to vary considerably. Within the private treatment sector the prevailing attitude seems to match that present in Great Britain prior to 1984, i.e., that the optimal and safest treatment for individuals indicating suicide risk is accomplished within the hospital. Among private practitioners there probably exists a wide range of willingness or unwillingness to attempt ambulatory care for such individuals.

Within the public sector, as is emphasized in the following discussion, deci-sions are made on a more pragmatic basis: the availability or unavailability of acute care hospital beds. In either case the functions performed by hospitals for individuals presenting with suicide risk need to be examined. Three such functions can be identified. Psychiatric hospitals or psychiatric sections in gen-eral hospitals first and foremost provide therapy specific to individual psycho-pathology. Patients admitted to the hospital undergo extensive assessment pro-cedures with elaboration of individualized treatment—plans related to their diagnosis and intended to provide relief from symptoms specific to the diag-nosis. Included among those symptoms is the propensity to self-destructive behavior.

Second, the fact of hospitalization provides the hospitalized patient a time-delay mechanism irrespective of the type of treatment provided. Especially for those individuals suffering impulse disorders, this delay mechanism may prove life-saving. The impulse to self-destruction tends to change rapidly over time such that repeated measures of suicide intention change hour by hour in a hospital setting.

Third, psychiatric hospitalization provides protective custody. This factor is institutionalized in the law wherein hospitalization is deemed warranted for individuals at risk for harm to themselves by virtue of mental illness, without designation of any required hospital function other than protection from self-destructive impulses. It must be noted that the protective functions of hospitals are imperfect at best. One example may suffice. I saw in consultation in an intensive care unit a man with homicide charges pending. He had killed a police officer in a gun battle and had himself sustained serious gunshot wounds, which in fact he seemed able to survive. In addition to placement in intensive care beside the nursing station he was assigned a police officer guard 24 hours a day. Despite these precautions a friend was allowed a 5-minute visit during which a pistol was placed under the patient's pillow and subsequently became the instrument of his death by suicide. Neither constant one-on-one nursing observation nor the policeman at his elbow were sufficient to prevent this in-hospital death. Although the courts behave as if they believe the protective custody function of hospitals should be infallible, practitioners know that it is not.

Case Volume Considerations

Extrapolations from experience in Houston, Texas, suggest that there are approximately two suicide attempts per hospital bed per day in the Houston area. Although hospital admissions for a suicide attempter may have an average length of stay less than those for medical admissions, it is clear that were all suicide attempters admitted to the hospital the medical-care community promptly would be overwhelmed. Obviously, only a few suicide attempters reach medical attention, typically for medical evaluation in an emergency center. In Great Britain, where there existed in the past a mandate for admission of all para-suicides, the number of acute admissions accounted for 15% of total hospital admissions (Willcox, 1985). Even at this 15% level, suicide attempts easily could compromise available medical care for other problems. If those at suicide risk associated with suicide ideation and suicide threat are added to the ranks of those already having attempted suicide, the magnitude of the problem is greater. The stark conclusion is that no simplistic rule for admission in the event of identified suicide risk is realistic. Thus refinement of the judgment of suicide risk is essential.

Risk Assessment

Issues in risk assessment in relation to potential suicide are covered elsewhere in this book (see Chapters 2, 8 and 9). Several levels of data must be considered when assessing the degree of risk for a given individual presenting with suicide ideation, threat, or attempt. In rank order, these levels involve (1) an overall assessment of the degree of risk by medical health professionals headed by the psychiatrist, who may be the hospital attending psychiatrist; (2) the individual's own judgment of the level of intentionality and of subsequent short-term risk plus the professional judgment of the reliability of that individual's account; (3) the availability of needed support systems outside the hospital; (4) the individual's ability and willingness to accept help and personal motivation for change with respect to suicide risk; and (5) formal identified risk factors for suicide.

Mental health professionals called on to state the degree of risk of suicide in an individual case must make coherent statements of their judgment. For legal reasons the documentation of the judgment process is important. Not every patient at risk of suicide warrants hospitalization, but when a decision is made for alternative therapy the rationale for this treatment and the awareness of risk-benefit ratios must be stated clearly. The judgment about clinical risk must take into account the factors discussed in subsequent paragraphs but should go beyond it to a global assessment that is not an intuitive guess but involves integration of all the available levels of data that contribute to the final assessment, as suggested by Pokorny (1983), incorporating clinical experience with more formal factors. A similar position has been advocated, for example, by Buie and Maltsberger (1983), who stated "formal elements . . . are secondary . . . the urge to suicide is largely independent of the observable mental status."

In a similar vein, assessing the reliability of individuals reporting on suicidal inclinations is a matter of clinical judgment that goes beyond codified criteria. A major issue in suicide assessment involves the impulsivity of the individual, which best can be evaluated in terms of that individual's history of translating intrapsychic impulse into disturbed behavior. For individuals whose impulse control has seemed relatively good, their own statement of future risk may be elicited by such questions as "What do you think you are going to do?" "Do you think you are going to be safe over the next several weeks?" "In your judgment are you able to resist these impulses?" In cases where individuals are not able to give encouraging answers to these questions, the reliability of individual judgment seems substantially in doubt, or there is a history of impulsive behavior, the need for hospitalization is greater.

When alternatives to hospital care are considered, the available support systems for the individuals in their day-to-day social functioning reach paramount importance. Is the family prepared to accept the burden of relating with and caring for a family member at risk for suicide? Is the employer, the school, the neighborhood willing to rally behind this individual to provide needed support?

When external supports are to be relied on, the therapist must accept the obligation of informing and rallying the appropriate individuals. It requires, first, the consent of that individual for the outreach efforts. J. P. Soubrier (personal communication) has emphasized the utility of contacting support systems through interviews with family and conversations with employers, school personnel, and even with friends and neighbors to achieve an extended milieu within the community as a means to reduce the suicide risk. If the individual at risk is unwilling for these contacts to be made, the need for hospitalization may seem greater.

A number of personal factors, some of which are detailed below, contribute to risk assessment. Foremost among them is the degree of therapeutic alliance that can be established in the emergency presentation and the degree of acceptance of need for help expressed by the individual. When substantial denial exists, the degree of turbulence in the life situation is denied, and the seriousness of suicide ideation or threat or attempt is minimized, the need for hospitalization may be considered greater (Kiev, 1976). Beck, Steer, Kovacs, and Garrison (1985) emphasized that the sense of hopelessness, pervasive pessimism, and lack of future orientation are key issues during assessment of risk. Long-term follow-up has suggested that individuals with a high score on measurements of hopelessness are especially in danger for subsequent suicide, although admittedly it is over the long term and not during the period that might be covered by an acute hospital admission. The crucial question of how much these factors of denial and hopelessness influence the short-term outcome for suicide risk has not been well researched.

The bulk of suicide risk research has focused on what can be referred to as formal measures of risk. First among them are demographic considerations detailed during the 1960s by Tuckman and Youngman (1963, 1968). These authors reported two follow-up studies after a suicide attempt and enumerated 11 risk factors indicating the likelihood of subsequent suicide.

1. Age over 45 years
2. Male sex
3. Unemployment
4. Marital status of separation, divorce, or widowhood
5. Living alone
6. Physical health problems
7. Medical treatment within the prior 6 months
8. Diagnosed psychopathology including alcoholism
9. Attempt accomplished through violent means in contrast to overdose
10. Having left a suicide note
11. History of suicide attempts

Those individuals, for example, who scored on 10 or 11 of these factors were found to have a tenfold increase in suicide compared to all other suicide attempters.

Numerous clinical reports have dealt with the psychiatric diagnosis of patients who committed suicide. For the most part, diagnoses were reached through retrospective analysis based on interviews with close survivors of suicidal individuals and have obvious disadvantages over prospective studies. Because suicide is a rare event prospective studies are difficult to accomplish. One confounding issue in both retrospective and prospective studies has been the absence of appropriate attention to issues of co-morbidity, for example, attention to the interactions between Axis I and Axis II diagnoses.

Hawton (1987) provided a tabulation from three retrospective studies indicating that 48% of suicides had a primary diagnosis of affective disorder, 23% of alcoholism, 6% of schizophrenia, and 19% of other psychiatric disorders. The striking conclusion from this tabulation is that almost all individuals dying from suicide can be evaluated as having been psychiatrically ill.

Prospective studies of death due to suicide among various categories of psychiatric illnesses suggest a range of 5 to 15% lifetime incidence, in contrast to the general population rate, which is estimated at 0.5 to 1.0%. Those with a primary diagnosis of affective disorders were estimated (Guze & Robins, 1970) as having a 15% lifetime suicide rate. Among affective disorders, manic-depressive disorder generally is viewed as having the highest rate (i.e., closer to 15%) in contrast to other affective disorders (Retterstol, 1974). A similar lifetime suicide rate is attributed to individuals with a primary diagnosis of alcoholism (Miles, 1977), although the importance of other diagnoses accompanying alcoholism is not indicated. Murphy and Robins (1967) emphasized that alcoholism-related suicides pertain especially to those individuals who have life circumstances involving a recent loss, rather than co-morbidity of other diagnoses such as affective disorder, bipolar disorder, and schizophrenia. The lifetime incidence of suicide among patients diagnosed as schizophrenic varies from 5% (Hawton, 1987) to 10% (Miles, 1977), with specific attention given to relatively young males with relapsing illness, high IQ, and high personal expectations. Such suicides are most apt to occur during nonpsychotic intervals, when depressive symptoms can be noted. More recently, attention has been drawn to the increased risk of suicide in individuals with panic disorder as well as in those with qualities of affective disturbance or propensity to violence who demonstrate cerebrospinal fluid reduction in 5-hydroxyindoleacetic acid (5-HIAA), suggesting a correlation between reduced central nervous system serotonin and destructive action in general (Markowitz, Weissman, Oullette, Lish & Klerman, 1989).

Specific psychological factors already have been identified that may contribute to suicide risk. They involve four properties primarily: (1) denial of importance of suicide risk; (2) personal sense of hopelessness or pessimism or lack of future orientation; (3) a tendency toward impulsive behavior; and (4) agitation or anxiety at the level of panic. Elements of personal history also rank among formal risk factors, which include a history of suicide attempt, a family history of suicide attempt, significant medical illnesses, and situational stresses.

Primary Indications for Hospitalization

It is established that not all individuals at risk for suicidal behavior can be hospitalized. Therefore it may be helpful to enumerate those conditions and circumstances that can be considered baseline indications for hospitalization with the recognition that additional indications may be relevant from time to time. First among them must be the failure of response to crisis intervention efforts. In the situation of an emergency presentation with suicide risk, the therapist is obliged to institute crisis intervention technology as discussed below. When this technology fails, the immediate or short-term risk for a suicide must be seen as great; and hospitalization therefore is indicated. Crisis intervention techniques can be judged to have failed when either of two circumstances obtains. First, if it has not been possible to establish a reasonable treatment alliance with an individual at risk, crisis intervention has failed. Second, if despite a reasonable therapeutic alliance the individual continues to express overwhelming immediate intent to kill himself or herself, crisis intervention techniques have failed. There exist no standard guidelines about how long to persist with someone whose suicidal ideation is refractory to one's best efforts. One to two hours in session with an individual maintaining an immediate intent probably is sufficient to discourage most therapists, who then resort to hospitalization on a voluntary or an involuntary basis.

Perhaps the most reasonable basis for hospitalization for individuals expressing suicide intent is the diagnosis of a disorder that optimally is treated with hospitalization regardless of whether suicide risk exists. This decision is referable to the primary function of psychiatric units in hospitals or to psychiatric hospitals, that is, the treatment of acute psychopathology. Among these conditions certainly should be included major depression, acute manic exacerbations of manic-depressive disorders, and acute psychotic presentations (whether of a functional or organic etiology) that include delusions relating to death. With all these conditions hospital treatment provides the most intense and most efficacious approach and is indicated irrespective of the degree of suicide risk.

A third indication for hospitalization involves those patients with manic-depressive disorder more typically in the depressive phase but possibly with mixed-depressive and manic-depressive features who express direct suicidal ideation. It is precisely because this diagnostic group is at highest risk for suicide that hospital care should be invoked for its protective functions.

The fourth and final category where hospitalization is indicated is somewhat more complex and involves all those presentations where it is judged that the advantages of hospitalization outweigh the disadvantages and, conversely, the disadvantages of ambulatory care outweigh the advantages.

The *advantages of hospital care* include the following: (1) There is a relative degree of safety from suicidal behavior inherent in the close observation available with hospitalization. (2) The highest possible intensity of treatment is available on a 24-hour basis. (3) Treatment is provided with the lowest morbidity. An example is the use of electroconvulsive therapy, where the total duration of illness can be minimized. (4) There is the possibility of lowered mortality, although admittedly this advantage has not been well researched.

The *disadvantages of hospital care* have been given less attention and have not been well researched. Among them, however, are the following. (1) Removal of the individual from suicidogenic life circumstances may hinder treatment. When an individual is hospitalized, the immediacy of life stresses is minimized. As a result it is difficult to bring into day-to-day treatment the importance of these circumstances. The unfortunate result is that often important psychological work involving these factors is postponed until the individual is returned to more normal living circumstances. (2) Hospitalization inevitably affords a degree of gratification for the patient's dependency yearning. This gratification is especially important in individuals who experience a great deal of conflict between autonomy seeking and dependency seeking. When dependency conflict is paramount in the patient's psychopathology, its gratification removes from the conflict area, at least for the term of hospitalization, the very issues that are central to restoration of the individual to health. (3) A related issue is the degree of overall ego regression experienced by the hospitalized individual. All nursing personnel can attest to the fact that everyone regresses to a certain extent when hospitalized. The basic experience of having one's day-to-day needs attended to by hospital personnel induces regression but ordinarily not to a pathological extent. Some individuals have the capacity to observe and utilize this regression in a psychologically constructive way. Most individuals with severe psychopathology do not have this capacity, and the hospital experience can become an interference when dealing with self-destructive impulses rather than being an asset. (4) Hospital care givers may respond to self-destructive behavior in ways that reinforce it as an adaptive method during subsequent life stresses.

In contrast with the above, a number of *advantages of outpatient therapy* for individuals with suicide risk must be emphasized. (1) The decision for ambulatory care stresses the responsibility placed on the suicidal individual for self-determination. This decision stands in contrast to the dependency gratification and tendency to regression that are inevitable in a hospitalization experience. It weighs heavily on the side of restoration of the suicidal individual to responsible adult functioning. (2) Ambulatory care provides a situation that minimizes disruption in the personal life of the individual. Frequently, occupational and educational responsibilities can continue to be met while therapy is ongoing. Interpersonal contacts continue and can be brought immediately into the therapy situation in a way that either is not possible or is substantially distorted if the individual is hospitalized. (3) Ambulatory care minimizes the stigma attached to mental illness that often is amplified through psychiatric hospitalization. This stigma may extend through family dynamics into the neighborhood or into the educational or occupational workplace. A requirement in ambulatory care that an individual continue in his or her ordinary life circumstances ensures that situational issues are experienced on a daily basis and are brought immediately into the therapy, whereas with hospitalization these issues often are delayed or obscured. (4) Ambulatory care, when appropriately structured, mobilizes support systems that otherwise might not be utilized for helping the individual resist self-destructive impulses.

The *disadvantages of ambulatory care* for suicidal individuals were referred to previously as self-evident but are enumerated here. (1) Foremost among them is the increased burden on the family and other members of the support system for the individual. The anxiety occasioned within families when a suicidal individual is not hospitalized often is extreme. It is important to communicate clearly with family and other members of the support system the importance of their role and sympathy for the increased burden they bear. (2) For the therapist, there is the added burden of uncertainty about the welfare of the suicidal individual, which often is seen as the determining factor in a decision for hospitalization. Compassion for individuals who pose some risk of suicide may, however, determine the decision for ambulatory care and may include as a consequence the need for the therapist to tolerate this uncertainty. (3) The final disadvantage of ambulatory care is the increase in dangerousness when a person at risk for suicide is allowed all the freedom of ambulatory care with the very real possibility of self-harm or even of harm to others. When a therapist elects to accept this degree of dangerousness, it is important that it be recognized that the protective capacity of hospitals is limited and, at best, time-limited. Moreover, it must be accepted that the risk for ultimate death from suicide in general is a long-term rather than a short-term issue that sooner or later must be dealt with in an ambulatory setting.

Alternatives to Hospitalization

A number of alternatives to hospitalization exist for an individual at risk for suicide. They include initial crisis intervention efforts, partial hospitalization programs, outpatient individual office treatment, and group therapy.

Crisis intervention is the first order of business for any individual presenting a suicide risk. The intervention may be preliminary to the decision to hospitalize, or it may be the definitive management of a suicide crisis. Although there are a variety of theoretical orientations to crisis intervention, four specific interventions are discussed here: (1) cognitive problem-solving interventions; (2) dysphoria-relieving interventions; (3) self-psychology interventions; and (4) referral as crisis intervention.

The ultimate goal of any crisis technology is to convert a crisis presentation to a manageable situation that may or may not require longer-term psychotherapy. Although how one handles oneself on the front line may blur across theoretical constructs, an awareness of these constructs informs and strengthens the work to be done.

Most crisis intervention manuals rest on a theoretical structure of cognitive behavioral therapy. This approach is not the only one to crisis intervention, but over time it has been demonstrated to be a technology with considerable efficacy. During crisis it is assumed that one party, the patient or client, suffers a reduction in adaptive mechanisms, and another party, the intervener, has ego strengths to lend in a situation where problems of living have become overwhelming.

The basic paradigm for intervention involves, first, the establishment of a therapeutic alliance. If the individual in distress and the intervener are not "on the same side," progress cannot be expected; hence considerable time may have to be devoted to establishing a working alliance. Once it is obtained, interventions can progress through a process of problem identification and selection of problem resolutions.

It is useful to establish a decision tree whereby problems are enumerated and are rank-ordered in terms of the urgency for solution. Every problem must be seen as potentially responsive to a variety of solutions, and each solution can be weighed in terms of advantages and disadvantages. The weighing process makes a statement against all-or-nothing thinking patterns. A patient or client in distress tends to see problems in terms of either–or possibilities. The therapist's role is to lend ego resources to the weighing of possibilities with various advantages and disadvantages in order to guide the person in distress to decision-making that is positive in its outcome. Such technology can be applied to serial problems until a plan of action is obtained. It is essential for such therapies that decisions be reinforced not as the decisions of the therapist but as decisions of the individual and that follow-up be planned and implemented so actualization of problem solving can be ensured.

Most suicide-related crisis presentations are highly emotionally charged, although a few individuals contemplate death calmly (and these individuals are likely to make plans deliberately and effectively). Most suicide ideation, though, is associated with feelings of desperation. Ventilation of painful affect regularly is needed before a distressed individual can engage in a useful search for alternatives to suicide. The task for the therapist is to tolerate the expression of distress and at the same time establish a controlled environment so the individual does not feel overwhelmed by the awfulness of his or her own turmoil. Reassurances generally are counterproductive. A desperate person is not comforted by such expressions as "It will be all right" or "Don't cry. We'll work something out." Rather, the therapist must identify what is being felt, legitimize the emotion, and communicate tolerance of emotional expression. It is not difficult to determine whether a person feels better or worse through the course of such ventilation. If worse, this realization can be communicated: "It is natural to be angry but you can see that you feel more out of control when you rage about this way," or "You must feel there is no end to your tears, but they will stop and we will be able to talk then." Emotional ventilation is not the endpoint of an intervention, but it can be effective in relieving severe dysphoria.

A contrasting program for crisis intervention rests on the self-psychology theory originated by Heinz Kohut. It involves the concept that an individual undergoing overwhelming stress loses the sense of personal cohesion and is "falling apart." For individuals to recover from the sense of falling apart it becomes necessary for them to experience themselves as restored to a sense of personal integration.

In the interpersonal field this goal is accomplished through another important individual responding appropriately to perceived needs. Those needs may be expressed in one of two ways, which at least initially are highly polarized. The individual may be seeking an alliance with another, the therapist, in a mirroring form where the sense of personal worth is reflected back in interactions with the therapist. The therapist in simplistic terms is called on to mirror the basic worth of the individual in distress. The term "mirror" derives from the efficacy of reflecting back virtually what is offered by the individual in distress. For example, if someone says "I'm overwhelmed; I cannot cope," the therapist in turn may say "I understand that you are so overwhelmed by current stresses that you have difficulty coping." This response embodies a validation of the stress that is being experienced and reassures the individual that his or her experience is both understood and greeted empathically.

At the opposite pole the individual in distress may feel a need to ally with an individual of substantial power and expertise as a means of incorporating into his or her personhood that very power. In this instance the role of the therapist is to validate that, "Yes, we have met in an experience where I have a great deal to offer you, where I have a great deal of power and expertise and you are able to benefit from this alliance with me." In this situation, crisis intervention demands of the therapist a different stance, one of assurance and competence, compared to the former position of empathic mirroring.

The latter position may feel foreign to the therapist but may be the only position that can reduce the affective distress of the suicidal individual. This crisis response is more difficult for most therapists working within the self-psychology model because it activates defenses against the grandiose strivings of the therapist. The selection of an effective stance requires careful listening and identification of the needs of the patient or client.

Finally, many crisis intervention situations involve, first and foremost, the need for referral of a suicidal individual into channels for intervention by others. The therapist undertaking crisis intervention must assess the range of needs of the individual in crisis and must have access to all possible community resources. Generally, these resources do not reside in one individual or in one agency. Thus multiple referrals are required with all the demands inherent in any referral process. The referral process is a four-pronged one involving (1) anticipation on behalf of the individual of what will happen in connection with referral, (2) explanation of what is to be expected, (3) dealing in advance with resistances that will be encountered, and (4) following up to be sure that referrals are completed.

After crisis intervention a useful alternative to hospitalization is partial hospitalization, that is, utilization of day-hospital programs. There does not yet exist specific research on the benefit of partial hospitalization versus 24-hour hospitalization with respect to risk for committing suicide. There do exist substantial data about the efficacy of partial hospitalization for psychiatric treatment overall (Comstock, Kamilar, Thornby, Ramirez, & Kaplan, 1985).

Partial hospitalization offers most of the advantages of inpatient therapy in terms of thorough assessment and intensity of therapeutic endeavor without the need for 24-hour institutionalization. Thus many of the disadvantages of 24-hour hospital care can be avoided and the advantages of ambulatory care can be realized. Patients treated with partial hospitalization demonstrate advantages equivalent to those undergoing inpatient treatment. Their subsequent requirements for inpatient treatment are reduced relative to those experiencing inpatient treatment and their level of stigmatization in the community and in the workplace is markedly reduced.

Outpatient office treatment is the most commonly chosen alternative to hospitalization for individuals at risk for suicide. The advantages of this treatment have already been enumerated. The choice for office treatment involves a high level of responsibility for the therapist. A number of legal issues are invoked here: It is essential that the outpatient therapist make a documented statement regarding assessment of risk versus benefit of outpatient therapy. Statements such as "In my judgment outpatient therapy is warranted despite recognition of significant risk for suicide" are important. In addition, considerable reinforcement of the wisdom of the choice for outpatient therapy may be obtained through consultation with a fellow professional. This consultation, again, should address assessment of the risks and benefits to be derived for the individual. Outpatient therapy where significant short-term risk for suicide exists should involve mobilization of support systems that extend beyond the therapist. When the individual at risk is resistant to such mobilization of others, hospitalization should be reconsidered. Despite the risks, outpatient therapy often should be considered the preferred means of treatment because of the substantial advantages of ongoing ambulatory function and reduction of stigma for the individual at risk.

Group therapy for self-destructive individuals has been insufficiently emphasized among therapies dealing with such individuals. Homogeneous groups have been demonstrated to be efficacious for treatment of self-destructive individuals (Comstock & McDermott, 1975). Such groups are feasible only when they are associated with emergency centers that provide a consistent flow of referrals. Heterogeneous groups accepting self-destructive individuals probably also are efficacious but have not been well researched. The level of therapist stress involved when working with homogeneous groups of suicidal individuals is considerable. Co-therapy functions are crucial to the successful conduct of such groups. External supervision or consultation also provides an important outlet for reduction of the level of stress for the therapists. Open-ended groups probably are preferable to closed groups purely for logistic reasons. A variety of theoretical orientations to group therapy can be considered. With cognitive therapy, following the model of Beck, Rusk, Shaw, and Emery (1979), the reduction in affective distress is impressive. Similar results have been obtained during interpersonal therapy (Klerman, Weissman, Rounsaville, & Chevron, 1984). It is my impression that psychodynamic mixed supportive and exploratory therapy groups similarly are effective, although they have not been as well codified through published manuals. Such therapies endeavor to translate between the here and now phenomena of group interaction and the individual dynamics of group members to explicate underlying conflicts and to eliminate from problem resolution the archaic aspects of internal and interpersonal conflict.

Guidelines for Practice: Chapter Summary

The foregoing analysis of suicide risk assessment and the decision for hospitalization hopefully presents a basis for realistic weighing of alternatives. An important aspect of this analysis is the role of professional liability and judicial attitudes toward decisions about suicidality. The existing court precedents provide a backdrop against which all clinical decisions must be made. It is important for individual therapists to document their decision-making process in order to avoid liability in those circumstances where suicide may occur. Suicide should be seen as an unwanted but inevitable aspect of dealing with severely mentally ill individuals. The survivors (family and friends) of the tragedy of suicide deserve assurance that diligent care was provided and that full awareness of associated risks was present. The psychiatrist has specific responsibility with respect to the decision to hospitalize. Other mental health professionals function during the initial assessment before hospitalization and the evaluation during hospitalization. At whatever level the decision-making process is viewed, all mental health professionals participate in the responsibility for reasonable documentation of how each decision is reached and each treatment intervention is implemented.

REFERENCES

Beck, A. T., Rusk, A. J., Shaw, B. F., and Emery, G. (1979). *Cognitive therapy of depression.* New York: Guilford.
Beck, A. T., Steer, R. A., Kovacs, M., & Garrison, B. (1985). Hopelessness and eventual suicide: a 10-year prospective study of patients hospitalized with suicidal ideation. *American Journal of Psychiatry, 142,* 559–563.
Buie, D. H., & Maltsberger, J. T. (1983). *The practical formulation of suicide risk.* Cambridge, MA: Firefly Press.
Comstock, B. S., Kamilar, S. M., Thornby, J. I., Ramirez, J. P., & Kaplan, H. B. (1985). Crisis treatment in a day hospital: impact on medical care-seeking. *Psychiatric Clinics of North America, 8,* 483–500.
Comstock, B. S., & McDermott, M. (1975). Group therapy for patients who attempt suicide. *International Journal of Group Psychotherapy, 25,* 44–49.
Farmer, R. (1987). Hostility and deliberate self-poisoning: the role of depression. *British Journal of Psychiatry, 150,* 609–614.

Guze, S. B., & Robins, E. (1970). Suicide and primary affective disorders. *British Journal of Psychiatry, 117*, 437–38.

Hawton, K. (1987). Assessment of suicide risk. *British Journal of Psychiatry. 150*, 145–153.

Klerman, G. L., Weissman, M. M., Rounsaville, B. J., Chevron, E. S. (1984). *Interpersonal psychotherapy of depression*. New York: Basic Books.

Kiev, A. (1976). Cluster analysis profiles of suicide attempters. *American Journal of Psychiatry, 133*, 150–153.

Markowitz, J. S., Weissman, M. M., Ouellette, R., Lish, J. D., & Klerman, G. L. (1989). Quality of life in panic disorder. *Archives of General Psychiatry, 46*, 984–992.

Miles, C. P. (1977). Conditions predisposing to suicide: a review. *Journal of Nervous and Mental Disease, 164*, 231–246.

Murphy, G. E., Armstrong, J. W., Jr., Hermele, S. L., Fischer, J. R., & Clendenin, W. W. (1979). Suicide and alcoholism. *Archives of General Psychiatry, 36*, 65–69.

Murphy, G. E., & Robins, E. (1967). Social factors in suicide. *JAMA, 199*, 303–308.

Pokorny, A. D. (1983). Prediction of suicide in psychiatric patients: report of a prospective study. *Archives of General Psychiatry, 40*, 249–257.

Retterstol, N. (1974). The future fate of suicide attempters. *Suicide and Life-Threatening Behavior, 4*, 203–211.

Tuckman, J., & Youngman, W. F. (1963). Identifying suicide risk groups among attempted suicides. *Public Health Reports, 78*, 763–766.

Tuckman, J., & Youngman, W. F. (1968). Assessment of suicide risk in attempted suicide. In H. L. P. Resnik (Ed.), *Suicidal behavior* (pp. 190–197). Boston: Little, Brown.

Waterhouse, J., & Platt, S. (1990) General hospital admission in the management of parasuicide: a randomised controlled trial. *British Journal of Psychiatry, 156*, 236–242.

PART V
Postvention, Training, and Legal Issues

13

Following a Suicide: Postvention

EDWARD J. DUNNE

The prevention of suicide is an endeavor that justifiably absorbs a great deal of the energy and time of mental health professionals. We view suicide and suicidal behaviors as negative situations, and we seek to eliminate them or at least substantially reduce their prevalence in our society. Deaths by suicide are almost always seen as preventable and therefore unnecessary. As professionals we study the demographics of suicide and the dynamics of suicidal behavior in an effort to learn better ways of predicting just who will be suicidal under what circumstances. Determining a patient's potential suicidal risk consumes a great deal of our professional energy. Unfortunately, these efforts are not always successful. In the United States alone, nearly 30,000 people annually complete suicide, despite our sincere efforts at prevention.

Every completed suicide represents the end of the psychological crisis for the victim. Most often, however, it is the beginning of a new crisis for those who were associated with them. Shneidman (in McIntosh, 1989) has estimated that each suicide directly and immediately affects an average of six people. In the United States alone, this estimate translates into at least 360,000 affected individuals on an annual basis. Because these people are likely to remain in the population for many years after such an event, conservative estimates suggest that at any one time there are several million people in the United States who have experienced the suicide of a relative or someone close to them (Andress & Corey, 1978). In recent years the term "survivor" has been used to describe these individuals. The term first gained acceptance during the 1970s through the developing self-help movement for relatives and friends of suicide victims. It is inaccurate, however, to assume that the term carries the same meaning as a "survivor" of some disaster, such as a plane crash or automobile accident. In the latter cases, the threat to well-being or safety diminishes rapidly after the event, whereas survivors of suicides become increasingly at risk for physical and psychological difficulties, even long after the event.

It is only recently that the mental health profession has begun looking at survivors of suicide in a systematic way. Farberow's (1972) comprehensive bibliography of all professional literature on suicide listed only eight instances of direct studies of survivors among the total of all work on suicide published

between 1896 and 1971. Although the proportion of works dealing with survivors relative to all studies of suicide is increasing, it remains only a tiny fraction
of all the published work on suicide itself. To date, for instance, there are fewer
than five professional books on the subject.

Those few professional studies that do exist suggest that the experience of
losing a loved one to suicide is among the greatest burdens individuals and
families may endure. The psychological and social ramifications can be profound. Unresolved and morbid grief, physical illness, depression, anxiety, familial and social disorganization, and interpersonal disturbances have been reported (Hauser, 1987). These sequelae arise from a variety of causes that are
specific to suicide death: the usually sudden and unexpected nature of the loss,
the fact that the death is frequently violent, the trauma associated with discovering the victim, the inexorable feelings of guilt and anger that surround the
death, the fact that these deaths frequently occur in systems already experiencing considerable amounts of stress, the compromising of mourning rituals,
the loss of accustomed social support, and the experience of real or imagined
stigma associated with suicide and mental illness. Most distressing is the evidence (Roy, 1986) of an increase in suicide risk for survivors themselves.

Individuals with a potential for an assortment of physical, psychological and
social problems as well as an increased danger of taking their own lives seem
to constitute a group for whom there is natural concern by the mental health
profession. Conversely, it could be expected that these survivors would turn
to mental health experts when seeking professional help. Regrettably, both
assumptions have proved false. The dearth of professional literature on survivors amply attests to the lack of clinical and scientific interest this group has
generated in the mental health community. In addition, professionals often
fear lawsuits and so may be reluctant to reach out to the family survivors of
their clients or patients. Correspondingly, survivors themselves frequently attest
to their avoidance of the professional community because of fears they will be
further stigmatized and blamed. Their attitudes toward professionals are often
colored by the fact that the person they are grieving was undergoing psychiatric
or psychological treatment at the time the death occurred.

Obviously, these issues are complex and deserving of serious, thorough evaluation by the professional community. It is to be hoped that the profession
itself will begin to address these issues more conscientiously in the near future.
This chapter concentrates on providing fundamental and essential information
and guidelines for practitioners responding to survivors after a suicide as a
step in the process of encouraging greater responsiveness on the part of individual professionals.

Who Are Survivors of Suicide?

In general, a survivor of suicide is any individual who was in relatively close
association with a person who ended his or her own life. Demographically, it

could be anyone, regardless of age, sex, race, socioeconomic, or marital status. This simple definition requires immediate clarification, however, because there are a great number of individuals who consider themselves "survivors" or who would be so considered by others, even though their association with the deceased individual would not be characterized as close. For instance, when celebrities such as movie stars or politicians kill themselves, the general public could be considered a "survivor" in a broad sense; and a small number of members of the public may be considered survivors in a more narrow sense because they are for some reason uniquely and profoundly affected by the death. Usually, however, the term "survivor" is reserved for individuals who have had a personal relationship with the deceased, although the apparent closeness of that relationship may not always be a good predictor of the severity of the individual's reaction. Some families' reaction to a suicide is such that even children who are unborn at the time of the event are affected by it, particularly if the family tries to keep the facts about the death a secret from children, or they become overly cautious and fearful during their subsequent child-rearing practices.

The greatest number of individuals who are survivors are members of the deceased's immediate network of family and friends and associates. For the purposes of this chapter, survivors are generally referred to as "family" unless clarity requires greater specificity, although the reader should keep in mind that the term "family" is used in a rather imprecise manner.

To some extent, survivor status is self-imposed by the individual, although even this differentiation is unreliable, as some people who are greatly affected by a suicide do not know the label "survivor" or misidentify their reactions as common to other grief experiences; moreover, those who are not personally distressed may be perceived and treated as survivors by others. The problem of definition has been addressed by McIntosh (1987) but as yet has no satisfactory resolution, complicating both research efforts and clinical interventions. Thus the clinician must be alert to all the possibilities in each case. Although such awareness appears to be a reasonable element of good clinical practice, it is not as standard as one might assume. During most intake interviews it is still rare for the clinician to ask directly about suicide deaths in the client's history. This oversight is made even more glaring by the significant relation between suicidal ideation, suicide, and prior status as a survivor.

What Is Special About This Experience?

The differences between grief experiences when the deceased ended his or her own life relative to the experiences of grief following a death from any other cause have only just begun to be studied (Barrett & Scott, 1990; Demi, 1984; McNiel, Hatcher, & Reubin, 1988). Even if these experiences turn out not to be statistically differentiable, the social and psychological experiences are not equatable in any simple way, and clinicians must be able to respond to the specifics of this situation with knowledge and compassion.

When a suicide occurs, the survivors generally find themselves on a path different from what they expect with respect to response from the community at large. As Andress and Corey (1978) suggested, suicides frequently occur at or near home and are discovered by a family member. They are always a police matter. Depending on the circumstances, the police presence is more or less prominent, which is consonant with our usual expectations that as "first responders" to accidents involving injury or death the police are on the scene and generally supportive. In the instance of a suicide, however, one role of the police is to rule out other possible causes of death, including homicide, and so the police take a decidedly more antagonistic position with the surviving family. This occurs at precisely the time when the family expects extra help from the police. In some instances family members may be placed under direct suspicion, particularly as a result of actions they took during the course of discovering the death (i.e., cutting the body down, handling the gun, holding suicide notes). Thus almost from the first moments of discovery, the family's experience of events surrounding the death begins to diverge from what they had always believed happens (may even have experienced before) when a death occurs. This divergence continues with time as more and more people and institutions come into contact with the family and behave in unexpected and uncharacteristic ways. Neighbors, perhaps put off by the police activity, do not rush to offer help. Relatives are not called immediately and do not gather quickly at the home. News reporters, anxious to get an "angle" on the death, ask insensitive and often accusatory questions. Contacts with clergy are strained and may result in a distortion, or denial, of funeral rites as a consequence of negative religious attitudes about suicide. The funeral director may counsel a small "private service" to allow the family to avoid the pain of having to deal with the morbidly curious. The family, seeking to learn more of the deceased's last few hours, may call the therapist, who, fearing recriminations or lawsuits refuses to talk with them or is guarded and aloof, sometimes openly blaming the family.

These events occur even while the family is attempting to deal with its own flood of emotions, which range from shock to grief to anger to despair. Members within the same family may react differently, causing further anguish. Some family members refuse to accept the official verdict of suicide and seek to "prove" some other cause. Some may want to publicly acknowledge the cause of death, whereas others, fearing stigma and stereotyping, press to keep the cause a secret. Some family members deliberately exclude others (typically children and grandparents) from full information and construct elaborate stories to explain the deaths as "natural." Some believe it is incorrect to grieve openly for a suicide, and others are unable to hide their anguish.

Because suicide almost always occurs within the context of a family, family members may blame each other for the death, compromising the support they can mutually give and receive. In many families the divergence of experience from expectation continues for many years. Under these circumstances social networks become smaller and more interdependent. Marital and intergenerational difficulties blossom. Families may feel increasingly stigmatized (Rudestam, 1987) and become more and more isolated from mechanisms of social support. As they seek to defend themselves from the blame they feel others place on them, they close themselves off from opportunities to work through their anguish. This cycle may continue uninterrupted for many years and in some instances is passed on from generation to generation as a conspiracy of silence and shame.

Individual and Family-Based Postvention

Shneidman (1973) described a series of interventions with a family following a suicide as a "postvention," thereby giving emphasis to the prevention aspects as well. Although postvention now includes a wide variety of techniques, the term remains appropriate for dealing with survivors after a suicide death.

In the past, survivors did not often seek out the services of a mental health practitioner following suicide, perhaps because they thought they would be blamed and stigmatized, or because they had lost confidence in the efficacy of the profession. Clinicians have become increasingly knowledgeable about working with this particular group. A greater understanding of the survivor experience as well as a diminution in the stigma attached to suicide deaths has started to change this situation.

Clinicians working with survivors find that survivors do not always seek outside help immediately. Although some do seek assistance immediately after the death, others do so only after many months or years of living with a host of unpleasant conditions and feelings derived from this experience. Having exhausted most of the other resources around them (family members, physician, clergy, friends) they finally turn to the mental health professional. A third subgroup of survivors are the individuals and families who are unaware of their survivor status or who have apparently "adjusted" to the death but find they have other issues with which to deal. Frequently, the unfinished nature of their grief or their unconscious fear of suicide becomes more evident during treatment and can finally be worked on in the open. No matter how long it takes them to come to treatment, however, survivors bring a unique constellation of symptoms, conditions, and defenses that must be addressed. This unique presentation poses challenges for clinicians who most often must overcome their own personal and professional biases to deal with these issues.

As a starting point, it is useful to conceptualize the task of the clinician in this situation. Resnik (1972), theorizing about the nature of the work in responding to the needs of survivors of a death by suicide, saw it as requiring both a reconstruction of the events and a therapeutic experience he called "psychological resynthesis." The therapeutic quality of psychological resynthesis stems from its combination of: (1) psychological first aid, offered at the time of the death; (2) psychological rehabilitation, offered within the first few weeks following the death; and (3) psychological renewal, offered approximately 6 months after the loss. Each intervention strategy is timed to coincide with the stages of grief described by Bowlby and Parkes (1970): shock, disorganization, and reorganization. For many survivors, however, grief is not as orderly as Resnik and Bowlby and Parkes suggested. These individuals often do not progress through these stages within the time frames prescribed. Frequently, stages persist for long periods. Many survivors find themselves returning to earlier stages long after subsequent stages have been entered. The clinician's first responsibility therefore is to carefully assess the survivor's current status with respect to resolution of grief and to attempt to focus the interventions to coincide with the actual condition of the survivors.

It is rare for mental health professionals to be among the "first responders" to a suicide. It is regrettable, as effective postvention within the first few hours after the death can serve to keep the survivors on a path that can lead them eventually to recovery and resolution. The first 48 hours after a suicide represent a "dangerous intersection" for survivors. Clinicians with the opportunity to work immediately with survivors should try to help the family remain open to the community and in touch with most systems of support. Families should be encouraged to treat the death much like any other death in the family, usually by participating in mourning rituals that allow full expression of grief. The availability of a knowledgeable, nonblaming clinician at this point in the family's experience can serve to reduce the subsequent potential for difficulty.

In most situations the clinician sees survivors or families who are referred some time after the death. Some of these survivors require no more than time-limited grief or crisis counseling, which is most likely the case for survivors whose prior functioning showed resilience in the face of setbacks and a sort of healthy flexibility in handling most of life's vicissitudes. The death of a significant other by suicide is a stressor of unparalleled magnitude in most people's lives, and even the most psychologically mature individual may encounter difficulty responding to it. Such individuals are often greatly helped as well by attending a self-help group composed of other survivors.

Others may find the experience overwhelming and a threat to whatever personal integration they have achieved. More intense intervention is appropriate in this case. The familial context of a suicide is usually difficult to ignore. Restoring the family to its role as immediate nurturer and support for its members is of prime strategic significance for heading off some of the more distressing aspects of this experience, such as scapegoating family members, cutoffs, and the like. This work may involve active problem-solving with the family to deal with such issues as who, what, and how to tell other members of the network or what to do about missed school days. Many cases challenge the skill of the clinician. Suicide destroys the original fabric of the family, forcing reintegration of the survivors. The pace at which individual family members are ready and able to do it varies, sometimes making individually aimed interventions a necessity. Placing pressure on those family members who are not yet prepared to deal with this highly charged issue probably does more harm than good. The clinician may need to work through one client to reach other family members in an indirect manner.

It is not possible to present a typical symptom picture of survivors as they enter treatment. The manifestations of the distress they evidence are varied. Depression is almost always a fairly standard feature, with symptoms such as appetite loss, tearfulness, sleep disturbances, and feelings of hopelessness and despair. Some individuals report extreme agitation, incorporated into searches for the causes of the suicide or for the "culprit" or manifested as a chronic sense of irritation with others in their social network. The behavioral expressions of grief are common to most individuals but may differ in intensity or duration (and are sometimes more notable by their absence). In addition, disturbing dreams and nightmares may be reported by both adults and children. Alternatively, blocking of dreams may be present. The client may experience disturbing visual recall of the events of the death, particularly if the survivor was a witness to the event or saw the deceased's body before removal by the coroner or funeral director. Obsessive thoughts centering around guilt or anger or even the last conversations with the deceased are frequently reported. Invasive and morbid images may also be described. Young children may act out the suicide in play. Parents may report a loss of developmental accomplishments in their children, poor performance in school, and school phobias.

A variety of minor physical symptoms such as weight loss or gain, tachycardia, arthritis, migraine, allergies, asthma, and tics may also be described. They may either be preexisting conditions exacerbated by the loss or appear within the first 6 months afterward.

The psychological lives these individuals and families report may seem dominated by a number of themes directly related to their status as survivors: the obsessive search for the "why" of the suicide, the legacy of guilt, the sense of stigmatization, an incomplete or unusual grieving pattern, the invasion of conscious thought by the idea of suicide as an acceptable solution, a pervasive sense of helplessness, chronic low self-esteem, reduction in the size and complexity of the social network, and finally and most troubling for the clinician an erosion of basic trust (Dunne, 1987b). Some of the families are in conflict over issues relating to scapegoating either other family members or outsiders. Families may be divided on questions about proper mourning behavior, or they may present a conflict about burial rites. The intensity of these experiences is different for each individual and family and may depend on such factors as the length of time since the death, the extent of support in the social network, the sturdiness of other social relationships, the nature of the relationship between the survivor and the deceased, and age, gender, and other sociocultural phenomena. These themes do not constitute the entire psychological landscape for these individuals and families, and no one evidences all. However, they do appear to be characteristic of many survivors and pose special challenges to the clinician.

When working with survivors whose loss is fairly recent, it can be helpful to take an educative approach (Anderson, Reiss, & Hogarty, 1986), particularly at the beginning, regardless of whether the work is with a family or an individual. As a basic principle, the clinician should assume the least pathology in the family, explaining that whatever dysfunction is evident is a consequence of the experience, rather than a result of underlying morbid psychological processes. It is important to take this powerful therapeutic stance with survivors, as it helps to diminish the excessive guilt that often plagues them. Furthermore, this approach can serve to reduce the fear of being or becoming mentally ill, a fairly common concern in the families of completed suicides, particularly those families where the deceased was manifestly psychiatrically disturbed prior to death.

The clinician informs the family of what is currently known about suicide and survivors, which should include a mention of the social history of suicide as it relates to the survivor's experience, as well as an indication of some of the sequelae of the experience of another's suicide that have been identified by clinicians working in this field. Care must be taken not to predict responses or to suggest that the survivors *should* be feeling or behaving in any way. A wide variety of responses are simply affirmed as "to be expected" in the present circumstance. It can be helpful to describe the generally poor public and professional understanding of the survivor experience, as it allows the clinician to explain any unfamiliarity with the phenomenon the survivors themselves might experience in the context of widespread ignorance.

The initial task of treatment is characterized to the survivors as one of re-establishing the balance between grief and functioning that is threatened by the overwhelming nature of deaths of this sort. A contract to work to resolve some of the issues raised by this ordeal is offered. It is important for the clinician to be forthright about disclosing his or her own experiences with suicide, both professionally and personally. Doing so, particularly when the clinician is a survivor, results in a sense of connection that greatly enhances the helping relationship. Families and individuals should be encouraged to ask questions as a means of allowing them the opportunity to express their immediate and long-term concerns. Most often these questions deal with the causes of suicide and some of the myths associated with it (e.g., there are *always* warning signs). The questions should be responded to in a frank, honest manner with the aim of reducing anxiety.

Once the essential information is given to the family, the clinician should carefully inquire into its current circumstances. The goal of this inquiry is threefold: to determine (1) the presence or absence of factors that either enhance or retard the resolution of grief; (2) whether any family members are experiencing suicidal thoughts or behaviors; and (3) the extent and nature of the social network. Regarding the resolution of grief, the clinician specifically needs to be informed about such things as the way the funeral service was conducted (if indeed there was one), the extent to which the larger familial and social network participated in this activity, and any other indications that the death was responded to in a manner different from the way death is ordinarily handled by the family. In some instances survivors, either acting on their own or following advice, dramatically alter the formal mourning rituals that have served to help them grieve in the past. Depending on whether it has taken place, the clinician may give specific instructions on the restoration of the ritual aspects of mourning consonant with the family's usual practice.

The second concern—evidence of suicidal thoughts or actions—is also directly approached. The survivors are helped to understand that such ideas tend to come to many people in their situation during this time, but they are rarely acted on and diminish with time. Anxiety about this experience is often intense, and some extra precautions may need to be established until everyone feels reasonably secure. It is useful to talk openly and frankly about this phenomenon, particularly to adolescents; who may be especially concerned about their ability to resist the idea of suicide. Being fearful that someone else may commit suicide is also addressed. It is helpful for families to acknowledge everyone's need to feel free from this worry. The clinician may need to help the family construct a system that serves to assure all the members of the family of each other's safety. When working with an individual who expresses this fear, it can be beneficial to encourage them to engage in an open discussion with the individual(s) involved, seeking a temporary behavior change that serves to reassure them.

The final avenue of inquiry concerns the extent to which the social network of the survivors has been affected by the suicide through withdrawal or isolation. Casual remarks offered in the course of gathering this information, such as: "We haven't heard from Aunt Ruth since the funeral" often yield important information about diminished support systems that must be rebuilt. Some individuals and families require coaching to help reapproach relatives and friends who are currently unavailable to them. This coaching can be done during the session, after an introduction that establishes this "pulling back" as a fairly routine occurrence following a suicide.

The remaining issues presented by survivors can usually be handled in a straightforward manner. The clinician should expect to encounter such responses as an exaggerated sense of responsibility, interrupted family life, marital discord, families and individuals stuck on certain issues surrounding the death, a sense of helplessness that accompanies adoption of a stigmatized identity, and anger, despair, and guilt. Techniques that avoid placing responsibility or blame on any individual and that focus on the consequences of behaviors rather than on the "underlying issue" are most useful at this time.

Survivors whose loss is at some distance from the time of initial contact with the clinician may present what appears to be more standard treatment issues, such as chronic depression, alienation, and a fear of intimacy. In this instance it may be necessary for the clinician to conduct a search into the client's experience following the suicide to determine what (if any) unfinished business remains. Some survivors minimize the suicide as "having happened years ago" and therefore having no current significance. Others appear involved with it even several years later. To some extent, however, most survivors are involved with the central themes of this concern such as unresolved grief, a search for the "why," the erosion of basic trust, anger and scapegoating, survivor guilt, and thoughts of suicide as an acceptable solution to problems. These areas may be interwoven with the fabric of ordinary life difficulties, or they may stand out in relief as the central psychological experience of the individual. The clinician may find the treatment dialogue returning again and again to these topics, even as progress is made in other areas.

It is not yet known what factors contribute to a better outcome for some individuals and families than others. Research into this area is only just beginning (McIntosh, 1987). The following elements, however, are presumed to be implicated in outcome: being a direct witness or discoverer of the suicide; the age of the victim and the age of the survivor; the nature of the relationship between survivor and victim; the method of the suicide; the social context of the suicide; the clarity of a psychiatric diagnosis for the deceased; the prior health of the survivor; the degree of stigma and burden experienced; the extent, intensity, and resilience of the social network and the degree to which the survivor is able to rely on regular sources of support such as the church, the mental health profession, and the extended family.

Group-Based Postvention: Self-Help Groups

One of the earliest forms of organized assistance offered to survivors came through the self-help movement during the early 1960s. Currently there are more than 300 mutual-aid or self-help groups within the United States that offer assistance specifically to survivors of a suicide; some of the groups are led by professionals in mental health settings. By bringing survivors into contact with other survivors, the groups can have a powerful impact on recovery by reducing the stigma often associated with these deaths. They also serve practical needs as well, such as providing information and offering models for recovery. These groups can serve as a valuable adjunct to more formal treatment, and clients should be encouraged to participate in them whenever feasible.

School-Based Postvention Programs

Suicides that occur in the context of schools severely burden the organization and threaten disruption. They require interventions that are both sensitive and professional, conducted by experts who are able to work with faculty, students, and parent groups (Lamb & Dunne-Maxim, 1987). Interventions such as closing the school so everyone could attend the funeral, holding large assemblies or special meetings, or constructing permanent memorials to the deceased within the school grounds have yielded to a more cautious approach that seeks to avoid glamorizing the death, which might provoke copying. We are only just beginning to examine the special needs of children who are survivors—needs at the immediate time of the death and during the ensuing years. Although children may manifest many of the same symptoms as adult survivors, their reactions are just as likely to be idiosyncratic. Recovery is complicated by the fact that the adults in their lives are usually grieving as well and are therefore less able to be as responsive as necessary. Any intervention with children, either at home or in the school, must be carefully thought out and conducted in a way to avoid dramatizing the event or leaving the children with a sense of hopelessness.

Special Considerations for the Clinician

The clinicians' responsibility to be educated about survivors is severely hampered by the lack of professional literature dealing with survivors in general and with therapeutic approaches to them. Some nonpublished sources, such as conferences, meetings, and seminars, are available but on a limited basis. The clinician should become familiar with professional organizations, such as the American Association of Suicidology and the International Association for the Prevention of Suicide, that routinely include survivor issues on the agendas of their annual meetings. Attendance at survivor group meetings as a trainee, whether conducted by a professional or run as a mutual aid group, provides an opportunity to meet survivors and become familiar with their issues.

Beyond the question of information deficiency, the clinician should also be alert to countertransference or attitudes that may contaminate the postvention. The mental health profession has had a long history of blaming families when a suicide occurs. This tradition is as old as, and similar to, the one that characterized the profession's treatment of families with a relative with schizophrenia for most of this century. Blaming the family in the case of suicide, however, is a centuries-old tradition of the nonprofessional community and therefore is more difficult to eradicate. The clinician accordingly must be especially careful to avoid this tendency. Work with survivors brings to the fore a number of disquieting questions about one's own mortality and the limits of one's professional capacity to keep someone from suicide. These concerns permeate the work with survivors as they do in few other therapeutic endeavors and can easily overwhelm the clinician, leading to feelings of hopelessness and acts of desperation. For this reason, the clinician should be sure to have backup help available and must learn to recognize his or her own personal limits.

It is common for the clinician to become frustrated and confused during the course of the treatment with the survivor because of the reappearance of issues such as the search for reasons for the suicide or because of the return of deep grief long after the clinician believed these issues had been mastered. It appears to be characteristic of survivors to suffer these returns to earlier stages in their grieving even as they are getting progressively better. The clinician must make frequent adjustments in the treatment to accommodate these fluctuations, which are probably more indicative of the lack of social support for grief resolution for survivors than of severe psychopathology.

It is important to state clearly some of the things that should be avoided when working with survivors. It has been suggested elsewhere (Dunne, 1987a) that the clinician who has been treating an individual who completes suicide should not treat additional family members. To do so ignores the clinician's own status as a survivor and enhances the risk of encouraging the development of countertransference phenomena based on either blaming of the family member or overly identifying with them and their loss.

As a survivor, the clinician must recognize that he or she is susceptible to the impulse to place blame on someone in order to avoid blaming themselves. It would be an extraordinary (and somewhat overly detached) clinician who did not experience some feelings of responsibility in the suicide death of a client. Because the mental health profession in general both blames and stigmatizes clinicians who lose clients to suicide, the clinician is motivated to shift responsibility to the new client or other family members. It can be overt and dramatic, as when the clinician encourages "working through the guilt" the client does not feel, or it can be much more subtle. Either way serves to further burden the family member.

Clinicians also must acknowledge their own feelings of loss and grief when a client dies. This feeling is likely to be aroused by the presence of the deceased's family members and thereby limit the clinician's freedom in responding to them. The clinician may also, consciously or unconsciously, attempt to avoid labeling efforts with the deceased patient as a failure and seek to "fix" things with this new family member. Further complicating the issue, the family member may seek out the therapist as part of their own search for the "why" of the suicide and therefore approach treatment with this hidden and potentially destructive agenda. Finally, the clinician is often suspicious of the family member's motives and may take a defensive position, thereby again limiting freedom of action. All of these factors point to the inadvisability of the surviving clinician serving as the therapist for any of the deceased client's family.

Clinicians ought likewise to excuse themselves from the treatment of survivors if they hold views about suicide that overly simplify the issue. Thus the notion that "everyone has a right to decide their own fate" often rings false with survivors, most of whom would have gladly sacrificed, at least momentarily, their relative's "rights" for their life. Conversely, the naive belief that "all suicides are preventable" serves to blame the family further because this particular suicide was not prevented.

The themes and interventions presented here are not applicable in all cases, as each survivor and surviving family present unique features that require special skill on the part of the clinician. Yet, in broad outline, these observations serve as a foundation for an approach to the treatment of survivors in formal postvention situations. The challenge for the clinician is twofold: (1) to examine his or her own perspective about suicide and survivors as a means of eliminating any vestiges of attitudes and beliefs that could seriously undermine the therapeutic work; and (2) to become aware of the need survivors have for competent yet compassionate mental health care.

Guidelines for Practice: Chapter Summary

Although the nearly 30,000 annual suicides in the United States directly affect more than 360,000 persons, we are only now beginning to examine the survivors' needs systematically. Our research suggests that many fall victim to a variety of mental and physical health problems during the years to come. Many are at greater risk of ending their own lives following the example of their loved one. Widespread ignorance about suicide and its aftermath has complicated the treatment of survivors. The mental health profession has only recently taken an active look at the needs of survivors, and survivors are only beginning to turn to the profession for assistance. The following guidelines for practice summarize the material presented in this chapter.

1. Presenting symptom pictures vary with the individual and the circumstances, although they most frequently involve depression, helplessness, and thoughts of suicide.

2. Interventions following a suicide ("postvention") should be timed to coincide with the stage of grief being experienced by the survivor.

3. Early intervention increases the likelihood of a good outcome.

4. The clinician should be well informed about the particular aspects of survivor grief that differentiate it from other losses.

5. Family postvention can be an efficient way to address the special issues of survivors.

6. An information-sharing, nonjudgmental approach serves to reduce the stigma associated with survivors of suicide.

7. Mutual-aid or self-help groups specifically for survivors of suicide can be a useful adjunct to more formal therapeutic approaches and are often all that is required.

8. Child survivors have special needs that are frequently unmet by the grieving adults in the family.

9. School-based postvention programs must be carefully planned and carried out by trained professionals.

10. The general lack of professional literature in this area, as well as the relatively high potential for the emergence of countertransference phenomena, suggest the importance of proper supervision for the clinician.

11. Clinicians who survive the loss of a client should refer the surviving family members to colleagues.

REFERENCES

Anderson, C. M., Reiss, D. J., & Hogarty, G. E. (1986). *Schizophrenia and the family.* New York: Guilford.

Andress, V. R., & Corey, D. M. (1978). Survivor-victims: who discovers or witnesses suicide. *Psychological Reports, 42,* 759–764.

Barrett, T. W., & Scott, T. B. (1990). Suicide bereavement and recovery patterns compared with nonsuicide bereavement patterns. *Suicide and Life-Threatening Behavior, 20*(1), 1.

Bowlby, J., & Parkes, C. M. (1970). Separation and loss within the family. In E. J. Anthony & C. Koupernick (Eds.), *The child in his family*, (Vol. 1; pp. 197–216). New York: Wiley Interscience.

Demi, A. S. (1984). Social adjustment of widows after a sudden death: suicide and non-suicide survivors compared. *Death Education, 8*(suppl.), 91–111.

Dunne, E. J. (1987a). A response to suicide in the mental health setting. In E. J. Dunne, J. L. McIntosh, & K. Dunne-Maxim (Eds.), *Suicide and its aftermath: understanding and counselling the survivors* (pp. 182–190). New York: Norton.

Dunne, E. J. (1987b). Special needs of suicide survivors in therapy. In E. J. Dunne, J. L. McIntosh, & K. Dunne-Maxim (Eds.), *Suicide and its aftermath: understanding and counselling the survivors* (pp. 193–207). New York: Norton.

Farberow, N. L. (1972). *Bibliography on suicide and suicide prevention, 1897–1957, 1958–1970.* DHEW Publ. No. (HSM) 72-9080. Washington, DC: U.S. Government Printing Office.

Hauser, M. J. (1987). Special aspects of grief after a suicide. In E. J. Dunne, J. L. McIntosh, & K. Dunne-Maxim (Eds.), *Suicide and its aftermath: understanding and counselling the survivors* (pp. 57–70). New York: Norton.

Lamb, F., & Dunne-Maxim, K. (1987). Postvention in schools: policy and process. In E. J. Dunne, J. L. McIntosh, & K. Dunne-Maxim (Eds.), *Suicide and its aftermath: understanding and counselling the survivors* (pp. 245–260). New York: Norton.

McIntosh, J. (1987). Research, therapy, and educational needs. In E. J. Dunne, J. L. McIntosh, & K. Dunne-Maxim (Eds.), *Suicide and its aftermath: understanding and counselling the survivors* (pp. 263–277). New York: Norton.

McIntosh, J. L. (1989). How many survivors of suicide are there? *Surviving Suicide, 1*, 1.

McNiel, D. E., Hatcher, C., & Reubin, R. (1988). Family survivors of suicide and accidental death: consequences for widows. *Suicide and Life-Threatening Behavior, 18*(2), 137.

Resnick, H. L. P. (1972). Psychological resynthesis: a clinical approach to the survivors of a death by suicide. In A. C. Cain (Ed), *Survivors of Suicide* (pp. 167–177). Springfield, IL: Charles C Thomas [reprinted from E. S. Shneidman & M. J. Ortega (Eds.) (1969). *International Psychiatry Clinics, 6*(2), 213–224.

Roy, A. (1986). Genetic factors in suicide. *Psychopharmacology Bulletin, 22*, 666–668.

Rudestam, K. (1987). Public perceptions of survivors of suicide. In E. J. Dunne, J. L. McIntosh, & K. Dunne-Maxim (Eds.), *Suicide and its aftermath: understanding and counselling the survivors* (pp. 31–44). New York: Norton.

Shneidman, E. S. (1973). *Deaths of man.* New York: Quadrangle Books.

14

Forensic Suicidology: Litigation of Suicide Cases and Equivocal Deaths

RONALD W. MARIS

What happens when an individual, especially a therapy client commits suicide? Naively, it may appear that the therapist's or hospital's work is finished when in fact it may be just beginning. When a patient dies, often litigation follows, sometimes for several painful and anxiety-producing years. *Forensic suicidology* is an aborning discipline that attempts to consider just such circumstances in which a suicidal death involves legal action. Suicidology is the scientific study of suicide, and forensic suicidology is that branch of suicidology dealing with legal issues, principles, or court cases. Forensic suicidology is akin to the better known terms "forensic psychiatry" and "forensic pathology" in that it teaches the application of suicidological knowledge to the purposes of the law.

Suicide issues that involve litigation usually focus on four broad areas: contested life insurance claims that argue the mode of plaintiff's death (Nolan, 1988) (was it an accident, suicide, homicide, or natural death?); malpractice claims (Litman, Maris, & Perlin, 1990; Perr, 1985) (including especially drug issues); jail and prison suicides (Danto, 1987; Hayes & Kajdan, 1981) and workmen's compensation cases (e.g., for employment-related accidents or deaths). When discussing these issues I refer to actual legal cases from my forensic practice (workmen's compensation cases are not considered). In the remainder of the chapter I examine (1) basic forensic concepts; (2) types of suicide litigation; (3) role of the suicide expert; (4) common "mistakes" that therapists and organizations (such as psychiatric hospitals) make when treating suicidal individuals; (5) retrospective assessment and prediction of suicides (especially the advantage of knowing the patient is dead by suicide); and (6) ethical issues and practical matters.

The approach to what follows is pragmatic and case centered. It is important not to confine one's analysis to suicides and their therapists but to consider also the treating organizations, such as hospitals, jails, city/county/state government, hotels, and schools, as organizations are frequently sued and carry large amounts of insurance.

Basic Concepts

One of the first problems that confronts nonlawyer suicidologists is a bewildering array of technical legal concepts that they must understand minimally in order to do their expert work effectively. Of course lawyers in the firm that retains you for your testimony act as your "interpreter," but the suicide expert still needs to be savvy with some basic legal concepts. It should also be noted that concepts vary from state to state and in state versus federal court.

Suicide is intentional, voluntary self-killing. It can be defined by elimination (although this definition is not the preferred one) as *not* an accident, homicide, or natural death. Suicide is also *not* a nonfatal suicide attempt, self-destruction where intent cannot be established (called "parasuicide") (Kreitman, 1977), gestures, ideation, suicide talk, or self-mutilation, although all of these traits can help indicate a completed suicide. In 27 states of the United States the law has a strong presumption against suicide, and suicide often requires a difficult burden of proof (Ring, 1988).* Of course, a suicide note would be a powerful indicator of suicide intent, but unfortunately suicide notes are absent in 75 to 85% of all suicides (Leenaars, 1988). The actual percentage of suicide note leavers may be as high as 30% of all suicides, but some of the notes are never discovered or are lost or hidden (R. Litman, personal communication). Usually the "manner of death" (suicide, accident, homicide, natural death, or undetermined) is the crucial issue in contested life insurance. We note in passing that the fact that a medical examiner or coroner classifies a death as suicide on the deceased's death certificate is often inadmissible or inconclusive evidence of suicide in court.

In malpractice or negligence cases the concept of the "standard of care" comes into play. The *standard of care* is "that degree of care which a reasonably prudent person or professional should exercise in same or similar circumstances" (Black, 1979, p. 1260; Bongar, Berman, Litman, & Maris, 1992). Thus generally only other, similar professionals (sometimes in similar geographical regions or professional practices) are entitled to express opinions about the standard of care. For example, psychologists may not be allowed to express opinions about the standard of care of psychiatrists or nonphysicians about drug treatment standards. However, it should be noted that psychologists and other nonphysician suicidologists are often allowed to give opinions about the standard of care.

Deviations from the standard of care are usually referred to as "negligence"

---

*States with some form of presumption against suicide include Alabama, Arizona, Arkansas, Colorado, Florida, Georgia, Idaho, Illinois, Iowa, Kentucky, Louisiana, Maine, Massachusetts, Michigan, Minnesota, Missouri, Nebraska, New Jersey, New York, North Carolina, Ohio, Oregon, Pennsylvania, South Carolina, Texas, Virginia, and Washington (Archin, Wettstein, & Roth, 1990). Dr. Robert Litman informed me that the presumption against suicide is derived from English Common Law. In Common Law, suicide was a crime, and therefore there was a presumption of innocence until proved guilty.

or "gross negligence" (although many other legal concepts are relevant). *Negligence* is "the failure to use ordinary care as a reasonably prudent and careful person or professional would use under same or similar circumstances" (Black, 1979, p. 930). Although negligence seems fairly straightforward, in fact it is a slippery concept that is difficult to illustrate out of context. Examples of negligence in my forensic practice include not taking a complete and prompt medical history of a patient, failure to secure previous medical records, or failure to do a mental status examination.

Gross negligence, on the other hand, means "such an entire want of care to indicate that the act of omission in question was the result of conscious indifference to the rights, welfare, life, property, or safety of the persons affected by it" (Black, 1979, p. 931). In my experience, the court has found the following behaviors to be grossly negligent: not evaluating a mentally disordered felon for more than 2.5 months after his being imprisoned, failing to administer a major tranquilizer to a psychotic inmate with a known history of a serious suicide attempt, and prescribing numerous psychotropic drugs without adequate assessment of their interaction effects or adequate follow-up monitoring.

When attempting to prove that a treatment or drug caused a client's suicide, the criterion is usually "proximate cause." A *proximate cause* is "that which in a natural and continuous sequence, unbroken by any efficient intervening cause, produces injury (including death) and without which the result would not have occurred" (Black, 1979, p. 1103). There may be more than one proximate cause. Some lawyers have objected that the proximate cause criterion is too stringent, akin to one billiard ball hitting another. Seldom is real life that clean or simple.

With jail or prison suicides, violations of the U.S. Constitution are often argued in federal court. For example, the Eighth Amendment to the U.S. Constitution guarantees that an inmate not to be subjected to cruel and unusual punishment. For example, deliberate indifference to serious medical needs of prisoners violates the Eighth Amendment (*Greason*, v. *Kemp*, 1990). Sometimes the Fourteenth Amendment is also involved. It stipulates, of course, that one can not deprive any person of life, liberty, or property without due process of law.

When investigating equivocal death in addition to a physical autopsy a "psychological autopsy" is often useful or even necessary. The *psychological autopsy* is an intensive retrospective interview(s) and data collection designed to reconstruct the psychosocial features and circumstances surrounding the death (concerning especially the mode of nonnatural deaths) of an individual (Brent et al., 1988; Clark and Horton, 1992; Litman, in Nolan, 1988, p. 70). Contrary to popular belief, the psychological autopsy is not standardized and is not any one questionnaire or form. It is especially helpful for providing evidence of a deceased person's intent to suicide after they are obviously unavailable to ask or interview.

Finally, we must briefly discuss *policy and procedure manuals or protocols*. These documents help to specify the standard of care in particular contexts. Most hospitals, jails, prisons, clinics, and schools have manuals stating their standards of treatment. For example, various suicide precautions (e.g., logged 15-minute checks or constant supervision of an inmate or patient within arm's reach 24 hours a day) or suicide attempt procedures are usually written down and coded. Such manuals are advisable (among other things) to help meet the requirements of the Joint Commission on Accreditation of Hospitals (JCAH), although after 1985 JCAH apparently removed their requirement for written suicide precautions as part of a hospital policy and procedure manual—as far as I can determine—(R. Litman, personal correspondence); and if the manual's procedures are not followed, it may indicate falling below the standard of care of suicidal individuals.

Types of Suicide Litigation

Most suicide litigation involves either life insurance, malpractice, jail or prison suicides, or workmen's compensation cases. Life insurance policies routinely contain a clause stipulating that if the policy holder commits suicide within 2 years of the date of issue of the policy payment shall be limited to premiums paid. A typical suicide clause in life insurance policies reads as follows.

If the insured, whether sane or insane, shall within two years from the date of issue hereof die as a result of suicide, the company's liability shall be limited to payment in one sum of the amount of premiums paid on the policy, less any debt owed to the company.

Sometimes an insurance company pays a policy even if the insured committed suicide. Insurance companies know the suicide rate of their policy holders and adjust the premiums of all policy holders to cover their expected losses from suicide. Still, insurance companies cannot afford to throw money away. If the insurance policy or policies are large and the policy holder's death equivocal, insurance companies often deny payment and claim a suicide defense. Usually the family sues the insurance company, especially if some of the deceased's insurance policies paid off and the family can afford to litigate or risk the remaining insurance proceeds. The insurance company must then prove by a preponderance of evidence that suicide did in fact occur. It is a difficult burden of proof and often (after some sparring and haggling) insurance companies settle the case rather than be exposed for large sums of money in a capricious jury trial (Nolan, 1988). Because the law usually presumes that suicide did not occur, it makes defending insurance claims on the grounds of suicide difficult.

To illustrate the major type of suicide litigation, one case for each type is presented in some detail from my own forensic practice. Although the cases are part of public record, identities are disguised to protect the privacy of their families.

S.F., a 49-year-old owner of a paper/graphics company in a large southern city, was found shot to death in the bedroom of his home Sunday afternoon,

September 9, 1984. He had recently (May 1984) taken out $750,000 in additional life insurance, made a will, established a trust fund, and spoken about his death with his lawyer. S.F.'s wife claimed that her husband's death was not a suicide (she believed it was a homicide by a disgruntled employee who had been fired recently) and sued the life insurance companies, who refused to pay on the grounds of suicide. Both plaintiff (wife) and the defendant insurance companies retained lawyers, experts, and death investigators. The case ended up in court for a jury decision in 1987.

Other pertinent data were as follows: After having had a successful career with the DuPont Company, S.F. quit his job in 1982; his newly formed company lost $300,000 in 1983, and S.F. had to borrow $1 million to keep the company afloat in 1984, although he had recently received a $464,000 inheritance from his mother. In August 1984 S.F. went to a psychiatrist because he was depressed; the psychiatrist testified that S.F. had terminal insomnia, suicidal thoughts, and decreased libido; he was withdrawn and anxious, had lost self-confidence, and feared losing control.

A brother of S.F. had committed suicide with a gun at age 35 in 1965, as had an uncle. S.F.'s mother died in June 1984 of pneumonia. S.F.'s daughter had been hospitalized for manic-depressive illness and was taking lithium carbonate at the time of her father's death. S.F. occasionally shouted, and he was openly irritable and angry. S.F.'s mother's sister had been hospitalized for depressive illness. At the time of death, S.F. was taking antidepressant medication (Pamelor), and he took Xanax and Restoril as well. S.F.'s internist testified that S.F. had prostate problems (no cancer), a kidney stone in 1982, nocturia, and decreased libido. He was taking Corgard for his blood pressure, had bad eyesight, and had recently lost weight. No suicide note was left. S.F. had started an MBA program but dropped out. Two weeks before his death S.F. had spoken to his brother (a physician) about suicidal thoughts.

The revolver that had killed S.F. had been placed in his mouth. There was no food in S.F.'s stomach, although he died Sunday afternoon. S.F.'s brother told the police at the time of S.F.'s death, "I was afraid this was going to happen."

The readers should ask themselves what their opinion about S.F.'s manner of death would have been had they been on the jury. What more would you like to know? In this case the jury found for the defendants and against the wife. Apparently the evidence for suicide was persuasive.

Malpractice arises when a patient attempts or completes suicide and the family believes it can be proved that the standard of care was below that expected of prudent, licensed professionals in same or similar circumstances (Gutheil, Bursztajn, & Brodsky, 1984; Litman, Maris, & Perlin, 1990). It is one of the most difficult forensic situations, as suicide can occur under the best of treatment conditions and because one understandably may be reluctant to testify against one's colleagues (Ruben, 1990). Many professionals contend that some legal protection should be available for mental health workers attempting to treat difficult, volatile, highly suicidal patients.

In *L.K. (wife)* v. *Dr. L. et al.* (1990) a 41-year-old geotechnical engineer (M.K.) with a long history of severe headaches and diabetes had gone to a neurologist for relief. Dr. L. prescribed Inderal, prednisone, and Meclomen (on April 21, 1987) and referred M.K. to biofeedback therapy. M.K. was married (to another diabetic) and had two young daughters (ages 8 and 13). There had been several prior incidents of angry outbursts and yelling at his children. There were no previous suicide attempts, but M.K. had lost his job in August 1981 and was out of work for 2 months. He took Serax at that time. Dr. L. did not advise M.K. about any possible drug interaction effects. In fact, when M.K. came back to see Dr. L. about drug reactions, Dr. L. told M.K.: "It's your diabetes, not the medications."

On May 4 M.K. awoke with a severe headache and stayed home from work in the morning. On the evening of May 5 M.K. passed out in his kitchen after coming inside from working in his garden. Earlier that same day M.K. had become confused and disoriented at lunch time and was literally lost for over an hour. On May 6 M.K. got up, made love to his wife, went to a biofeedback appointment, bought his wife a ring, then went home and shot himself (in the chest) with a .38 revolver he had had since 1979. He put garbage bags under his body before shooting himself and held a bible and a picture of his wife and daughters in his free hand. M.K. left a suicide note saying among other things:

My dearest L. I could see no other solution. . . . My diabetes and medications were out of control. . . .

Unfortunately, both his wife and 8-year-old daughter ignored a note on the front door not to come in and to go next door to call the police. They found him dead in a closet just off the kitchen. A jury found in favor of the plaintiff, Mrs. M.K., and awarded her $2.8 million.

Jail and prison suicides typically involve young sociopathic white men arrested on drunk and disorderly charges often with many prior such arrests (Bland, Newman, Dyck, & Orn, 1990; Oliver & Roberts, 1990). Their suicides or attempts usually occur within the first few hours of incarceration by hanging or other asphyxiation. Before the fact of suicide, it is difficult to determine which inmate to put on suicide precautions (e.g., to place in a holding cell or in restraints). It is easy to be righteously indignant after the fact of suicide, as hindsight is 20:20. Note, too, that often jail suicides can be prevented only at the cost of severe measures, which deny basic human freedoms or dignities guaranteed by the U.S. Constitution.

F.F. was a 28-year-old black homosexual man with a history of schizophrenia and a serious prior suicide attempt by stabbing while experiencing a command hallucination. As a result he was hospitalized at the state psychiatric hospital (diagnosis: chronic, paranoid schizophrenia) in South Carolina and treated successfully with a major tranquilizer (Haldol). F.F. was discharged in mid-1983. After discharge he took his Haldol irregularly and again became aggressive and disoriented. In a florid psychotic condition he hitchhiked on Interstate 26 and was arrested by a state trooper. F.F. fought with the state patrolman, who had to radio for help to subdue him.

After handcuffing, F.F. was taken to a nearby rural South Carolina county jail, arrested, and booked for resisting arrest. He was put in a hold cell 10 feet from the jailer, in which the bars were covered with a fine-meshed wire to prevent hanging. Except for a small panel in front of the cell toilet, F.F. was in constant view of the jailer. After identifying F.F. from information in his wallet, the jailer called F.F.'s mother at home. She told the jailer, "He ain't right, he was a mental patient." A teletype message was also received by the jail from the state hospital that F.F. was mentally disturbed and probably needed Haldol intramuscularly from the nearest hospital emergency room. The arrest took place late on a Friday afternoon. Despite warnings from F.F.'s mother and the South Carolina State Hospital, F.F. was not treated. He paced the cell all weekend, slept little, ran back and forth into his cell bars, and muttered: "I am tired of being in this world" and "Everything is evil and should be destroyed."

After the teletype message, F.F. was put on constant suicide watch orders. He became increasingly agitated and psychotic. About 8 a.m. Monday morning F.F. suddenly became quiet. The jailer had left his desk to talk on the phone for about 5 minutes, during which time F.F. had crouched behind the toilet screen, swallowed the leg of his jogging pants (it took two large men pulling together to get it out of his throat), and stuck his head in the toilet bowl. The mother sued the county, the jail, and various officials and was awarded funeral expenses, court costs, and damages.

Nonmedical Suicide Expert

In most instances the legal norm is that a suicide *expert* is a physician, often a psychiatrist or a pathologist (Litman, 1990). If the issues involve psychotropic drugs or physical autopsies, the suicide expert probably should be a medical doctor or have special training in forensics and psychopharmacology. On the other hand, suicidology is rapidly emerging as a separate professional specialty. For example, it is not uncommon for psychiatrists to have never had a completed suicide in their practice and to be unable to cite a single scientific journal reference on suicide; it is likely that they themselves have done no research and have published nothing on suicide.

Nonmedical experts must *qualify* themselves to be able to give a *DSM-III-R* or *DSM-IV* mental diagnosis, to discuss the appropriateness of various treatments, or to have an opinion about psychoactive drugs. Unfortunately, there is currently no board certification in suicidology. However, clinical experience with suicidal individuals, research and publication on suicide topics, supervised training programs and postdoctoral fellowships, and experience in jails or correctional facilities or with medical examiners in forensic settings are useful and relevant experiences. Often a nonmedical expert is attacked as being "merely academic." Here it is especially important to document one's (supervised) clinical experience with suicidal individuals.

It may be necessary to educate lawyers, judges, and juries about suicidology, one's qualifications, and the panorama of suicidal situations. Most nonmedical suicidologists are usually qualifiable as suicide experts but often only after considerable insults to one's professional status. The court tends to be conservative and traditional, and any experts tend to be distrusted and even resented. For example, at a murder trial in rural North Carolina, when considering my qualifications, the judge informed the jury that an "expert" was "just a 'pert' from out of town."

Suicidology is a multidisciplinary science, with even dimensions of art (Perlin, 1975). To be a suicide expert requires knowledge in psychiatry, psychology, sociology, pathology, pharmacology, biostatistics and epidemiology, law, philosophy, forensics, criminology, nursing, social work, and much more. The Johns Hopkins Medical School postdoctoral suicidology training program, where I trained initially (Perlin, 1975), included a supervised acute treatment/emergency room program, service as a deputy medical examiner and assistant in death investigations, a moot court for various suicide cases, clinical field placements at suicide prevention centers, individual suicide case conferences, and formal courses in the psychiatric (particularly affective disorders), psychological, sociological, and public health aspects of suicide and suicide prevention. One major suicide training problem is that there are currently no postdoctoral programs in suicidology in the United States. Thus it is even more difficult to demonstrate competence as a suicide expert to the court.

An expert is qualified to have an *opinion*. Some judges insist that experts limit their testimony to stating and defending a specific opinion (rather than, for example, reviewing options and alternatives for the jury). During litigation, facts are seldom self-evident, nor are they subject to only one interpretation. Still, the expert must make every effort to know the relevant facts cold. It never ceases to amaze me how many experts do a sloppy job of preparing a foundation in fact for their opinions. One should carefully gather information from both sides of a suicide case, although "your side" typically feeds you somewhat biased, one-sided, incomplete evidence. It is useful to ask for the complete list of evidence to determine if anything important is being withheld from you. In the last analysis most experts are paid to form and defend an opinion. For example, was the death a suicide? Was there malpractice? Could the adolescent jail suicide have been anticipated and prevented? Was a jailer grossly negligent? Suicide (or any) expert witnessing is thus a bit like wine-tasting: You, as expert, are entitled to your opinion.

Investigating a Suicide Case

How does an expert form a defensible opinion (Litman, Curphey, Shneidman, Farberow, & Tabachnik, 1963)? Much of what suicide experts do is carefully read materials provided to them by lawyers—who, of course, want the expert to form an opinion helpful to their clients and who typically provide limited, somewhat biased evidence. *Evidence* can be defined as "all the means by which any alleged matter of fact, the truth of which is submitted to investigation, is established or disproved" (Black, 1979, p. 498). It may include depositions of key witnesses, informal interviews, medical records, opinions of other experts, death certificates, autopsy and police reports, private investigators' reports, letters, notes (sometimes suicide notes), photographs, diaries, even desk calendars. I always ask for a complete list of evidence (technically, evidence is presented at the trial). It is a good idea to take notes (with page, paragraph, line, date, and so on specified) in order to document your opinion in a deposition or in court room testimony, although your notes are discoverable by the opposing side. You can count on being "grilled" by opposing counsel(s) so be prepared. Some favorite deposition techniques include changing topics rapidly, trying to force you speak off the top of your head without benefit of notes or records, and provoking you to try to make you angry. This strategy can make you look disorganized, unprepared, and irrational.

It may be necessary to do some of your own interviewing or investigating, although lawyers typically do not like it (because it is expensive and difficult for the lawyers to control what you do). Remember that your work products are discoverable, so you may not want to tape interviews or take notes you would not want someone else to read. In a large case it is not unusual or inappropriate for a suicide expert to do some of their own investigation and to gather specific data more relevant to the expert's opinion (e.g., a depression inventory or a suicide prediction scale). It is also important to ask if you can speak to key witnesses from the other side, although you will routinely be denied access to them. At least you will have tried to gather more complete, impartial information if asked that question at trial. If possible, the expert should be present for as much of the actual trial as is permitted (although others recommend the expert leave after their own testimony).

Procedures that are often useful for file preparation are making a list of key persons involved in the case (a *dramatis personae*) and creating a chronology of key events complete with appropriate dates (e.g., the deceased's birth and death dates, marriage and divorce dates, birth dates and names of the deceased's children, hospitalization, drug, and treatment dates, work history, and so on). It is also helpful to outline all the materials you read and then to put index tabs in the margins of your notes by key subject matters (e.g., the major points of your expert opinion). It is also useful to keep up-to-date financial records, with total hours and a summary of fees you billed. You need a current curriculum vitae (two copies) in your file as well. Finally, your tentative expert opinion and supporting citations probably should be written out in a easily readable form. All notes, correspondence, telephone messages (sometimes with a billing record from the telephone company) are useful to keep in the file, especially by whom and when you were first contacted with regard to the case. Much of this information should *not* be taken to a deposition, but rather to the trial itself.

Remember, usually you are an expert, not a lawyer. It may be wise for you to ask the lawyers to explain the essential points of the law at issue and to provide you with definitions of key terms (such as were listed above). Otherwise, you may not consider appropriate facts. Most lawyers keep their experts too uninformed about the relevant law and about other aspects and information about of the case. For example, it is helpful for the expert to know who has the burden of proof, what kind of evidence counts and does not count (and has been admitted at trial), what the law says about suicide, what are the pertinent precedents, what sections of the U.S. Constitution were argued to have been violated in a jail suicide, and so on.

Finally, realize that you will never be fully knowledgeable about all aspects of a suicide case. The lawyers who have lived with the case (often for several years) are always better informed than the expert. Do not by afraid to admit that you do not know some facts or have no expertise about some matters. Your role as an expert is limited. Rely on your side's lawyers to help you. Always pause a moment after each question you are asked to give your side's lawyers time to object.

Prediction of Individual Suicides

An important controversy in suicide litigation is whether individual cases of suicide can be predicted with any degree of accuracy. For example, Motto argued that we can only predict the probability of suicide *risk*, not the suicides themselves (1992). Elsewhere we have addressed this complex subject at length from a variety of professional perspectives (Maris, Berman, Maltsberger, & Yufit, 1992). If individual suicides cannot be predicted, the questions of whether a suicide occurred and if anyone is responsible for it are clearly in doubt. A related issue is whether a specific case was in fact a suicide, homicide, accident, or natural death (Jobes, Berman, & Josselson, 1987). Actually, predictive questions vary by type of suicide litigation. For insurance cases we usually ask: Is it a suicide? but for malpractice cases the questions may be: What is the standard of care, did treatment fall below the standard of care, and could the suicide have been prevented? For jail or prison suicides, the issue may be: Can a jail ever be fully suicide-proofed or were constitutional rights violated?

Suicide outcomes are seldom, if every, 100% certain, or else there would be no litigation. However, just because a suicide outcome is not certain, it does not mean prediction is not possible. Usually a high probability of suicide (significantly higher than most other competing options) with a known error variance is what is required for reasonable prediction. Decisions about life insurance, malpractice, and negligence must be made despite unclear facts.

I do not wish to appear cavalier when discussing the enormously complex issue of suicide prediction. For example, we often do not *know* that a suicide outcome has a higher probability than, for instance, the outcome of continuing to live another year. Litman argued (personal communication, 1990) that even with a high suicide risk it is almost always a higher probability that there will *not* be a suicide, especially within a short period of time. Moreover, even if we do know suicide is highly probable, we seldom know *when* it will occur.

One way out of this dilemma is to posit that the person most responsible for and knowledgeable about suicide is the would-be suicide himself or herself (although, of course, it is not always the case). The clinician guarantees care at or above the standard, *not* suicide prevention. It is even conceivable that (in rare instances) "good care" might include a suicide outcome. Another resolution to the dilemma is that suicide can be "postdicted," if not predicted.

One should not lose sight of the fact that in forensic cases we often *know* the suicide outcome, that is, a particular person did actually commit suicide. In such cases plaintiff lawyers tend to become indignant and may argue that surely any fool could have seen the suicide coming and should have prevented it. Before the fact, however, suicide is obviously difficult to predict. For example, many people become clinically depressed (perhaps 2–3% of all Americans), abuse alcohol (about 10% of the general population), and even contemplate suicide at some time in their life (15–20%), but only about 1 to 3 in 10,000 (0.01%) complete suicide during any one year.

Because suicide is a rare event, it is common to have problems with false positives when predicting suicide. A false positive would be to predict a suicide that in fact was not a suicide. Other relevant concepts are true positives (sensitivity) and true negatives (specificity). For example, psychiatrist Alex Pokorny claimed (1983): "We do not possess any item(s) of information . . . that permit us to identify to a useful degree the particular persons who will commit suicide." In Pokorny's research on 4800 psychiatric inpatients with the best of predictors, 1209 patients were predicted to be future suicides, but only 67 actually were. Admittedly, suicide prediction is a difficult problem, but Pokorny's lack of success may mean that his predictors were still too crude, that he did not have a high enough risk population, or that even 4800 cases were too few for suicide prediction. That is, his results may be more of a comment on Pokorny's research procedures than on the impossibility of suicide prediction.

There is no one type of suicide, and therefore there is no one set of predictors. Models of prediction must vary with type of suicide being predicted. Different types of suicide have different sets of predictor variables, similar to different types of physical illness. However, it does not mean that there are so many predictors of suicide present in such complex mixes for each suicide type that suicide cannot practically be predicted or can be predicted only by experts. In many instances suicides can be predicted fairly well utilizing a dozen or two predictors, especially over a long enough period. In other instances we may have to settle for predicting suicide risk, such as "there is a 1% group risk of suicide in the next year."

One such list of predictors is offered in Table 14.1. Minimally, every suicide investigation ought to gather information on all 15 basic predictor types listed in Table 14.1. In the space we have available here it is not possible to do justice to different suicide types and their specific predictors (however, see Maris et al. [1992] Chs. 1 and 4). It must suffice for now to give a brief overview.

Almost all suicides have diagnosable mental disorders, usually one of the *affective* or *mood disorders*, although suicide risk is also high among schizophrenics and some personality and panic disorders. About 15% of persons with primary depressive illness eventually commit suicide. *Alcoholism* is another major predictor of suicide. In Robins' St. Louis research (Robins, 1981) 72% of completed suicides were either depressed (47%) or alcoholic (25%). No other single predictor was present in more than 5% of all suicides. Roy (1986) estimated in a large series of studies that 18% of all alcoholics eventually commit suicide.

Ideas about suicide, *talk* of suicide or dying, and *preparation* for suicide occur in most suicide cases. Sometimes the best predictor of suicide is simply to ask people if they are feeling like killing themselves and, if so, what method or plan they have. Partially self-destructive behavior may also indicate ideas about suicide (e.g., drinking heavily, sexual promiscuity, missing work or school, failing to take medications). Of course, before one kills oneself there is a *suicide attempt*. Any individual with a history of one or more suicide attempts is at much greater risk to complete suicide than those who have never attempted suicide. About 15% of nonfatal suicide attempters eventually kill themselves.

When someone attempts or contemplates suicide, clearly the *lethality of the method* is a crucial factor. One of the reasons men have higher suicide rates than women is that they are more likely to use firearms or hanging than are women. *Social isolation* or loss of social support increases the risk of suicide. In my Chicago suicide surveys (Maris, 1969, 1981) 50% of suicide completers had no close friends compared to only 20% of nonfatal suicide atempters. When one lives alone obviously the possibility of rescue after a suicide attempt is lower

Table 14.1. Common Single Predictors of Suicide

 1. Depressive illness, mental disorder
 2. Alcoholism, drug abuse
 3. Suicide ideation, talk, preparation, religious ideas
 4. Prior suicide attempts
 5. Lethal methods
 6. Isolation, living alone, loss of support
 7. Hopelessness, cognitive rigidity
 8. Older white men
 9. Modeling, suicide in the family, genetics
10. Work problems, economics, occupation
11. Marital problems, family pathology
12. Stress, life events
13. Anger, aggression, irritability, 5-HIAA
14. Physical illness
15. Repetition and "comorbidity" of factors 1 through 14, suicidal careers

Source: Assessment and Prediction of Suicide (Ch. 1) by R. W. Maris, A. L. Berman, J. T. Maltsberger, and R. Yufit (Eds.), 1992, New York: Guilford Publications. By permission.

than if others were around. Beck, Kovaks, and Weisman (1986) concluded that *hopelessness* is a better predictor of suicide than is depression. Suicide completers also tend to be rigid or dichotomous in their thinking and are less likely to conceive of alternatives to suicide to resolve their life problems.

Demographically, we know that suicide is particularly likely among *elderly white men*. Ninety-two percent of all suicides in the United States are committed by whites. Suicide rates of men are three to four times those of women. Although the suicide rates among young men have increased (until about 1977), it is still true that suicide rates tend to increase with age for men and that recently they seem to be going up among older men (McIntosh, 1990).

Suicides tend to run in families. A *history of suicide* in one's first-degree relatives increases the probability of one's own suicide. Some of the influence is probably genetic, and some may be modeling or copying. *Work problems* are common among suicides. As many as one-third of all suicide completers are unemployed at the time of their death (Maris, 1981), and they typically have erratic job histories.

Marriage and *having children*, other things being equal, are usually associated with lower rates of suicide. Suicide rates are almost always highest among widowed and divorced people. *Stress*, especially chronic stress, and negative life events tend to exacerbate suicide. Remember, too, that suicide is normally an aggressive, violent act. Suicide completers typically are *angry*, irritated, and dissatisfied. Freud argued that often suicides were disguised or displaced murders of internalized other people (e.g., parents or lovers). Biological research suggests that suicide is more common among those with low levels of brain serotonin and especially its metabolite *5-hydroxyindoleacetic acid* (Winchel, Stanley, & Stanley, 1990).

Although *physical illness* in and of itself is not a reliable predictor of suicide, about 35 to 40% of all suicide completers have significant physical illnesses. Illnesses that have been found to be related to suicide include epilepsy, musculoskeletal disorders, malignant neoplasms, and gastrointestinal disorders. Finally, *repetition* and *interaction* (see related concept of *comorbidity*) of the above predictors are critical factors in what has been called "suicidal careers" (Maris, 1981). Most people bear up under single or first insults to their adaptive repertoire. Typically, suicide completers display a gradual, chronic pattern of entropy.

Common Therapy Mistakes That Prompt Litigation

Anytime someone dies a nonnatural death, especially by suicide or an accident, there may be a lawsuit. One way to minimize the probability of suit is for clinicians to *monitor their professional involvements* carefully. If there is a choice of screening clients, it would be wise to anticipate suicidal outcomes and perhaps limit the number of depressed or alcoholic (and so on) male patients treated. However, if the clinician works in a jail or hospital, he or she may have no choice but to treat or manage dangerously suicidal people. Even here a hospital can be judicious about who has admitting privileges, and clinicians can be careful about whom you take on as a partner in the practice. Thus the initial therapy mistake is not monitoring professional involvements carefully enough.

In a sense, "mistake" is a misleading concept. It implies that any time a suicide occurs someone has "fouled up." Although that may be generally true, most therapists eventually encounter suicides in their practice if they work long enough. Even with competent therapists or care that is above the standard, there are some suicides. Suicide is not *prima facie* evidence of a "mistake." There may even be rational suicides (Maris, 1981) that under certain limited circumstances are the preferred outcome—heretical and dangerous as this idea is.

Once therapists become involved with suicidal clients or inmates, other treatment mistakes come into play. There should be a treatment manual, protocol, or plan, and it should be followed scrupulously. A prompt, thorough history is essential (obtained as quickly as the protocol states). For example, if the protocol calls for suicide precautions and logged 15-minute checks in progress notes, be sure to make them. Write or call promptly for prior treatment records, including prior medication records. Keep good written records and notes (preferably typed) that an outsider could read and understand. I was involved in a malpractice case where the psychologist being sued made sloppy, cryptic, fragmentary notes that were not even complete sentences—much to his legal disadvantage. Sloppy note-taking often reflects or suggests a sloppy clinical practice.

Be sure to ask clients about suicide. Probe for self-destructive thoughts or preparations. Avoid superficial, insincere, perfunctory inquiries about suicide. Use indirection and sometimes confront inconsistencies in patients' statements and behaviors. Be involved with your clients; avoid fragmentation in their care. Few warm, supportive, likable, caring therapists are sued, even if their clients commit suicide. Almost always, in my experience, the physicians, jailers, and psychologists who are sued were arrogant, aloof, inattentive to details, overly busy, delegated too much of their patients' care to others, and so on. One can be a caring, concerned therapist without being seductive or inappropriately personal. Among other things, call the family after the death and go to the deceased's funeral.

Training of personnel is often crucial. Staff should be aware of high suicide risk profiles in their own areas (e.g., jails, schools, psychiatric clinics). Other critical mistakes are (1) not to know what times clients or inmates are at greatest risk to suicide; (2) not to control methods of suicide; (3) not to suicide-proof a treating environment; and (4) not to place patients in the correct environment (e.g., with another patient or inmate, on 24-hour constant watch, or in rooms where bathroom doors cannot be locked).

Of course, extreme caution is needed regarding the prescription and monitoring of psychiatric drugs. Many tranquilizing drugs make depression worse. Several of my forensic cases have involved possible side effects of the benzodiazepine drugs, especially Halcion (triazolam) (APA, 1990). Antidepressant medications are also commonly used to make explicit overdoses. There have also been special concerns recently with Prozac (fluoxetine hydrochloride) and suicide. Be sure to obtain complete drug histories, ask what other nonpsychiatric drugs (e.g., insulin, β-blockers) the patient takes, and be careful to assess drug interaction effects.

Ethical Issues and Practical Matters

A recurring issue is whether a suicide expert's role is more like that of a lawyer or that of an objective scientist (Amchin et al., 1990; Battin & Mayo, 1980). Surely, the suicidologist expert is more like a scientist than a lawyer or advocate. How, then is it that both sides always have their experts? Is it just a matter of money? Of course, one should believe in their opinion and marshal evidence to support it. There are some cases an expert should simply refuse to take. There is little worse than agreeing to defend a position of which you are not convinced. Even if the facts are not clear, they are usually clear enough. One cannot leave everything up to a jury and certainly not to lawyers.

How much should you charge for your expert work? Fees for experts usually vary between roughly $100 and $300 an hour (but can go as high as $400 to $500 per hour). Usually slightly more is charged for giving depositions and testifying in court (and you will earn it). It may seem inappropriate to be making so many money from someone else's misfortune and misery, and expert's fees are probably too high. Ultimately, expert's costs are passed on to the general public in the form of increases in insurance rates. Some experts ask what the senior law partners involved in the case charge and then charge that (or $1 more) themselves.

You may wish to ask for a retainer (of $1,000 to $2,000) to guarantee that you will be paid. Long delays in payment are common, especially with plaintiff firms. Do not underestimate the value of your expert opinion in forcing the settlement of a case. If a case is settled, experts do not receive a percentage. In many cases experts prepare for a trial over a period of years and are never given the chance to testify. Settlements are common (sometimes at the last moment before trial or even during trial). Lawyers and their clients are often unwilling to risk losing, even if their cases are good. Suicide experts report feeling like "elegant prostitutes." Occasionally lawyers do not even tell their experts a case was settled.

It is difficult to market your expert services, given the stigma against advertising in the medical and legal professions. Most cases come by referral from colleagues and a few via publications or speeches at conferences. You might consider preparing papers for trial lawyer or forensic psychiatry meetings. There are "expert-at-law" and forensic sciences professional journals (e.g., *Experts at Law, Journal of Forensic Sciences*) and societies, as well as agencies for listing your expert services (although many professionals avoid them). Despite the problems involved, expert witnessing and death investigations are rewarding and exciting work. There is a definite outcome that affects the lives of particular human beings immensely.

Guidelines for Practice: Chapter Summary

Forensic suicidology is that branch of suicidology dealing with legal issues, principles, and court cases. As a therapy practitioner you are expected by law to exercise the same degree of care a reasonably prudent professional would in the same or similar circumstances. If you do not and your patient or client dies, you may be liable for negligence or malpractice suits.

Ordinarily, the standard of care (which differs some for individuals and organizations) includes having clear treatment procedures or plans (often written in a manual or protocol) that are followed carefully. This treatment procedure protocol for suicide clients often includes obtaining a prompt and thorough psychosocial history (that includes securing prior medical, treatment, and drug records); having a physical examination done; giving a coded (*DSM-III-R*) intake diagnosis; formulating a treatment plan; making clear, prompt, thorough, and legible therapy or progress notes; being familiar with the basic predictors of suicide; assessing the client for suicide potential (including openly asking about suicide intentions, plans, and methods); being trained in suicide prevention; carefully monitoring psychoactive drugs and anticipating possible drug interactions and side effects; avoiding fragmentation of the care of the patient; and being a concerned, involved, and supportive therapist.

The chapter also considered the behavioral scientist (as well as the suicide therapist) as a nonmedical suicide expert. As a nonmedical expert the psychologist can become involved primarily in contested life insurance claims, malpractice cases of their colleagues, and jail and prison suicides. Being qualifiable as a suicide expert requires relevant clinical experience with suicidal patients and a research and publication record on suicide. Expert opinion is based on evidence (usually medical or therapy records and testimony of witnesses) you have reviewed and occasionally on your own death investigations. An expert needs to be well versed in the major predictors and epidemiology of suicide.

REFERENCES

American Psychiatric Association (1990). *Benzodiazepine dependence, toxicity, and abuse*. Washington, DC, American Psychiatric Association.
Amchin, J., Wettstein, R. M., & Roth, L. H. (1990). Suicide, ethics, and the law. In S. J. Blumenthal & D. J. Kupfer (Eds.), *Suicide over the life-cycle* (pp. 637–664). Washington DC: American Psychiatric Press.
Battin, M. P., & Mayo D. J. (1980). *Suicide: the philosophical issues*. New York: St. Martins Press.
Beck, A. T., Kovacs, M., & Weisman, A. (1986). Hopelessness: an indicator of suicidal risk. *Suicide and Life-Threatening Behavior, 5*, 98–103.

Black, H. C. (1979). *Black's law dictionary*. St. Paul, MN: West Publishing.

Bland, R. C., Newman, S. C., Dyck, R. J., & Orn, H. (1990). Prevalence of psychiatric disorders and suicide attempts in a prison population. *Canadian Journal of Psychiatry, 35*, 407– 413.

Bonger, B., Berman, A. L., Litman, R. E., & Maris, R. W. (1992). Suicide and the standard of care. *Suicide and Life. Threatening Behavior 22*, (3) in press.

Brent, D. A., Joshua, J. A., Kolko, D. J., & Zelenak, J. P. (1988). The psychological autopsy: methodological considerations for the study of adolescent suicide. *American Academy of Child and Adolescent Psychiatry, 7*, 362–366.

Clark, D. C., & Horton, S. L. (1992). Assessment in absentia: the value of the psychological autopsy method for studying antecedents of suicide and preventing future suicides. In R. W. Maris, A. L. Berman, J. T. Maltsberger, & R. Yufit. (Eds.), *Assessment and prediction of suicide*. New York: Guilford.

Danto, B. L. (1987). Suicide following arrest and lockup. *American Journal of Forensic Psychiatry, 8*(2), 51–60

Experts-at-Law (national magazine for trial attorneys and legal consultants). North Hollywood, CA

Greason et al. v. *Kemp et al.* U.S. Court of Appeals No. 88-8563 (pp. 1211–1227) (Georgia, 11th Cir., 1990).

Gutheil, T. G., & Appelbaum, P. S. (1982). *Clinical handbook of psychiatry and the law*. New York: McGraw-Hill.

Gutheil, T. g., Bursztajn, H., and Brodsky, A. (1984). Malpractice prevention through the sharing of uncertainty. *New England Journal of Medicine, 311*, 49–51.

Hayes, L. M., & Kajdan, B. (1981). *And darkness closes in . . . national study of jail suicides*. Washington, DC: National Center on Institutions and Alternatives.

Jobes, D. A., Berman, A. L., & Josselson, A. R. (1987). Improving the validity and reliability of medical-legal certification of suicide. *Suicide and Life-Threatening Behavior, 17*, 310– 325.

Journal of Forensic Sciences. Philadelphia: American Academy of Forensic Sciences.

Leenaars, A. A. (1988). *Suicide notes: predictive clues and patterns*. New York: Human Sciences Press.

Litman, N. (Ed.). (1977). *Parasuicide*. London: John Wiley & Sons.

Litman, R. E., Curphey, T. J., Shneidman, E. S., Farberow, N. L., & Tabachnick, N. D. (1963). Investigation of equivocal suicides. *Journal of the American Medical Association, 184*, 924– 929.

Litman, R. E. (1990). Psycholegal aspects of suicide. In W. Curran, A. L. McGarry, C. S. Petty. (Eds.), *Modern legal medicine, psychiatry, and forensic science* (pp. 841–853). Philadelphia: Davis.

Litman, R. E., Maris, R. W., and Perlin, S. (1990). Malpractice. In A. L. Berman (ed.), *Suicide prevention: case consultation*. New York; Springer-Verlag.

Maris, R. W. (1969). *Social forces in urban suicide*. Chicago: Dorsey Press.

Maris, R. W. (1981). *Pathways to suicide: Survey of self-destructive behaviors*. Baltimore: Johns Hopkins University Press.

Maris, R. W., Berman, A. L., Maltsberger, J. T., and Yufit, R. (Eds.). (1992). *Assessment and prediction of suicide*. New York: Guilford.

McIntosh, J. (1990). Older adults: the next suicide epidemic? Presented at the annual conference of the American Association of Suicidology, New Orleans.

Motto, J. A. (1992). An integrated approach to estimating suicide risk. In R. W. Maris, A. L. Berman, J. T. Maltsberger, & R. Yufit (Eds.), *Assessment and prediction of suicide*. New York: Guilford.

Nolan, J. L. (1988). The suicide case: investigation and trail of insurance claims. Washington, DC: American Bar Association.

Oliver, J. M., & Roberts, J. B. (1990). Jail suicide and legal redress. *Suicide and Life-Threatening Behavior, 20*(2), 138–147.

Perlin, S. (Ed.) (1975). *Handbook for the study of suicide*. New York: Oxford University Press.

Perr, I. N. (1985). Suicide litigation and risk management; a review of 32 cases. *Bulletin of American Academy of Psychiatry and Law, 13*, 209–219.

Pokorny, A. D. (1983). Prediction of suicide in psychiatric patients. *Archives of General Psychiatry, 40*, 249–257.

Ring, L. M. (1988). Obtaining insurance proceeds over a suicide defense. In J. L. Nolan (Ed.), *The suicide case: investigation and trial of insurance claims* (pp. 1–12). Washington DC: American Bar Association.

Robins, E. (1981). *The final months: a study of the lives of 134 persons who committed suicide*. New York: Oxford University Press.

Roy, A. (1986). Alcoholism and suicide. In R. W. Maris, (Ed.), *Biology of suicide*. New York: Guilford.

Ruben, H. L. (1990). Surviving a suicide in your practice. In S. J. Blumenthal & D. J. Kupfer (Eds.), *Suicide over the life cycle* (pp. 619–636). Washington, DC: American Psychiatric Press.

Winchel, R. M., Stanley, B., & Stanley, M. (1990). In S. J. Blumenthal & D. J. Kupfer (Eds.), *Suicide over the life cycle* (pp. 97–126). Washington, DC: American Psychiatric Press.

15

Training and Supervisory Issues in the Assessment and Management of the Suicidal Patient*

BRUCE BONGAR, JAMES W. LOMAX, AND MORT MARMATZ

Completed and attempted suicide represents a serious public health problem in the United States today. Each year almost 30,000 individuals take their own lives, making suicide the eighth leading cause of death in this country (Alcohol, Drug Abuse, and Mental Health Administration 1989; Hirschfeld & Davidson, 1988). The data on completed suicide become all the more disturbing when one considers that completed suicides arguably represent only a small percentage of the number of attempts; that suicide may be statistically underreported; and that the rates of suicide in most industrialized countries are increasing, particularly among the age group 15–24 and among persons over age 65 (Maris, 1988).

Suicide has also been found to be the most frequently encountered emergency situation for mental health professionals (Schein, 1976); and clinicians have consistently ranked work with suicidal patients as the most stressful of all clinical endeavors (Deutsch, 1984). Patient suicide has a significant acute impact on the professional lives of a substantial number of therapists (Chemtob, Bauer, Hamada, Pelowski, & Muraoka, 1989), and it is alarmingly clear from recent empirical findings that the average professional psychologist involved in direct patient care has better than a 20% chance of losing a patient to suicide at some time during his or her professional career. The odds climb to better than 50% for psychiatrists (Chemtob, Hamada, Bauer, Kinney & Torigoe, 1988a; Chemtob, Hamada, Bauer, Torigoe & Kinney 1988b). Even a psychologist in training, according to one study, has a one in seven chance of losing a patient to suicide (Brown, 1987).

Another study (Kleespies, Smith, & Becker, 1990), investigating the incidence, impact, and methods of coping with patient suicide during the training years of psychology graduate students, found that one of every six students had experienced a patient's suicide at some time during their training. Trainees with patient suicides reported levels of stress that were equivalent to that found

*Portions of this chapter were adapted from *The Suicidal Patient: Clinical and Legal Standards of Care*, by B. Bongar, 1991, Washington, DC: American Psychological Association. By permission.

in "patient samples with bereavement and higher than that found with professional clinicians who had patient suicides" (Kleespies et al., 1990, p. 257). This study found also that trainees who had lost a patient to suicide responded (in order of frequency) with feelings of shock, guilt or shame, denial or disbelief, feelings of incompetence, anger, and depression, and a sense of being blamed. After the suicide, trainees most frequently turned to their supervisors for both emotional support and help in understanding the suicide.

Therefore it is particularly important that training programs in psychology and psychiatry

not convey, either explicitly or implicitly, the impression that patient suicides are an unlikely event (Brown, 1987). . . . It may be time that psychiatrists and psychologists explicitly acknowledge patient suicide as an important occupational hazard. [Chemtob et al., 1989, p. 299]

Despite these obvious and grim realities in our field, it also seems that we may not be training clinicians adequately to manage suicidal patients. In a comprehensive study Bongar and Harmatz (1989) found that training directors of the traditional programs in clinical psychology (i.e., member programs of the Council of University Directors of Clinical Psychology) rated the study of suicide as important to the graduate education of psychologists and indicated that graduate training was the most appropriate place for this training to occur. Unfortunately, only 35% of the programs offered any formal training in the management of suicidal patients as part of their curriculum. Even when all the efforts of the traditional scientist-practitioner and professional school training programs are combined, only 40% of all graduate programs in clinical psychology offer formal training in the study of suicide (Bongar & Harmatz, 1990). The issue of what constitutes adequate training is a particularly thorny issue for the core mental health disciplines (e.g., psychiatry, psychology, nursing, and social work) as well as for the growing number of other mental health counselors, marital and family therapists, and so forth.

Even though such formal training in the study of suicide does not commonly occur, resources for this training are available. For example, there exists a professional organization, the American Association of Suicidology (AAS), devoted to the scientific study of suicide and life-threatening behaviors (Berman, 1986). The AAS Training Committee has in recent years developed a comprehensive bibliography on the study of suicide (AAS, 1987), and the organization's journal has published recommendations for conducting high quality research in suicidology (Smith & Maris, 1986).

Specialists who study suicide have compiled an impressive body of research on clinical risk factors, epidemiology, psychiatric diagnoses, clinical assessment and management strategies, psychotherapy, pharmacology, risk-benefit analysis, hospitalization, postvention, and law and ethics (Beck, 1967; Beck, Kovacs, & Weissman, 1979; Fawcett, Scheftner, Clark, Hedeker, Gibbons & Coryell, 1987; Farberow, 1957; Frederick, 1978; Gutheil & Appelbaum, 1982; Jacobs, 1984; Klerman, 1987; Litman, 1982; Maltsberger, 1986; Maris, 1981; Motto,

1986; Murphy, 1984; Pokorny, 1964; Robins, 1985; Sainsbury, 1986; Shneidman, 1981, 1985, 1986; Simon, 1987; K. Smith, 1988; Wekstein, 1979). It is clear that sound procedures and techniques for the management of suicidal patients are clinical skills that can be taught (Berman, 1986; Bongar, 1991; Bongar, Peterson, Harris, & Aissis, 1989; Roy, 1986; Shneidman, 1986; Simon, 1988).

As we have already noted, patient suicide is not a rare event in clinical practice and should be considered a very real personal and occupational hazard by those mental health professionals involved in direct patient care (Chemtob et al., 1988a,b). To meet the challenge of the suicidal patient, clinicians need to approach this population with confidence—not with fear (Berman & Cohen-Sandler, 1983). Confidence building begins with demanding professional standards aimed at moving clinical practice beyond merely defensive care.

Although there are relatively few true emergency situations in mental health practice, of these emergencies suicide is probably the most compelling; and its unsuccessful assessment or management can be a source of great anguish to the practitioner (Lomax, 1986). Most practicing mental health professionals see patients in their professional practice activities who meet the profile of an elevated risk for attempted or completed suicide. *Clinical wisdom among mental health practitioners "admonishes that it is not a matter of whether one of their clients will someday commit suicide, but of when"* (Fremouw, de Perczel, & Ellis, 1990, p. 129). Patient suicide is not a tragedy that exclusively confronts the mental health professions, but "the incidence of its occurrence is such a frequent issue, both professionally and sometimes legally, that it demands special consideration" from mental health professionals (J. Smith, 1986, p. 62).

Further complicating the clinical situation is the fact that although investigators have identified many clinical indicators of suicide risk there are "no pathognomonic predictors of suicide" (Simon, 1988, p. 89). Authorities who investigate suicidal phenomena have not reached consensus on the key risk factors, both short term and long term, that distinguish suicide completers.

Lomax (1986) has called for training programs where suicide care is recognized as a "core curriculum" item in the training sequence. This statement calls for trainees to demonstrate "basic competence in suicide care" (Lomax, 1986, p. 56). That is, there are basic standards that can be promulgated in a training sequence. For example, it is critical for the clinician who assesses and treats suicidal outpatients to know the resources that are available for emergencies and outpatient crises. As a specific standard-setting example of this position, the Board of Professional Affairs' Committee on Professional Standards of the American Psychological Association has stated that psychologists must be well prepared to handle psychological emergencies, and that it behooves all psychologists to

make certain that they are prepared to develop intervention strategies that are appropriate to the demands of a psychological emergency. Thus, if adequate training was not obtained at the graduate level, then the educational and experience requirements will have to be obtained through postdoctoral or continuing education work. [Board of Professional Affairs, Committee on Professional Standards, 1987, p. 708]

Lomax (1986) also pointed out the importance of using Shneidman's (1975) plan for learning about the phases of prevention, intervention, and postvention.

Shneidman emphasizes the importance of teaching psychiatry residents what is known about the recognition of suicidal risk, including epidemiological factors, the presence of psychopathology that increases lethality and perturbation, the medical and psychological evaluation of patients who have made a suicide attempt, and the knowledge of the implications of suicide attempts and the completions on the families of the primary patients. . . . the resident must learn his or her own characteristic responses to the suicidal patient in each phase of prevention (the suicidal statement or clue), intervention (the response to the patient at the time of the suicide attempt) and postvention (the response to a person who has attempted or completed suicide). [Lomax, 1986, p. 58]

Whether the mental health professional works in a college counseling center, a community mental health agency, inpatient setting, or outpatient private practice, he or she is likely to see patients who present with an elevated risk for suicide. The clinician must have readily at hand the crisis intervention and emergency management tools necessary to deal with the problem of patient suicidality. There is a consensus that crisis management principally entails therapeutic activism, the delaying of the patient's suicidal impulses, the restoration of hope, environmental intervention, and consideration of hospitalization (Fremouw et al., 1990). Regarding the last point, particular problems may emerge regarding the question of hospitalization if the clinician becomes overinvested in outpatient care or overly dependent on the use of the hospital. A particular danger here is that, in an attempt to avoid this interruption, some psychotherapists may inadvertently expose their patients to even greater danger by avoiding hospitalization (Farberow, 1957).

The issues of changes of supervisory settings and therapist absences are worthy of a further cautionary note. Motto (1979) commented that the period surrounding a loss or change of a therapist is an especially vulnerable time for the suicidal patient. He cited the example where at one institution an examination of four recent suicides revealed that three had occurred in the context of vacation interruptions. This problem is common in training institutions because trainees, interns, and residents often move in and out of clinical assignments and often leave the area after completion of training.

Bongar et al. (1989), in a report on integrative approaches to the assessment and management of the suicidal patient, pointed out that one of the most significant factors when assessing suicide risk and determining the prognosis for the success of subsequent treatment is the quality of the therapeutic alliance (see also Bongar, 1991; Jacobs 1989; Peterson & Bongar, 1989; Simon, 1987, 1988; K. Smith, 1988). The therapeutic alliance has been suggested by Simon (1988) as the bedrock indicator of the patient's willingness to seek help and sustenance through personal relationships during serious emotional crises. Maltsberger (1986) specifically addressed the issue of the absence of sustaining interpersonal relationships as a key factor in the clinical formulation of suicide risk. In short, the importance of a good therapeutic alliance can be emphasized to trainees and through clinical supervision demonstrated to be an ongoing and robust measure of the treatment's effect on the patient's vulnerability to suicide.

So important are these therapeutic alliance skills that it may be helpful to note the more general comment by London (1986) that it is wise to consider training novice clinicians in the fundamental skill of establishing good interpersonal relationships as a prerequisite to the later development of specialized and advanced therapeutic proficiencies.

Training institutions can benefit from self-knowledge. Hospitals, clinics, and other training sites should ask:

"Do we treat persons at special risk for suicide?" If yes, then there must be a security area and policies for special management of suicidal persons. [Litman, 1982, p. 220]

Such policies are best determined by a training institution's suicide prevention committee that represents the staff and administration (Litman, 1982). Such a committee can establish written guidelines after a survey of the security areas and after talking with staff and patients. It is critical that these policies be incorporated routinely into the training and supervision of all staff, students, and supervisors. Litman also stated that a reasonable performance requires that the "patient be evaluated for suicide risk, that a treatment plan be formulated, and that the staff follow the treatment plan according to the hospital's policies" (Litman, 1982, p. 220).

In this regard, Pope (1986) stressed the importance of staying within one's area of competence and of knowing one's personal limits, observing "that working with suicidal patients can be demanding, draining, crisis filled activity. It is literally life or death work" (Pope, 1986, p. 19). In addition to obtaining adequate training and knowledge of the literature on suicide, mental health professionals must become familiar with the legal standards involving rights to treatment and to refuse treatment, as well as the rules regarding confidentiality, involuntary hospitalization, and so forth. He noted that a standard of care must involve screening for suicide risk during the initial contact and ongoing alertness to this issue throughout the course of treatment. There also should be frequent consultation and ready access to facilities needed to implement appropriate affirmative precautions (e.g., emergency teams, hospitals, crisis intervention centers, and day treatment).

As an example of the specific technical proficiencies that can be taught via the supervisory process, we can examine the treatment of depression in an outpatient setting as the kind of specific technical proficiency that needs to be brought to bear in a suicidal crisis. Ideally, those clinicians who undertake the treatment of severely depressed patients should have broad-spectrum training, including an understanding of the limitations and benefits of the various psychosocial and organic therapies. Moreover, any clinician who undertakes the management of severely depressed suicidal outpatients must have ready access to appropriate inpatient facilities in case a voluntary or involuntary hospitalization is indicated. The implicit requirement is that mental health professionals know the specific clinical and research literature on all recommended psychosocial interventions for depression, as well as the appropriateness of routinely requesting a consultation for psychotropic medication for the treatment of patients with affective disorders (Lesse, 1989) or to consider the potential utility of such specialized interventions as electroconvulsive therapy in cases of patients with psychotic features in a depressive illness (J. W. Lomax, personal communication, 1991).

Technical proficiency also means that the mental health professional who sees a suicidal patient in an outpatient setting must learn to distinguish carefully between acute suicidal states related to *DSM-III-R* Axis I clinical syndromes and to chronic suicidal behavior as part of an Axis II personality disorder; the care and management of these patients may differ dramatically. Of course, patients with Axis II personality disorders also could develop an Axis I condition at any time. An additional and obvious consideration when treating the suicidal patient is whether the assessment and treatment take place in an inpatient or outpatient setting (Simon, 1987, 1988).

Clinicians who treat depressed suicidal patients must be knowledgeable about current clinical and research developments. For example, the supervisor might point out to the trainee the changes in our empirical understanding of acute and chronic predictor variables (Beck & Steer, 1989; Fawcett et al., 1987), new models for examining attempter behavior (Clark, Gibbons, Fawcett, & Scheftner, 1989), findings on high-risk populations (Weissman, Klerman, Markowitz, & Ouelette, 1989), and current comprehensive reviews on depression (Keller, 1988) and suicide (Mann & Stanley, 1989).

However much or little one opts immediately to incorporate such new developments in one's own treatment planning (e.g., to adopt recent directive and prescriptive treatment recommendations for certain psychopathological conditions), one effect of examining this information in our training programs is to gain better understanding of the current applied and theoretical approaches to intervention and to allow supervisors and trainees to reassess competencies within the context of the changing theoretical and empirical knowledge base. This matter is of vital importance. For example, for the treatment of patients with severe depression, Lesse (1989, p. 195) pointed out that not all mental health professionals "are equipped by personality or training to manage severely depressed patients, let alone those who have suicidal preoccupations or drives. This pertains to senior personnel and trainees alike." Senior personnel should exercise caution when placing residents or trainees in charge of a suicidal patient's care. Here it is critical that both supervisors and trainees recognize the limitations of the trainee's education and experience level. Good training programs expose the trainee to a graduated level of responsibility when working with high-risk patients (J. W. Lomax, personal communication, 1991). Lesse contended that therapists with their own depressive propensities, those who are threatened by aggressive patients, or those who are unable to handle crisis or emergency situations should not treat depressed patients. He also recommended that severely depressed patients not be cared for by clinicians who do not have the capacity for intensive psychopathological investigation.

In coming to terms with their own limitations, supervisors and trainees must remember, as Welch (1989) pointed out, that all mental health providers (psychiatrists, psychologists, social workers, psychiatric nurses) are limited to varying degrees as to their specific professional competencies. He continued:

One might further argue that the greatest threat to "quality of care" comes not from those with limited training but from those with a limited recognition of the limitations of their own training. [Welch, 1989, p. 28].

Therefore one of the initial tasks of the mental health professional, who is called on to treat the suicidal patient, is the need to learn in his or her basic training sequence how to evaluate a priori the strengths and limitations of his or her own training, education, and experience in the treatment of specific patient populations in specific clinical settings (e.g., an understanding of their own technical proficiencies as well as their emotional tolerance levels for the intense demands required for treating suicidal patients). Specifically, as part of their fundamental training as mental health professionals, clinicians must learn how to make the difficult and highly personal decision to conduct their own self-study of personal and professional competence to treat suicidal patients before the fact, not wait to assess this competence when suicidal thoughts or behaviors emerge in patients whom they are seeing in treatment. Pope (1986) has commented that the American Psychological Association's Specialty Guidelines for the Delivery of Services by Clinical Psychologists dictates that clinical psychologists limit their practice to demonstrated areas of professional competence, for example, "to ensure that you meet the legal, ethical and professional standards of competence in working with suicidal clients" (Pope, 1986, p. 19).

Responsibilities of Supervisors

Many clinicians serve as supervisors of trainees, interns, employees, and other colleagues. One estimate placed 64% of psychologists spending at least part of their professional time in supervision (VandeCreek & Harrar, 1988). Although to date few court cases in the mental health field are based solely on supervisor negligence,

Slovenko (1980) has warned that supervisor liability may be the "lawsuit of the future" . . . the principles under which supervisor liability can be claimed have been established in other fields such as medicine and could be applied to professional psychology. [VandeCreek & Harrar, 1988, p. 13]

In general, supervisor negligence can be found if three criteria are met.

1. A professional relationship of any duration exists between supervisor and supervisee

2. The behavior of the supervisor or supervisee fell below an accepted standard of care for the profession

3. A patient was injured; and the substandard care of the supervisee or supervisor was the proximate cause of the patient's injury [VandeCreek & Harrar, 1988]

Simon (1987, 1988) and VandeCreek and Harrar (1988) observed that under the doctrine of *respondeat superior* (let the master respond), psychiatrists, psychologists, and other mental health professionals may be held monetarily responsible for the negligent acts of others working under their supervision, control, or direction. Under a doctrine of vicarious liability, the one who controls the conduct of the treatment may be required to pay damages to the plaintiff. Simon (1987), however, cautioned that there is little case law on the specific subject of clinical supervision in the field of mental health and negligence.

Supervisors should be particularly careful when assigning what might be construed as high-risk activities to trainees or interns, as the supervisor is expected to exercise close supervision and control. They should carefully note the training and competencies of trainees and make certain they do not assign to a student a patient whose problems may be beyond the trainee's level of training, education, and experience. Lomax has warned of the importance that institutions carefully delineate the lines of supervisory responsibilities and the actual responsibilities of each supervisor. For example, it is critical that one understand the difference in responsibility and liability of a supervisor in a psychiatry residency program, wherein the supervisor is generally a paid faculty member of the department and a credentialed member of the clinical entity where the resident is assigned and is thus legally liable for all those that she or he supervises—in contrast to individuals who provide "individual supervision" (as educational consultants or consultee centered consultations" (J. W. Lomax, personal communication, 1991). Because there is little formal graduate training or residence training currently offered in the study of suicide (Bongar & Harmatz, 1989; Lomax, 1986), we recommend that all supervisors and training programs require that new trainees, interns, residents, and staff receive formal didactic and supervised experience in the management of both acute and chronic suicidal states. Such training should be part of the trainee orientation for training sites, psychiatric residencies, psychological service centers, hospitals, and so forth. (Such training must include close and early case supervision on this specific high-risk task for trainees, interns, residents, and staff.) Lomax (1986) has recommended that ideally each training program should have a designated expert in suicide care with a continuing clinical responsibility for suicidal individuals. "He or she should develop the local curriculum and have personal responsibility for the most critical elements of clinical supervision" (Lomax, 1986, p. 61). For additional guidelines and recommendations for the general supervision of clinical practice, the reader is directed to the work of Alonso (1985), Beutler (1988), Kaslow (1984), Langs (1979), Lomax (1986), Norcross (1984), Peterson (1985), Stoltenberg and Delworth (1987), and to a comprehensive bibliography on supervision (Robiner & Schofield, 1990).

Patients (and their families) have a right to an informed consent as to the trainee status of the psychotherapist. There is an ethical and legal obligation to tell patients of the therapist's trainee status and level of experience. Failure to provide patients with this information also may expose both trainee and supervisor to "possible lawsuits alleging fraud, deceit, misrepresentation, invasion of privacy, breach of confidentiality, and lack of informed consent" (VandeCreek & Harrar, 1988, p. 14).

In particular, the doctrine of informed consent would mandate that clinical supervisors share enough information with the patient regarding the trainee's level of education and experience, as well regarding the expected level of care, that a prudent patient can make an informed decision (VandeCreek & Harrar, 1988) about his or her own care.

While some psychologists have worried that sharing this information might scare off most clients, an equally persuasive argument can be made to the client that teaching facilities, with close supervision by experts, can provide exceptionally good care. [VandeCreek & Harrar, 1988, p. 14]

The above statements to patients about the expert quality of the supervision mandates that the supervisors themselves be soundly trained and highly knowledgeable when working with suicidal patients. It is essential to point out that even when a patient gives informed consent to care by a trainee the patient has not thereby consented to receive substandard care (Bongar, 1991; VandeCreek & Harrar, 1988). VandeCreek and Harrar pointed out the conclusion of some courts that trainees should have the "same standards of care as professionals who provide the same service (e.g., *Emory University* v. *Proubiansky*, 1981). It is critical to understand that if a patient is harmed the student, supervisor, employing hospital/agency/clinic, and possibly the sponsoring educational institution may all be made defendants, and liability can be alleged as either direct or vicarious (VandeCreek & Harrar, 1988).

Recommendations for Training Programs

When assessing and treating suicidal patients, we recommend that training institutions have policy guidelines for such management and that such guidelines be required in all institutions that offer supervised clinical experience. Furthermore, we concur with Bongar's (1991) position that any time a patient presents with even a mild-to-moderate level of elevated suicide risk the supervisor meet with the patient (along with the trainee) to ensure that all foreseeable elements of the elevated risk profile have been ascertained and that the management plan demonstrates affirmative precautions for safeguarding the patient. We believe that this procedure not only protects the patient but can serve as a valuable modeling experience for trainees to observe their supervisor at work in this high-risk endeavor. For example, Lomax (1986) has urged that, for psychiatry residency programs, each suicidal patient should be evaluated by a more senior resident or faculty member, and the resident should be observed

first-hand as he or she evaluates patients and initiates treatment. Particular emphasis should be placed on whether the resident is able to engage the suicidal patient in the interview process, whether he or she is able to frame questions and other interview responses in a way that facilitates the flow of communication, and whether he or she balances the need to obtain uncomfortable information with the avoidance of undue narcissistic injury to the patient. The resident's treatment plans for suicidal patients should be evaluated with respect to the clarity of the plans and the appropriate use of dispositions available, including hospitalization and psychotropic agents. [Lomax, 1986, p. 60]

Lastly, as for all nonsupervised clinical care, assessments and management precautions should be meticulously documented. Bongar (1991) noted that if the training site does not require that the supervisor sign each individual progress note after reviewing the trainee's work, at the very minimum the supervisor should sign (and if indicated amend) any risk-benefit analysis in the chart (e.g., where the assessment and management precautions for dangerousness are documented).

As practitioners, mental health professionals run the risk of losing a patient to suicide and of experiencing the traumatic impact of this event. As mentioned earlier, one study found that losing a patient to suicide was so personally and professionally impactful that almost half of the psychologists who lost a patient reported intrusive symptoms of stress during the weeks that followed the suicide—stress comparable to that of individuals who had suffered the death of a close family member (Chemtob et al., 1988a). This same study suggested that

patient suicide should be acknowledged as an occupational hazard for psychologists, not only because of its frequency but also because of its impact on psychologists' professional and personal lives . . . training programs do not currently have established protocols for helping trainees to deal with the aftermath of a patient's suicide; therefore trainees and their supervisors are left to their own devices. [Chemtob et al., 1988a, 419–420]

Goals and Objectives for the Training Sequence

The exact nature and duration of training in the study of suicide is clearly open to debate. Yet clinical competence in managing suicidal patients does not mean informational competence alone (Berman, 1986). There is research to support the contention that knowledge of risk factors and the capacity to respond in an effective way to those patients who present as an imminent risk of suicide may be independent areas of clinical competence (Inman, Bascue, Kahn, & Shaw, 1984). Thus clinical supervision as part of any formal training is of vital importance. A model curriculum in the study of suicide might include the setting of minimal standards of competence and the assessment of both knowledge and skill as a standard of competence (Berman & Cohen-Sandler, 1983).

Such a model program (incorporating research, theory, professional practice issues, and clinical supervision techniques) could be integrated into the typical psychiatric residency experience, graduate seminar on professional practice, and so forth. This model should include not only the basic data on research and clinical techniques but also the current clinical and empirical findings on the management of special high-risk populations (e.g., the problem of suicide among the young, the elderly, the chronically mentally ill, and the physically ill). This training model would also include essential knowledge of the necessary postvention strategies for suicide survivors, including protocols for helping the clinician who had treated the patient (Brown, 1987; Chemtob et al., 1988a,b; Peterson & Bongar, 1990).

Lomax (1986) has proposed that a core curriculum in suicide care include critical didactic and clinical experiences in the conceptual categories of knowledge, skills, and attitudes. The knowledge base could be covered in formal didactic settings such as seminars or lectures. Furthermore, knowledge also evolves from the supervisory process. This didactic component should occur early in the training sequence (e.g., during the first 6 months of psychiatry rotations). Skill development can be carefully addressed in the trainees' clinical assignments and individual supervision, and ideally assignments to working with high-risk patients would not occur until the trainee has mastered the basics of nonemergency clinical care. The training sequence must also help in the development of functional attitudes toward a patient's suicidal statements and behaviors. Here individual supervision is the

starting point for developing competence in recognizing awareness of one's attitudes about patients who talk about suicide or give other clues toward contemplated self-destruction—including patients whose behavior is more obviously influenced by wishes to produce an effect on others in their environment or remove them from an intolerable situation. . . . Individual supervision plays a particularly critical role in the emotionally charged event of postvention. [Lomax, 1986, p. 60]

Finally, it is important to understand that it is in our mental health training programs where we begin the process of becoming clinicians—that, by implication, educational and training credentials imply competence (Barron, 1987). Maris and colleagues (1973) warned that we must especially be careful to guard against the conception of training in the study of suicide as a mechanical dissemination of factual knowledge, and that those involved in teaching the subject understand the "crucial importance of consultation and supervision as methods of training and education in suicidology" (p. 34).

Each training program in the core mental health disciplines has its own unique features and therefore should develop curricula that are suitable to the specific character and realities of each training institution. Lomax (1986) noted that there must be an evaluation component to assess the trainees' mastery of suicidal care (both written and clinical). Furthermore, each program must decide what it considers to be the minimal practice competencies that students need to develop in the areas of suicide prevention, intervention, and postvention, and these goals and objectives should be operationalized and made explicit to trainees early in their clinical education.

Guidelines for Practice: Chapter Summary

It is useful to note that the problem of suicide risk assessment is basically the same as other evaluative procedures in medicine and psychology, that is,

[D]ata are gathered from the patient, from the family, and other collaterals, from the history, from specific tests, and from direct observation. All information is tinged with intuitive elaboration of a nature and degree that is unique to each clinician. Though the available information at any given time may be incomplete, ambiguous, contradictory, or of questionable validity, a decision is made, primarily on intuitive grounds, as to estimated level of risk. As new data become available, and circumstances change, the estimate of risk is modified accordingly. [Motto, 1989, p. 256]

In addition, Bongar (1991) cautioned that training in the study of suicide is more than the "cookbook" memorization of a few demographic and general clinical risk factors and that high quality training in the care of the suicidal patient provides a basic understanding of the complexities and controversies in our understanding of suicide, and how these affect clinical practice. Ultimately, the success of the training sequence can be judged on how well each trainee understands the necessity of treating each patient's risk profile and management as a unique set of variables to be comprehensively addressed in his or her clinical assessment and case management activities.

REFERENCES

Alcohol, Drug Abuse, and Mental Health Administration (1989). *Report of the Secretary's task force on youth suicide (Vols. I–IV)*. DHSS Publ. No. (ADM) 89-1621-1624. Washington, DC: U.S. Government Printing Office.

Alonso, A. (1985). *The quiet profession: supervisors of psychotherapy*. New York: Macmillan.

American Association of Suicidology (1987). AAS Training Committee Bibliographic Recommendations. Available from the American Association of Suicidology, Denver, CO.

Barron, J. (1987). Clinical psychology: racing into the future with old dreams and visions. *The Clinical Psychologist, 40*(4), 93–96.

Beck, A. T. (1967). *Depression: Clinical, experimental, and theoretical aspects*. New York: Harper & Row.

Beck, A. T., Kovacs, M., & Weissman, A. (1975). Hopelessness and suicidal behavior. *Journal of the American Medical Association, 234*, 1146–1149.

Beck, A. T., Kovacs, M., & Weissman, A. (1979). Assessment of suicidal intention: the scale for suicide ideation. *Journal of Consulting and Clinical Psychology, 47*, 343–352.

Beck, A. T., & Steer, R. A. (1989). Clinical predictors of eventual suicide: a 5- to 10-year prospective study of suicide attempters. *Journal of Affective Disorders, 17*, 203–209.

Berman, A. (1983). Training Committee Report. Unpublished manuscript. Available from the American Association of Suicidology, Denver, CO.

Berman, A. (1986). A critical look at our adolescence: Notes on turning 18 (and 75). *Suicide and Life-Threatening Behavior, 16*, 1–12.

Berman, A. L., & Cohen-Sandler, R. (1982). Suicide and the standard of care; optimal vs. acceptable. *Suicide and Life-Threatening Behavior, 12*, 114–122.

Berman, A. L., & Cohen-Sandler, R. (1983). Suicide and malpractice: expert testimony and the standard of care. *Professional Psychology: Research and Practice, 14*(1), 6–19.

Beutler, L. E. (1988). Introduction to the special series: training to competency in psycho-therapy. *Journal of Counsulting and Clinical Psychology, 56*, 651–652.

Board of Professional Affairs, Committee on Professional Standards (1987). Casebook for providers of psychological services. *American Psychologist, 42*, 704–711.

Bongar, B. (1991). *The suicidal patient: clinical and legal standards of care.* Washington, DC: American Psychological Association.

Bongar, B., & Harmatz, M. (1989). Graduate training in clinical psychology and the study of suicide. *Professional Psychology: Research and Practice, 20*, 209–213.

Bongar, B., & Harmatz, M. (1990). Clinical psychology graduate education in the study of suicide: availability, resources, and importance in NCSPP versus CUDCP programs. In B. Bongar and E. A. Harris (Chairmen), *Contemporary developments in the assessment and management of the suicidal patient.* Symposium conducted at the meeting of the American Psychological Association, Boston.

Bongar, B., Peterson, L. G., Harris, E. A., & Aissis, J. (1989). Clinical and legal considerations in the management of suicidal patients: an integrative overview. *Journal of Integrative and Eclectic Psychotherapy, 8*(1), 53–67.

Brown, H. N. (1987). The impact of suicides on therapists in training. *Comprehensive Psychiatry, 28*(2), 101–112.

Chemtob, C. M., Bauer, G. B., Hamada, R. S., Pelowski, S. R., & Muraoka, M. Y. (1989). Patient suicide: occupational hazard for psychologists and psychiatrists. *Professional Psychology: Research and Practice, 20*, 294–300.

Chemtob, C. M., Hamada, R. S., Bauer, G. B., Kinney, B., & Torigoe, R. Y. (1988a). Patient suicide: frequency and impact on psychiatrists. *American Journal of Psychiatry, 145*, 224–228.

Chemtob, C. M., Hamada, R. S., Bauer, G. B., Torigoe, R. Y., & Kinney, B. (1988b). Patient suicide: frequency and impact on psychologists. *Professional Psychology: Research and Practice, 19*, 421–425.

Clark, D. C., Gibbons, R. D., Fawcett, J., & Scheftner, W. A. (1989). What is the mechanism by which suicide attempts predispose to later suicide attempts? A mathematical model. *Journal of Abnormal Psychology, 98*(1), 42–49.

Deutsch, C. J. (1984). Self-report sources of stress among psychotherapists. *Professional Psychology: Research and Practice, 15*, 833–845.

Emory University v. *Porubiansky,* 282 S.E. 2nd 903 (1981).

Fawcett, J., Scheftner, W., Clark, D., Hedeker, D., Gibbons, R. & Coryell, W. (1987). Clinical predictors of suicide in patients with major affective disorders: A controlled prospective study. *American Journal of Psychiatry, 144*(1), 35–40.

Farberow, N. L. (1957). The suicidal crisis in psychotherapy. In E. S. Shneidman & N. L. Farberow (Eds.), *Clues to suicide* (pp. 119–130). New York: McGraw-Hill.

Frederick C. J. (1978). Current trends in suicidal behavior. *American Journal of Psychotherapy, 32*(2), 172–200.

Fremouw, W. J., de Perczel, M., & Ellis, T. E. (1990). *Suicide risk: assessment and response guidelines.* New York: Pergamon Press.

Gutheil, T. G., & Appelbaum, P. S. (1982). *Clinical handbook of psychiatry and the law*. New York: McGraw-Hill.

Hirschfeld, R., & Davidson, L. (1988). Risk factors for suicide. In A. J. Frances & R. E. Hales (Eds.) *American Psychiatric Press Review of Psychiatry Vol. 7* (pp. 307–333). Washington, DC.: American Psychiatric Press.

Inman, D., Bascue, J., Kahn, W., & Sharp, P. (1984). The relationship between suicide knowledge and suicide interviewing skills. *Death Education, 8*, 179–184.

Jacobs, D. G. (1984). Psychopharmacologic management of the psychiatric emergency patient. *General Hospital Psychiatry, 6*, 203–210.

Jacobs, D. G. (1989). Psychotherapy with suicidal patients: the empathic method. In D. G. Jacobs & H. N. Brown (Eds.), *Suicide: understanding and responding: Harvard Medical School perspectives on suicide* (pp. 329–342). Madison, CT: International Universities Press.

Kaslow, F. W. (Ed.) (1984). *Psychotherapy for psychotherapists*. New York: Haworth.

Keller, M. B. (1988). Afterword to section II: unipolar depression. In A. J. Frances & R. E. Hales (Eds.), *American Psychiatric Press review of psychiatry* Vol. 7; pp. 284–287). Washington, DC: American Psychiatric Press.

Kleespies, P. M., Smith, M. R., & Becker, B. R. (1990). Psychology interns as patient suicide survivors: Incidence, impact, and recovery. *Professional Psychology: Research and Practice, 21*(4), 257–263.

Klerman, G. L. (1987). Clinical epidemiology of suicide. *Journal of Clinical Psychiatry, 48* (Suppl.), 33–38.

Langs, R. (1979). *The supervisory experience*. New York: Jason Aronson.

Lesse, S. (1989). The range of therapies with severely depressed suicidal patients. In S. Lesse (Ed.), *What we know about suicidal behavior and how to treat it* (pp. ix–xiv). Northvale, NJ: Jason Aronson.

Litman, R. E. (1982). Hospital suicides: lawsuits and standards. *Suicide and Life-Threatening Behavior, 12*, 212–220.

Lomax, J. W. (1986). A proposed curriculum on suicide for psychiatric residency. *Suicide and Life-Threatening Behavior, 16*, 56–64.

London, P. (1986). Major issues in psychotherapy integration. *International Journal of Eclectic Psychotherapy, 5*, 211–216.

Maltsberger, J. T. (1986). *Suicide risk: the formulation of clinical judgement*. New York: New York University Press.

Mann, J. J., & Stanley, M. (1988). Afterword to section III: suicide. In A. J. Frances & R. E. Hales (Eds.), *American Psychiatric Press Review of Psychiatry* (Vol. 7; pp. 422–426). Washington, DC: American Psychiatric Press.

Maris, R. W. (1981). *Pathways to suicide: a survey of self-destructive behaviors*. Baltimore: Johns Hopkins University Press.

Maris, R. W. (1988). Preface: overview and discussion. In R. W. Maris (Ed.), *Understanding and preventing suicide: plenary papers of the first combined meeting of the AAS and IASP* (pp. vii–xxiii). New York: Guilford.

Maris, R. W., Dorpat, T. L., Hathorne, B. C., Heilig, S. M., Powell, W. J., Stone, H., & Ward, H. P. (1973). Education and training in suicidology for the seventies. In H. L. P. Resnick & B. C. Hathorne (Eds.), *Suicide prevention in the seventies* (pp. 23–44). Washington, DC: U.S. Government Printing Office.

Motto, J. A. (1979). Guidelines for the management of the suicidal patient. In *Weekly Psychiatry Update Series Lesson 20, 3*, 3–7. Available from Biomedia, Princeton, NJ.

Motto, J. A. (1986). Clinical considerations of biological correlates of suicide. *Suicide and Life-Threatening Behavior, 16*, 1–20.

Motto, J. A. (1989). Problems in suicide risk assessment. In D. G. Jacobs & H. N. Brown (Eds.), *Suicide: Understanding and responding: Harvard Medical School perspectives on suicide* (pp. 129–142). Madison, CT: International Universities Press.

Murphy, G. (1984). The prediction of suicide. *American Journal of Psychotherapy, 38*, 341–349.

Norcross, J. C. (1984). The training of clinical psychologists: some training predictions and recommendations. *Clinical Psychologist*, *37*(1), 23–24.

Peterson, D. R. (1985). Twenty years of practitioner training in psychology. *American Psychologist*, *40*, 441–451.

Peterson, L. G., & Bongar, B. (1989). The suicidal patient. In A. Lazare (Ed.), *Outpatient psychiatry: diagnosis and treatment* (2nd ed.; pp. 569–584). Baltimore: Williams & Wilkins.

Peterson, L. G., & Bongar, B. (1990). Training physicians in the clinical evaluation of the suicidal patient. In M. Hale (Ed.), *Teaching methods in consultation-liaison psychiatry* (pp. 89–108). Basel: Karger.

Pokorny, A. D. (1964). Suicide rates in various psychiatric disorders. *Journal of Nervous and Mental Disease*, *139*, 499–506.

Pope, K. (1986). Assessment and management of suicidal risks: clinical and legal standards of care. *Independent Practitioner*, January, 17–23.

Robiner, W. N., & Schofield, W. (1990). References on supervision in clinical and counseling psychology. *Professional Psychology: Research and Practice*, *21*, 297–312.

Robins, E. (1985). Suicide. In H. I. Kaplan & B. J. Sadock (Eds.), *Comprehensive textbook of psychiatry* (Vol. 4; pp. 311–1315). Baltimore: Williams & Wilkins.

Roy, A. (1986). Preface. In A. Roy (Ed.), *Suicide* (p. vii). Baltimore: Williams & Wilkins.

Sainsbury, P. (1986). The epidemiology of suicide. In A. Roy (Ed.), *Suicide* (pp. 17–40). Baltimore: Williams & Wilkins.

Schein, H. M. (1976). Obstacles in the education of psychiatric residents. *Omega*, *7*, 75–82.

Shneidman, E. S. (1975). Suicide. In A. E. M. Freedman, H. I. Kaplan, & B. J. Saddock (Eds.), *Comprehensive textbook of psychiatry* (Vol. 2). Baltimore: Williams & Wilkins.

Shneidman, E. S. (1981). Postvention: the care of the bereaved. *Suicide and Life-Threatening Behavior*, *11*, 349–359.

Shneidman, E. S. (1985). *Definition of suicide*. New York: John Wiley & Sons.

Shneidman, E. S. (1986). Some essentials of suicide and some implications for response. In A. Roy (Ed.), *Suicide* (pp. 1–16). Baltimore: Williams & Wilkins.

Simon, R. I. (1987). *Clinical psychiatry and the law*. Washington, DC: American Psychiatric Press.

Simon, R. I. (1988). *Concise guide to clinical psychiatry and the law*. Washington, DC: American Psychiatric Press.

Slovenko, R. (1985). Forensic psychiatry. In H. I. Kaplan & B. J. Sadock (Eds.), *Comprehensive textbook of psychiatry* (Vol. 4; pp. 1960–1990). Baltimore: Williams & Wilkins.

Smith, J. (1986). *Medical malpractice psychiatric care*. Colorado Springs, CO: Shepards/McGraw-Hill.

Smith, K. (1988). The psychotherapy of suicidal patients. Presented at the annual meeting of the American Association of Suicidology, Washington, DC.

Smith, K., & Maris, R. (1986). Suggested recommendations for the study of suicide and other life-threatening behaviors. *Suicide and Life-Threatening Behavior*, *16*, 67–69.

Stoltenberg, C. D., & Delworth, U. (1987). *Supervising counselors and therapists*. San Francisco: Jossey-Bass.

VandeCreek, L., & Harrar, W. (1988). The legal liability of supervisors. *Psychotherapy Bulletin*, *23*(3), 13–17.

Weissman, M. M., Klerman, G. L., Markowitz, J. S., & Ouelette, R. (1989). Suicidal ideation and suicide attempts in panic disorder and attacks. *New England Journal of Medicine*, *321*, 1209–1214.

Wekstein, L. (1979). *Handbook of suicidology: principles, problems and practice*. New York: Brunner/Mazel.

Welch, B. (1989). A collaborative model proposed. *American Psychological Association Monitor*, *20*(10), 28.

16

Guidelines for Risk Management in the Care of the Suicidal Patient*

BRUCE BONGAR

When a clinician is ignorant of the law, the courts and the legal system seem to hover as menacing specters around the practitioner's clinical endeavors. Yet a clinically useful understanding of how the legal system works may actually enhance clinicians' enjoyment of their practice activities by, in Simon's words, "making the law a working partner" (1988, p. xv). Although clinicians are not required to be lawyers, they are required to practice within the law and should attempt to "incorporate legal issues into their management of patients—turning the law to clinical account for the benefit of the patients" (Simon, 1988, p. xv). In addition, practitioners have an moral and ethical responsibility to provide care for patients, a duty that often transcends any minimal standard that may be imposed by the law or regulatory agencies, a duty to strive for an optimal standard of care in their own practices (Bongar, 1991).

It is further argued (Bongar, 1991) that in the current climate of increased malpractice actions against mental health practitioners it would be naive for the practicing clinician not to consider appropriate clinical and legal management issues when treating certain high-risk populations (e.g., patients who are dangerous to self or others). Indeed, Simon (1988) argued that it would be not merely naive but foolhardy to ignore risk management procedures in the course of treating such patients. Here the key factors are to know when to apply risk management practices and to make certain that patients are helped and not harmed by such practices (Simon, 1988).

A concern that supercedes the many specific suggestions on risk management in this chapter is my belief that the best overall risk management strategy remains a sensitive, caring therapeutic alliance within the context of the best possible clinical care. (I discuss this alliance specifically at the end of the chapter.)

As a starting point for our understanding risk management in the suicidal scenario, we turn now to what Harris (1990) has determined are the elementary techniques of successful risk management: techniques that, if understood and used effectively, significantly minimize the risk of being found negligent in a

---

*Portions of this chapter were adapted from *The Suicidal Patient: Clinical and Legal Standards of Care*, by B. Bongar, 1991, Washington, DC: American Psychological Association. By permission.

malpractice action, as well as being of substantial benefit and utility in defending oneself against complaints filed with professional ethics committees, licensing boards, and other bodies by which one is held responsible.

Harris (1990) noted that mental health professionals who wish to incorporate high-quality risk management activities as part of their professional practice activities must first and foremost understand their professions' ethical standards and combine this understanding with specific laws and regulations that govern practice in their state. Effective risk management includes the additional requirement of obtaining essential clinical assessment and management information on specific at risk populations, understanding the relation between the law and mental health practitioners, knowing the rules and limitations regarding confidentiality and informed consent, understanding how courts determine malpractice, and learning how professional liability insurance policies work. (See Bennett, Bryant, VandenBos, and Greenwood [1990] for a more extended discussion of these general issues and their relation to professional liability.) Finally, Harris commented on the critical relation between effective risk management and the important elements of documentation and consultation.

Guthiel (1990) claimed that the latter two elements are "the twin pillars" of liability prevention (p. 338). He went on to state that good documentation provides a durable contemporaneous record, not only of what happened but of the exercise of the mental health professional's judgment, the risk-benefit analysis, and the patient's ability to participate in planning his or her own treatment. The use of consultation, then, provides a "biopsy" of the standard of care, capturing in a practical way the reasoning of the

"average and reasonable practitioner," that mythical being who represents the reference standard for the determination of the standard of care and alleged deviations therefrom. [Guthiel, 1990, p. 338]

Before beginning our discussion of risk management, it is crucial to state that the information contained in this chapter is no substitute for a timely, formal consultation with a knowledgeable attorney and with one's professional colleagues. In particular, clinicians with specific questions or those who are threatened with a suit should follow the suggestions of Wright (1981) and VandeCreek and Knapp (1983) that the first step (when one has specific legal concerns or has reason to believe that a malpractice suit is imminent) is straightforward: Consult an attorney who is expert in matters of mental health and the law.

A crucial premise for understanding risk management and the suicidal patient is that clinicians have a duty to take steps to prevent suicide if they can reasonably anticipate the danger. "Therefore, the key issues in determining liability are whether the psychotherapist should have predicted that the patient was likely to attempt suicidal behavior, and (assuming there was an identifiable risk) whether the therapist did enough to protect the patient" (Stromberg et al., 1988, p. 467).

In this regard, Pope (1986, p. 19) stressed the importance of staying within one's area of competence and of knowing one's personal limits, observing "that working with suicidal patients can be demanding, draining, crisis filled activity. It is literally life or death work." In addition to obtaining adequate training and knowledge of the literature on suicide, mental health professionals must become familiar with the legal standards involving rights to treatment and to refuse treatment, as well as the rules regarding confidentiality, involuntary hospitalization, and so forth. Pope noted that a standard of care must involve a screening for suicide risk during the initial contact and ongoing alertness to this issue throughout the course of treatment. There also should be frequent consultation and ready access to facilities needed to implement appropriate affirmative precautions (e.g., emergency teams, hospitals, crisis intervention centers, day treatment).

The courts and the legal system have been sympathetic to the difficulties clinicians have when predicting suicides and, as a result, rarely have imposed liability in the absence of prior observable acts or verbal threats by the patient. As one example of this reality, in the case of *Bogust* v. *Iverson*, a college guidance counselor was held not liable when a student committed suicide 6 weeks after sessions with the counselor had ended. The student had not talked about suicide and had not exhibited behavior that would have prompted the counselor to initiate procedures for a civil commitment of the student (Stormberg et al., 1988).

Bongar (1991) noted that the general legal standard for patient care clearly includes a thorough understanding of the complexities of procedures for assessing elevated risk and specific clinical management techniques for the suicidal patient. Mental health professionals have been *held liable* when they have not taken adequate precautions to manage patients. The courts do not necessarily defer to a clinician's decisional process when they find that "due to a totally unreasonable professional judgment, he or she underestimated the need for special care, or failed to take the usual precautions" (Stromberg et al., 1988, p. 468).

The mental health professional's assessment and treatment efforts represent an opportunity to translate knowledge (albeit incomplete) of elevated risk factors into a plan of action (Bongar, 1991). The management plan for patients who are at elevated risk for suicide should ameliorate those risk factors that are most foreseeably likely to result in suicide or self-harm (Brent, Kupfer, Bromet, & Dew, 1988). Here there are several general principles that should guide the treatment of patients at elevated risk for suicide and that apply across broad diagnostic categories. The most basic principle is that, because most suicide victims take their own lives or harm themselves in the midst of a psychiatric episode (Barraclough, Bunch, Nelson, & Sainsbury, 1974; Dorpat & Ripley, 1960; Murphy, 1988; Robins, Murphy, Wilkinson, Gassner, & Kays, 1959; Shaffer, 1974; Shaffer, Gould, & Trautman, 1985; Shaffi, 1986), it is critical to understand that a proper diagnosis and careful management/treatment plan of the acute psychiatric disorder could dramatically alter the risk for suicide (Brent et al., 1988). Although there is a loose fit between diagnosis and suicide,

Simon (1988) noted that suicide rarely occurs in the absence of psychiatric illness; the data on adult suicides indicate that more than 90% of these suicide victims were mentally ill before their deaths.

In addition to the above requirements for acute management, Litman (1988), Simon (1987, 1988), and Gutheil (1990) noted the special precautions that clinicians must take when assessing and treating patients who present with chronic suicidal ideation and behavior (e.g., when the clinician takes repeated calculated risks in not hospitalizing). Gutheil (1990) noted that here the mental health clinician feels the tension between short-term solutions (e.g, a protected environment) and long-term solutions (e.g., actual treatment of the chronicity).

Other general principles include family involvement for support and improved compliance; diagnosis and treatment of any co-morbid medical and psychiatric condition; the provision of hope, particularly to new-onset patients; the restriction of the availability of lethal agents; and indications for psychiatric hospitalization (Brent et al., 1988). To this list a risk management perspective would add the critical necessity of assessing personal and professional competencies in order to treat at-risk patients, as well as meticulous documentation and the routine involvement of "a second opinion" through consultation (Bongar, 1991; Bongar, Peterson, Harris, & Aissis, 1989).

Our assessment and management activities should include a specific evaluation of the patient's competence to participate in management and treatment decisions, especially the patient's ability to form a therapeutic alliance (Bongar, 1991; Gutheil, 1984, 1990; Kahn, 1990; Luborsky, 1990). Bongar (1991) noted that an essential element in strengthening this alliance is the use of informed consent; that is, patients have the right to participate actively in making decisions about their psychological/psychiatric care. Clinicians need to evaluate the quality of this special relationship, directly and continuously, to understand that the quality of this collaborative alliance is inextricably part of any successful treatment/management plan (Bongar, 1991; Bongar et al., 1989; Gutheil, 1984, 1988, 1990; Kahn, 1990; Luborsky, 1990; Motto, 1979; Shneidman, 1981, 1984; Simon, 1987, 1988).

Simon (1988) noted that suicide results from the complex interplay of a number of diagnostic (psychiatric and medical), constitutional, environmental, occupational, sociocultural, existential, and chance casual elements. It is not simply the result of misdiagnosis or inadequate treatment. Courts sometimes have trouble understanding that psychotherapists are "ordinary mortals struggling with" this conundrum (Simon, 1987, p. 264) and that mental health professionals are not able to guarantee control over the behavior of their patients, particularly patients in outpatient treatment.

As a first step in risk management, clinicians must determine their own technical and personal competence to work with such high-risk patients. Although the law does not record a case to date of negligent psychotherapy where the basis is a failure to cure or to relieve a psychiatric symptom, verbal psychotherapies are not without risk (Simon, 1987). Patients who are improperly diagnosed or given an inappropriate type of psychotherapy may indeed regress and present with suicidal ideation or behavior (Simon, 1987; Stone, 1989a,b). Unusual or extreme therapies, as well as innovative or regressive therapies, in addition to sexual seduction by therapists and malignant countertransferences, can be taken as evidence of treatment gone awry. The presence of such factors may increase the risk of suicide significantly (Simon, 1987). Stone (1989b) has cautioned about the risks of forcing a patient to see that his or her reality situation is empty of possible gratification, as well as being cautious when the context of this malignant insight is such that the patient feels even more hopeless and helpless. At times therapy is less than adequate owing to ideological, theoretical, or technical prejudices (Lesse, 1989). "Not all psychiatrists and other psychotherapists are equipped emotionally or technically to manage suicidal patients" (Lesse, 1989, p. 215).

In coming to terms with their own limitations, one must remember, as Welch (1989) pointed out, that all mental health providers (psychiatrists, psychologists, social workers, psychiatric nurses) are limited to varying degrees as to their specific professional competencies. Welch continued:

One might further argue that the greatest threat to "quality of care" comes not from those with limited training but from those with limited recognition of the limitations of their own training." [Welch, 1989, p. 28]

Bongar (1991) observed that one of the initial tasks of the mental health professional called on to treat the suicidal patient is the need to have evaluated a priori the strengths and limitations of his or her own training, education, and experience in the treatment of specific patient populations in specific clinical settings (e.g., an understanding of their own technical proficiencies, as well as their emotional tolerance levels for the intense demands required for treating suicidal patients).

Risk Management and Documentation

Gutheil (1980) remarked that the prudent mental health practitioner might well use paranoia as a motivating force to make psychiatric records effective for forensic purposes, utilization review, and treatment planning. Gutheil's key points were "If it isn't written down, it didn't happen" and "What you see is what you've got." As a general rule, he continued, clinicians should write their chart notes as if a lawyer were sitting on their shoulders, reviewing every word. He stated that using paranoia as a guiding reality principle in these litigious times is a sound basis for effective record keeping, because:

We mental health professionals should face, with dispassionate resoluteness, the cold fact that certain people are out to get us. These people are called "lawyers," and the reason they are out to get us is simple: they are paid to do so. The plot is variously termed "malpractice litigation," "contemporary narcissistic entitlement" or the "American disease." . . . These facts are familiar to anyone able to read a newspaper and need not be belabored, but it is this reality-based paranoia that may serve as our stimulus in attempting to achieve records unassailable from the viewpoints of utilization review, forensic considerations, and treatment. [Gutheil, 1980, pp. 479–480]

As Gutheil (1980) further noted, honest error is separable from negligence in theory, but in practice juries often confound the distinction. There is no infallible protection against this fact of forensic life.

Thus an essential element in risk management is the maintenance of timely, meticulous records. However, the clinician should never lose sight of the most important purpose of clinical records and the rationale that properly underlies the keeping of such meticulous high-quality records—namely, such documentation is an organizing framework for focusing the mental health professional's attention on the making of sound clinical judgments (Bongar, 1991).

This ethos of meticulousness is of particular importance in clinical situations that are suffused with uncertainty (Gutheil, 1990). Such situations specifically require that the clinician "think out loud for the record."

For example, in the case of not hospitalizing a suicidal outpatient, such informed record-keeping would often include "thinking out loud for the record" regarding the dangers to which the patient might be exposed and the "careful articulation of the pros and cons, including known risks and disadvantages and the reasons for overriding them . . . specific dates, and names are included, showing that the treating professional did not operate alone and unchecked in making this difficult but commonly encountered situation" (Gutheil, 1980, p. 482). The treatment planning builds explicitly on past observed and recorded data and consultations. Gutheil noted that "as a general rule, the more uncertainty there is, the more one should think out loud in the record" (1980, p. 482). For a detailed discussion of the details of maintaining such meticulous records in clinical practice see Bongar (1991), Gutheil (1980), Simon (1988), and VandeCreek and Knapp (1989).

The lack of documentation can fatally cripple the defendant's case, even if the therapist had acted in a conscientious and professionally sound manner. Numerous case consultations have supported this conclusion, including a case where the consultant believed there was no negligence on the part of the treating staff, but "the almost complete lack of records left a legitimate issue as to the fact and so the settlement against the hospital and psychiatrist was made" (Perr, 1985, p. 217). The settlement in that case was for $500,000 [VandeCreek & Knapp, 1989, p. 30]

The power of documentation when retrospectively evaluating the quality of assessment and treatment is underscored by the observation that "clinicians who make bad decisions but whose justification for the intervention is well documented often come out better than clinicians who have made reasonable decisions but poor documentation leaves them vulnerable" (Gutheil, 1984, p. 3).

The typical good clinical record should be explicit about such treatment decisions as whether to hospitalize the patient as well as those concerning therapeutic impasses, pass/discharge privileges, any uncertainty about diagnosis, and evaluation of psychosocial supports. Furthermore, VandeCreek and

Knapp (1989) observed that the clinician should carefully document any decisions to reduce the frequency of observations of suicidal patients. Bongar (1991) pointed out that every significant decision point also must include a risk-benefit analysis that indicates all actions one considered, the reasons that led one to take an action, and the reasons that led one to reject action. The record must indicate specifically why consultation and supervision were or were not employed and include a written record of the consultant's recommendations. Also, Gutheil (1980, p. 482) noted that "malpractice suits, it must be obvious, have been won or lost on matters of timing. . . . For this reason alone, as well as for the clinical need to reconstruct events with accuracy, the use of time notations (as well as dates) is a useful habit to develop."

In summary, the clinical record should reflect which sources of information were consulted, what factors went into the clinical decision, and how the factors were balanced by the use of a risk-benefit assessment (Bongar, 1991). Such risk-benefit notes are the decisional roadmarks in a psychotherapist's clinical formulation of the management/treatment plan (Simon, 1988).

In addition, it is a grave error to ignore the written records from a patient's previous treatment (Simon, 1987, 1988). With patient permission, the clinician may wish to contact family members, who can help determine the gravity of past suicide attempts (Simon, 1987, 1988).

All of this historical information needs to be recorded in the chart and incorporated into the ongoing treatment plan. Bongar (1991) pointed out that the absence of efforts to obtain previous medical and psychotherapy records is a reliable channel marker for finding other signs of inadequate clinical care.

Hypothetical Risk-Benefit Note

Bongar (1991) has written that a model risk-benefit progress note would include a completed assessment of risk and incorporate precisely which information alerted the clinician to that risk and which high risk factors were present in that situation and the patient's background, as well as low-risk factors (e.g., reasons to live, care of a young child). What questions did one ask, and what answers did one get? How did this information, the patient's history, and one's own clinical/evaluative judgments lead to the actions taken and rejected? This analysis should include the specific pros and cons of each action from a clinical and a legal perspective. One should state with whom one formally consulted, what was communicated to them, the nature of their response, and the actions they recommended. Also, one should indicate whether the recommendations were clear-cut. If there were alternative recommendations from the consultant, they should be described in detail, together with the rationale for not exercising those alternatives. If the opinions of consultants differed from one's own or from each other, what were the sources of difference?

Whenever possible, the risk-benefit note should detail that the clinician understood the role of informed consent and the right of the competent patient to participate collaboratively in the decision-making process. Specifically, the chart should describe the mental health professional's efforts to involve the competent patient (and, when indicated their significant others) in an open discussion of the risks and benefits of a particular course of action. If there is any disagreement in this process, it is wise to advise the patient and family immediately that they have the right to obtain a "second opinion" and to facilitate such a consultation.

The clinical record should be as timely as possible, but it should not prevent one from including details at a later date that one neglected in the heat of the moment. However, altering or rewriting the record after one discovers that there may be questions about one's decisions (e.g., a suit) can destroy the possibility of an adequate defense. As Hoge and Appelbaum (1989, p. 620) noted, "no single act so destroys the clinician's credibility in court."

Obviously, no mental health professional can ever obtain all the information recommended for every forensically significant situation. The more information that is contained in the record, however, the more the record demonstrates that even though the result may have been unfortunate the practitioner behaved in a reasonable professional manner given the information she or he had at the time. The extra time and effort required to draft comprehensive records pays high dividends should the tragedy of a patient suicide occur.

Research indicates that it may be advisable to warn the support system and significant others of a patient's suicidal potential and generally to increase their involvement in management and treatment (VandeCreek & Knapp, 1989). Such involvement can be a strong factor in promoting the patient's recovery. Observing that suicide is often a highly charged dyadic process, Shneidman (1981) urged support group involvement in suicide prevention efforts. He also stated that at the very least every mental health professional who treats suicidal patients must carefully assess the interpersonal matrix for the role of significant others as either helpers or hinderers in the treatment process. If the patient does commit suicide, the clinician has established the communication channels and, ideally, good relations with the family that may facilitate a healthy resolution of ensuing sorrow and grief.

In understanding a malpractice action, Simon (1988) has modified Sadoff's "dereliction of duty directly causing damages" (Sadoff, 1975) to the "4Ds": (1) *duty* of care; (2) *deviation* from the standard of care; (3) *damage* to the patient; and (4) *directly* the result of the deviation from the standard of care. However, Gutheil contended that, in the real world, the determination of malpractice rests on the malignant synergy of

a bad outcome from whatever cause, in concert with "bad feelings" in the plaintiffs, including guilt, rage, grief, surprise, betrayal, and sense of being left alone with the bad outcome (psychological abandonment). [Gutheil, 1990, p. 335]

The suicide of a loved one is a catastrophic outcome to treatment and one that Gutheil believed is destined to leave in its wake a host of bad feelings toward the clinician. Litigation over suicide may provide a mechanism for displacement of the family's guilt onto the clinician, serving the grieving family as an "antidepressant" mechanism for grief avoidance.

Informed Consent and Confidentiality: Legal Considerations

When clinicians formulate a treatment plan, they face the essential task of involving the patient in the treatment process. However, the law of informed consent can be confusing to mental health professionals, who tend to see this task as an intrusion by the legal system into the treatment process and who reduce it "to a meaningless, mechanistic ritual of form signing" (Hoge & Appelbaum, 1989, p. 613). If the mental health professional, instead, views the process of informed consent as an opportunity to increase communication and collaboration between the psychotherapist and patient, this particular task "can have a powerful therapeutic influence of its own" (Hoge & Appelbaum, 1989, p. 613).

The legal and ethical rationale for informed consent is grounded in the principle that patients should have the right to participate actively in making decisions about their psychological care. Not only are patients likely to cooperate more in a treatment they have had an active role in selecting, but the likelihood is greater that the chosen treatment will specifically address the patient's real concerns (Hoge & Appelbaum, 1989). The clinician should note, however, Simon's (1988) four exceptions to the requirement for informed consent:

1. Emergencies, in which immediate treatment is needed to prevent imminent harm
2. A waiver, with which the patient knowingly and voluntarily waives his or her right to be informed
3. Therapeutic privilege, with which the psychologist determines that a complete disclosure might have deleterious effects on the patient's well being
4. Incompetence on the part of the patient to give consent (for a detailed discussion of informed consent and competence see Simon [1988, pp. 26–42])

Hoge and Appelbaum (1989) stated that when screening for incompetence legal definitions have tended to be vague. They recommended that clinicians conceptually use the following hierarchy (in order of increasing stringency).

The clinician must ask whether the patient can:

1. Show evidence of a choice concerning treatment
2. Achieve a factual understanding of the issues at hand
3. Rationally manipulate the information provided to him or her
4. Appreciate the nature of the situation and the consequences of the decision

Failure in any of the above standards may warrant a judicial determination of incompetence (Hoge & Appelbaum, 1989). Realistically, there are situations (e.g., emergencies when any delay of treatment could result in harm to the patient) in which the requirements of informed consent are suspended. "The law is willing to presume that a reasonable person would consent in such circumstances" (Hoge & Appelbaum, 1989, p. 614). (For a complete and in-depth discussion of the complexities of specific theories of informed consent, the reader is directed to the work of Appelbaum, Lidz, and Meisel [1987].)

By providing pertinent information to both patient and family over the course of treatment, the clinician not only allows active collaboration but also fosters close monitoring of the patient's and family's concerns. In fact, under the rules of informed consent the patient (and often the family) has a right to be told about the risks and benefits of the suggested course of action and of any reasonable alternative treatments (Bennett et al., 1990). All such information should be given in a "neutral dispassionate manner, utilizing to the extent possible a scientific approach to the pros and cons of alternative forms of treatment" (Sadoff, 1990, p. 332).

An open discussion of the risks and benefits facilitates cooperation, widens the protective net, and increases available sources of vital information. Moreover, such an open sharing of the risks with patients and family can lessen the experience of shock and surprise should the tragedy of a patient's suicide occur. Yet the delicate problem remains of how to involve the support group without violating patient confidentiality.

Confidentiality

Confidentiality is often referred to as the patient's right to have communications that are given in confidence not disclosed to outside parties without the patient's implied or express authorization (Simon, 1988). Once a doctor-patient relationship is established, the mental health professional assumes an "automatic duty to safeguard patients' disclosures" (Simon, 1988, p. 57). Yet the clinician's duty to maintain confidentiality is not absolute, and there are clearly situations where breaching confidentiality is both ethically and legally valid.

Simon (1988) pointed out that the competent patient's request for the maintenance of confidentiality must be honored unless the patient is a clear danger to himself or herself or to others. However, the legal duty to warn or inform third parties exists in some jurisdictions only if the danger of physical harm is threatened toward others (Simon, 1988).

Mental health professionals must understand laws and regulations related to breaching confidentiality when patients are a "danger to self." Pope, Tabachnick, and Keith-Spiegel (1988, p. 550) have noted that:

Apparently the argument made by the defense in many of the early "duty to protect" cases (e.g., *Tarasoff* v. *Regents of the University of California*, 1976), that absolute confidentiality is necessary for psychotherapy, is not persuasive for many psychotherapists: Breaking confidentiality is seen by a large number of practitioners as uniformly good practice in cases of homicidal risk, suicidal risk, and child abuse.

Shneidman (1981) has gone even further with this position and stated that confidentiality, when a patient has exhibited suicidal behavior, should not be an important issue between psychotherapists and their patients. He believed that the main goal of suicidal therapy is to defuse the potentially lethal situation. Thus to hold to the principle of confidentiality is contradictory to a basic tenet of an ethical psychotherapeutic relationship.

Gutheil (1984) pointed out that with patients for whom suicidal issues are predominant it is important to assess their ability to participate in therapeutic alliance with the clinician.

The patient who is collaborative, who sees the issue as a joint problem for both patient and clinician, is in a completely different position from the patient who sees himself as being acted upon. . . . The distinction between the patient who can cooperate but does not and the patient who is too sick to cooperate may mean the difference between success and failure in the litigation area. [Gutheil, 1984, p. 3]

Guidelines for Practice: Chapter Summary

Historically, mental health professionals have been reluctant to obtain "second opinions," mostly because of the issue of confidentiality. The basic argument has been that a psychotherapist could not disclose information to a second therapist because it would negatively affect the patient. As a tragic example of this attitude, one study found that only 27% of clinicians (psychologists, psychiatrists, and social workers) routinely seek consultation to assist in their assessment of suicide (Jobes, Eyman, & Yufit, 1990). A different pattern emerges in medical specialties other than psychiatry, where it is routine and customary to obtain consultation or a second opinion when conflicts or uncertainties arise.

Shneidman (1981) has cautioned that there is almost no instance in the therapist's professional life when consultation with a colleague is more important than when dealing with a highly suicidal patient. An additional perspective is essential if the clinician is to "keep in mind the total picture of the patient and not be blinded by his or her theoretical constructs" (Sadoff, 1990, p. 335).

Thus I argue that any time a clinician writes a risk-benefit progress note in a situation of uncertainty during assessment or management (even when only a low level of elevated risk is detected), she or he also should weigh the value of obtaining a formal consultation. In this regard there are two important points. Although a careful risk-benefit analysis that examines both sides of the risk equation (the hidden benefits and risks, as well as overt ones) can do much to refute an allegation of negligence, "The risk-benefit analysis may prove wrong in hindsight" (Gutheil, 1990, p. 337).

In the aftermath of a patient's suicide, there is often a tendency to trigger in involved observers what some decision analysts have called "the hindsight bias: the observer's perception that what happened was inevitable (and hence predictable or foreseeable) because, in fact, it happened" (Gutheil, 1990, p.

336). Gutheil further observed that hindsight bias is often accompanied by "magical thinking": a form of reasoning characterized by extremes of categorical thinking (e.g., the clinician must hospitalize or not hospitalize, medicate differently, should not have granted privileges) coupled with the perception that the clinician is the only active agent—the patient being inert or helpless, completely under the clinician's control (Gutheil, 1990).

A good risk management analysis does not preclude a poor outcome, however. Although a risk-benefit analysis (prior to any decision to take a calculated risk in the management plan) shows that the clinician has weighed both sides of the equation, a simultaneous consultation for any high-risk decision or in any clinical situation of uncertainty provides a useful "biopsy" of the standard of care (Gutheil, 1990). Such consultation, in conjunction with documentation of this decision-making process, can serve as durable proof that the clinician has not been negligent in confronting the decision in question. The consultation captures, in the most practical manner, the reasoning of the hypothetical "reasonable and average practitioner" and, more importantly,

frees the clinician from the accusation of ideologic insularity; the "second opinion" may thus provide invaluable input, especially in moments of crisis. [Gutheil, 1990, p. 338]

Furthermore, the use of consultation may aid in providing a shift in treatment, help in dealing with a therapeutic impasse, and further help to assuage the fears and subsequent guilt of the family and significant others if the treatment should happen to fail (Sadoff, 1990).

The following list is suggested (Bongar, 1991) as the sort of specific questions that could be discussed with a consultant when treating the suicidal patient. These questions include a review of:

1. The overall management of the case, specific treatment issues, uncertainties in the assessment of elevated risk or in diagnosis. It can include a review of the mental status examination, history, information from significant others, the results of any psychological tests and data from risk estimators, suicide lethality scales, and so on; also, a review of the psychologist's formulation of the patient's *DSM-III-R* diagnosis, together with any other specific psychotherapeutic formulations, clinical assessments, and evaluation of any special treatment and management issues (e.g., co-morbidity of alcohol/substance abuse, physical illness).

2. Issues of managing the patient with chronically suicidal behavior, patient dependency, patient hostility and manipulation, toxic interpersonal matrices, lack of psychosocial supports, patient's competence to participate in treatment decisions, and an assessment of the quality of the therapeutic alliance and the patient's particular response to the psychologist and to the course of treatment (e.g., intense negative or positive transference).

3. The psychologist's own feelings about the progress of treatment and feelings toward the patient (e.g., the psychologist's own feelings of fear, incompetence, anxiety, helplessness, or even anger) and any negative therapeutic

reactions (e.g., negative therapeutic reaction, countertransference, and "therapeutic burnout").

4. The advisability of using medication or the need for additional medical evaluation (e.g., any uncertainties as to organicity or neurological complications); also a request for a reevaluation of any current medications the patient is taking (e.g., effectiveness, compliance in taking medication, side effects, polypharmacy.)

5. The indications and contraindications for hospitalization; a review of available community crisis intervention resources for the patient with few psychosocial support; referral for day treatment; emergency and backup arrangements and resources, and planning for the psychologist's absences.

6. Indications and contraindications for family and group treatment; indications and contraindications for other types of psychotherapy and somatic interventions; questions on the status of and progress in the integration of multiple therapeutic techniques.

7. The psychologist's assessment criteria for evaluating dangerousness and imminence (e.g., does the consultant agree with the clinician's assessment of the level of perturbation and lethality); review of specifics of the patient's feelings of despair, depression, hopelessness, cognitive constriction, and impulses toward cessation.

8. The issues of informed consent and confidentiality; the adequacy of all current documentation on the case (e.g., intake notes, progress notes, utilization reviews, family meetings, supervisor notes, telephone contacts).

9. Whether the consultant agrees with the psychologist's current risk-benefit analysis and management plan in particular. Does the consultant agree that the dual issues of foreseeability and the need to take affirmative precautions have been adequately addressed?

Bongar (1991) also noted that the opinion of a reasonable colleague can be the best immediate "cross-validity check" on the standard of care. In summary, I agree with Shneidman's dictum that "Suicide prevention is not best done as a solo practice" (1981, p. 344), and that routine consultation is the wisest possible course of action for assessment and management of the suicidal patient.

REFERENCES

Appelbaum, P. S., Lidz, C. W., & Meisel, A. (1987). *Informed consent: legal theory and clinical practice*. New York: Oxford University Press.
Barraclough, B., Bunch, J., Nelson, B., & Sainsbury, P. (1974). A hundred cases of suicide: clinical aspects. *British Journal of Psychiatry, 125*, 355–373.
Bennett, B. E., Bryant, B. K., VandeBos, G. R., & Greenwood, A. (1990). *Professional liability and risk management*. Washington, DC: American Psychological Association.
Bogust v. *Iverson* (1960). 10 Wis. 3d 129, 102 N.W.2d 228.
Bongar, B. (1991). *The suicidal patient: clinical and legal standards of care*. Washington, DC: American Psychological Association.
Bongar, B., Peterson, L. G., Harris, E. A., & Aissis, J. (1989). Clinical and legal considerations

in the management of suicidal patients: an integrative overview. *Journal of Integrative and Eclectic Psychotherapy, 8*(1), 53–67.

Brent, D. A., Kupfer, D. J., Bromet, E. J., & Dew, M. A. (1988). The assessment and treatment of patients at risk for suicide. In A. J. Frances & R. E. Hales (Eds.), *American Psychiatric Press review of psychiatry* (Vol. 7; pp. 353–385). Washington, DC: American Psychiatric Press.

Dorpat, T. L., & Ripley, H. S. (1960). A study of suicide in the Seattle area. *Comprehensive Psychiatry, 1*, 349–359.

Gutheil, T. G. (1980). Paranoia and progress notes: a guide to forensically informed psychiatric record-keeping. *Hospital and Community Psychiatry, 31*, 479–482.

Gutheil, T. G. (1984). Malpractice liability in suicide. *Legal Aspects of Psychiatric Practice, 1*, 1–4.

Gutheil, T. G. (1988). Suicide and suit: liability and self-destruction. In D. G. Jacobs & J. Fawcett (Chairmen), Suicide and the psychiatrist: clinical challenges. Presented at a symposium sponsored by the Suicide Education Institute of Boston in collaboration with The Center of Suicide Research and Prevention, at the American Psychiatric Association Annual Meeting, Montreal.

Gutheil, T. G. (1990). Argument for the defendant-expert opinion: death in hindsight. In R. I. Simon (Ed.), *Review of clinical psychiatry and the law* (pp. 335–339). Washington, DC: American Psychiatric Association.

Harris, E. A. (1990). Risk management. Workshop sponsored by the American Psychological Association Insurance Trust, San Francisco.

Hoge, S. K., & Appelbaum, P. S. (1989). Legal issues in outpatient psychiatry. In A. Lazare (Ed.), *Outpatient psychiatry* (pp. 605–621). Baltimore: Williams & Wilkins.

Jobes, D. A., Eyman, J. R. & Yufit, R. I. (1990). Suicide risk assessment survey. Presented at the annual meeting of the American Association of Suicidology, New Orleans.

Kahn, A. (1990). Principles of psychotherapy with suicidal patients. In S. J. Blumenthal & D. J. Kupfer (Eds.), *Suicide over the life cycle: risk factors, assessment, and treatment of suicidal patients* (pp. 441–468). Washington, DC: American Psychological Press.

Lesse, S. (1989). The range of therapies with severely depressed suicidal patients. In S. Lesse (Ed.), *What we know about suicidal behavior and how to treat it* (pp. ix–xiv). Northvale, NJ: Jason Aronson.

Litman, R. E. (1988). Treating high-risk chronically suicidal patients. In D. G. Jacobs & J. Fawcett (Chairmen), Suicide and the psychiatrist: clinical challenges. Presented in a symposium sponsored by the Suicide Education Institute of Boston in collaboration with The Center of Suicide Research and Prevention, at the American Psychiatric Association Annual Meeting, Montreal.

Luborsky, L. (1990). Who is helped by psychotherapy? *Harvard Mental Health Letter, 7*(2), 4–5.

Motto, J. A. (1979). Guidelines for the management of the suicidal patient. *Weekly Psychiatry Update Series Lesson 20, 3*, 3–7. Available from Biomedia, Princeton, NJ.

Murphy, G. E. (1988). The prediction of suicide. In S. Lesse (Ed.), *What we know about suicidal behavior and how to treat it* (pp. 47–58). Northvale, NJ: Jason Aronson.

Perr, I. N. (1985). Psychiatric malpractice issues. In S. Rachlin (Ed.), *Legal encroachment on psychiatric practice*. San Francisco: Jossey-Bass.

Pope, K. (1986). Assessment and management of suicidal risks: clinical and legal standards of care. *Independent Practitioner*, January, 17–23.

Pope, K. S., Tabachnick, B. G., & Keith-Spiegel, P. (1988). Good and poor practices in psychotherapy: national survey of beliefs of psychotherapists. *Professional Psychology: Research and Practice, 19*, 547–552.

Robins, E., Murphy, G. E., Wilkinson, R. M., Gassner, S., & Kays, J. (1959). Some clinical considerations in the prevention of suicide based on a study of 134 successful suicides. *American Journal of Public Health, 49*, 888–898.

Sadoff, R. L. (1975). *Forensic psychiatry: a practical guide for lawyers and psychiatrists.* Springfield, IL: Charles C Thomas.

Sadoff, R. L. (1990). Argument for the plaintiff—expert opinion: death in hindsight. In R. I. Simon (Ed.), *Review of clinical psychiatry and the law* (pp. 331–335). Washington, DC: American Psychological Association.

Shaffer, D. (1974). Suicide in childhood and early adolescence. *Journal of Child Psychology and Psychiatry, 15,* 275–291.

Shaffer, D., Gould, M., & Trautman, P. (1985). Suicidal behavior in children and young adults. Presented at the Psychobiology of Suicidal Behavior Conference, New York.

Shaffi, M. (1986). Psychological autopsy study of suicide in adolescents. Presented at the Child Depression Consortium, St. Louis.

Shneidman, E. S. (1981). Psychotherapy with suicidal patients. *Suicide and Life-Threatening Behavior, 11,* 341–348.

Shneidman, E. S. (1984). Aphorisms of suicide and some implications for psychotherapy. *American Journal of Psychology, 38*(3), 319–328.

Simon, R. I. (1987). *Clinical psychiatry and the law.* Washington, DC: American Psychiatric Press.

Simon, R. I. (1988). *Concise guide to clinical psychiatry and the law.* Washington, DC: American Psychiatric Press.

Stone, A. A. (1989a). A response to Dr. Klerman. *Harvard Medical School Mental Health Letter, 6*(1), 3–4.

Stone, A. A. (1989b). Suicide precipitated by psychotherapy. In S. Lesse (Ed.), *What we know about suicidal behavior and how to treat it* (pp. 307–319). Northvalue, NJ: Jason Aronson.

Stromberg, C. D., Haggarty, D. J., Leibenluft, R. F., McMillan, M. H., Mishkin, B., Rubin, B. L., & Trilling, H. R. (1988). *The psychologist's legal handbook.* Washington, DC: Council for the National Register of Health Service Providers in Psychology.

Tarasoff v. *Regents of the University of California et al.* (1976). 551 P2d.334, 131 Cal Rptr. 14, CA Sup. Ct.

VandeCreek, L., & Knapp, S. (1983). Malpractice risks with suicidal patients. *Psychotherapy: Theory, Research and Practice, 20,* 274–280.

VandeCreek, L., & Knapp, S. (1989). *Tarasoff and beyond: legal and clinical considerations in the treatment of life-endangering patients.* Sarasota, FL: Professional Resource Exchange.

Welch, B. (1989). A collaborative model proposed. *American Psychological Association Monitor, 20*(10), 28.

Wright, R. (1981). What to do until the malpractice lawyer comes. *American Psychologist, 36,* 1535–1541.

Index

Note: Page numbers followed by f refer to illustrations, page numbers followed by t refer to tables.